Conference Proceedings of the Society for Experimental Mechanics Series

Series Editor

Kristin B. Zimmerman, Ph.D.
Society for Experimental Mechanics, Inc.
Bethel, CT, USA

More information about this series at http://www.springer.com/series/8922

Simon Quinn • Xavier Balandraud
Editors

Residual Stress, Thermomechanics & Infrared Imaging, Hybrid Techniques and Inverse Problems, Volume 9

Proceedings of the 2016 Annual Conference on Experimental and Applied Mechanics

Springer

Editors
Simon Quinn
University of Southampton
Southampton, UK

Xavier Balandraud
CNRS, UMR 6602, Institut Pascal
Aubiére, France

Université Clermont Auvergne
Sigma Clermont, Institut Pascal
Clermont-Ferrand, France

ISSN 2191-5644 ISSN 2191-5652 (electronic)
Conference Proceedings of the Society for Experimental Mechanics Series
ISBN 978-3-319-42254-1 ISBN 978-3-319-42255-8 (eBook)
DOI 10.1007/978-3-319-42255-8

Library of Congress Control Number: 2016949067

Printed on acid-free paper

This Springer imprint is published by Springer Nature
The registered company is Springer International Publishing AG Switzerland

Preface

Residual Stress, Thermomechanics & Infrared Imaging, Hybrid Techniques and Inverse Problems represents one of ten volumes of technical papers presented at the 2016 SEM Annual Conference & Exposition on Experimental and Applied Mechanics organized by the Society for Experimental Mechanics and held in Orlando, FL, June 6–9, 2016. The complete Proceedings also includes volumes on *Dynamic Behavior of Materials*; *Challenges In Mechanics of Time-Dependent Materials*; *Advancement of Optical Methods in Experimental Mechanics*; *Experimental and Applied Mechanics*; *Micro- and Nanomechanics*; *Mechanics of Biological Systems and Materials*; *Mechanics of Composite & Multifunctional Materials*; *Fracture, Fatigue, Failure and Damage Evolution*; *and Joining Technologies for Composites and Dissimilar Materials.*

Each collection presents early findings from experimental and computational investigations on an important area within Experimental Mechanics, Residual Stress, Thermomechanics & Infrared Imaging, Hybrid Techniques and Inverse Problems being three of these areas.

Residual stresses have a great deal of importance in engineering systems and design. The hidden character of residual stresses often causes them to be underrated or overlooked. However, they profoundly influence structural design and substantially affect strength, fatigue life and dimensional stability. Since residual stresses are induced during almost all materials' processing procedures, for example welding/joining, casting, thermal conditioning and forming, they must be taken seriously and included in practical applications.

In recent years, the applications of infrared imaging techniques to the mechanics of materials and structures have grown considerably. The expansion is marked by the increased spatial and temporal resolution of the infrared detectors, faster processing times and a much greater temperature resolution. The improved sensitivity and more reliable temperature calibrations of the devices have meant that more accurate data can be obtained than were previously available.

Advances in inverse identification have been coupled with optical methods that provide surface deformation measurements and volumetric measurements of materials. In particular, inverse methodology was developed to more fully use the dense spatial data provided by optical methods to identify mechanical constitutive parameters of materials. Since its beginnings during the 1980s, creativity in inverse methods has led to applications in a wide range of materials, with many different constitutive relationships, across material heterogeneous interfaces. Complex test fixtures have been implemented to produce the necessary strain fields for identification. Force reconstruction has been developed for high strain rate testing. As developments in optical methods improve for both very large and very small length scales, applications of inverse identification have expanded to include geological and atomistic events.

Southampton, UK
Aubiére, France

Simon Quinn
Xavier Balandraud

Contents

Chapter 1
Fatigue Behaviour of Stainless Steels: A Multi-parametric Approach

R. De Finis, D. Palumbo, F. Ancona, and U. Galietti

Abstract In recent years different experimental methods have been experienced to enhance the fatigue characterisation of materials with the aim to overcome the Standard long-lasting tests, i.e. Wohler curve determination. Standard fatigue treatment requires at least 15 specimens being tested to get an estimation of material fatigue limit and it is worth noting that this kind of tests do not provide any information on damage phenomena occurring in the material. Thus, topic to be addressed in this research have to do with development of lock-in infrared measurement based thermal method for rapid evaluation of fatigue limit. By performing a single test , the adopted method leads to match different parameter information. The Assessed parameters are in number more than the ones provided by TSA, as well. Moreover, the adopted technique points to study damage by analysing the different phenomena involved in fatigue and in this regard, the aim of this paper is to show how a thermal technique can attain an early assessment of the failure processes during a cyclically loading test. The author is, also, focused on to illustrate the strong points of a method based on infrared measurements for assessing endurance limit for both austenitic and martensitic stainless steels while considering, as reference, the Standard Test methods.

Keywords Lock-in thermography • Fatigue • Austenitic/martensitic stainless steels

1.1 Introduction

Infrared thermography has been successfully exploited as an experimental, non-destructive, real-time and non-contact technique to observe physical processes of: damage, fatigue, and failure. The strong points of the technique concern the mechanical characterisation of metallic, composites and structural components [1–3]. Different works was meant to show that the surface temperature monitoring is a reliable technique to detect the damage phenomena and thus, the material intrinsic dissipation are evaluated. Following from the work of Luong [1] in to the use of thermal sources to assess fatigue limit, in literature, different approaches have been performed to study the fatigue damage with thermography based on: the monitoring of the surface temperature[2, 4] , the evaluation of "dissipative" thermal heat sources[5], the evaluation of the phase variation in thermographic signal by using lock-in thermography [6–8]. Referring to the temperature, it is worth noting that this parameter is very sensitive to the external influences and thus, room temperature and/or loading machine grip heating can affect the measurements and have to be considered in the analysis. In this framework [2], De Finis et All propose a robust technique to thermal data analysis in order to filter out all the noisy all the heat sources compromising the adiabatic condition of the sample during the test and for early detecting the dissipation processes. Despite the direct use of thermal parameter has opened-up the possibility of mechanical fatigue characterising with less-lasting and less expensive experimental campaign, temperature measurements are affected by thermal properties of the material [6], as well, i.e. thermal conductivity. Material with high conductivity (e.g. aluminium alloys or welded joints) experience low temperature during the test due to high percentage of reflected radiation. Another issue related to the use of temperature is represented by the lattice microstructures which cause extremely low temperature increments [2, 3]. To improve the analysing experimental dataset and to avoid the appearance of the 'external' influences another approach based on lock-in thermography is proposed. The technique, moreover, provides several parameters accounting for the study of damage phenomena. By demodulating thermal signal the harmonic analysis allows for achievement of 1° phase and 2° amplitude harmonic components: two significant parameters for assessing fatigue behaviour of material [7, 8], and all external heat source influence is eliminated. In this work lock-in thermography will support the Thermoelastic Stress Analysis by studying evolution of damage phenomena despite the loss of adiabatic conditions [6, 9]. In particular the phase shift of first

R. De Finis (✉) • D. Palumbo • F. Ancona • U. Galietti
Department of Mechanics, Mathematics and Management (DMMM), Politecnico di Bari, Bari, Italy
e-mail: rosa.definis@poliba.it; davide.palumbo@poliba.it; francesco.ancona@poliba.it; umberto.galietti@poliba.it

S. Quinn, X. Balandraud (eds.), *Residual Stress, Thermomechanics & Infrared Imaging, Hybrid Techniques and Inverse Problems, Volume 9*, Conference Proceedings of the Society for Experimental Mechanics Series, DOI 10.1007/978-3-319-42255-8_1

component synchronous at the mechanical frequency is related to the appearance of plastic zones or cracks in the material and so it is strictly related to damage [6]. Moreover, if thermal phenomena are present a double mechanical load frequency component arises related to dissipative heat sources in the material and thus it is associated to the temperature [8]. In this paper the phase shift and double frequency harmonic amplitude are jointly used to carry out a fatigue damage study for assessing endurance limit of both austenitic and martensitic stainless steels.

1.2 Theoretical Framework

The thermoelastic stress analysis is a full-filed, well-established and non-contacting technique for evaluating surface stresses related to the small temperature increments when during the fatigue experiment adiabatic conditions are achieved [6, 10, 11]. Under the hypothesis of a linear elastic, isotropic and homogeneous material the temperature change occurs isoentropically: the so called thermoelastic effect [12]. The classical thermoelastic equation (1.1) states:

$$\Delta T = -T_0 K \Delta \sigma_I \tag{1.1}$$

where K is the thermoelastic constant, ΔσI is the change in the stress invariant and T0 is the specimen environmental temperature. If thermoelastic effect is present the reversible conversion between mechanical and thermal energy is feasible [10, 11]. During the cyclically loading test, when the intrinsic stress level in the material exceed the yield strength, high stress gradients and local plasticity appear caused by dissipative heat sources. In these conditions the classical thermoelastic equation (1.1) lose validity and TSA cannot be used for assessing surface stress field of material [10]. Despite this limitation in applying TSA, the phase of thermoelastic signal still represents an important parameter to evaluate fatigue behaviour of material [6].

Considering a reference signal of load cell installed on loading machine, the thermoelastic signal and reference signal are considered as two vectors rotating at the same speed (the same mechanical frequency). Time after time the shift in phase values is constant and it varies only in case of loss of adiabatic condition. Hence, for studying the fatigue behaviour another parameter can be used in place of the TSA provided parameters, it is represented by phase shift between thermoelastic signal with respect reference signal and its amplitude [6]. By using a lock-in amplifier, the signal can be reported to reference signal and a suitable signal demodulation must be realized to obtain the parameters. The adopted mathematics model describing temperature change during the test, is [8, 13]:

$$T_m(t) = a + bt + \Delta T_1 \sin(\omega t + \varphi_1) + \Delta T_2 \sin(2\omega t + \varphi_2) \tag{1.2}$$

Where "a" coefficient is proportional to the environmental temperature, "b" coefficient depends on mechanical load frequency and time, "f" is the mechanical load frequency, T1,2 and φ1,2 are coefficient proportional to, respectively amplitude and phase shift of first and second harmonic component of Fourier Fast Transform [8, 9]. In this paper the assessment of fatigue behaviour of martensitic and austenitic stainless steel is shown by evaluating φ1 and T2 parameters of thermal signal demodulation.

1.3 Experimental Setup

A typical setup for assessing signal is drawn in and Fig. 1.1.

The tested materials are the austenitic AISI 316 and martensitic X4 Cr Ni 16-4 stainless steels used in Oil&Gas Industry.

Martensitic stainless steels have a higher mechanical strength obtained by a quenching heat treatment but limited corrosion resistance. The AISI 316 material is a well-known austenitic stainless steel while X4 Cr Ni Mo 16-4 deal with the addition of Chromium in the lattice (11–16 % in weight) that allows for [14]:

- improvement of corrosion resistance through the formation of oxides,
- avoidance of the depleting of Chromium from lattice.

Fig. 1.1 Loading machine and IR detector

During the fatigue tests the procedure is stepwise with incremental stress amplitude and mean stresses. Three 'dog bone' shaped specimen for both material have been tested in an insulated chamber covering the clamping area and minimizing environmental heat change. During fatigue testing, a thin sub-micron matt black coating was applied on the specimen gage-length section to decrease the surface-heat reflections.

Low-cycle fatigue experiments were performed on both steels at 17 Hz on a MTS (Material Test System) machine with a R-ratio of 0.5. A load-control mode was used. Maximum stress levels ranging from 80 to 440 MPa and 440 to 830 MPa, respectively for AISI 316 and X4 Cr Ni 16-4 were applied. Note that all of the stresses were applied in a sinusoidal waveform.

Thermography detection was conducted using a state-of the-art FLIR X 6540SC thermographic-IR-imaging system with a 640 × 512 pixels focal-plane-array cooled detector that is sensitive to a radiation wavelength of 1.5–5.1 μm. The temperature sensitivity is less than 25 mK. The system has a highest-speed data-acquisition capability of 125 Hz at a full frame of 640 × 512 pixels.

The procedure for processing phase data does not consider any external heat influences and consists in applying a Gaussian filter to all data matrix for each loading step, first. In the data matrix the pixel is a single value of phase shift or 2° order amplitude component, only the pixels referring to the gauge length of specimen are considered in the analysis and thus a windowing is needed. Moreover, a reference data matrix has been subtracted from the leftovers to eliminate the noise of early load steps (Fig. 1.2). For specimen 1 of AISI 316 the reference load step has been fixed to 35 MPa while for X4 Cr Ni 16-4 material the subtracted image refers to 167.50 MPa. For the T2 parameter during the data processing any subtraction has been made. In the next paragraph are shown the results for both materials.

1.4 Results

In this paper a comparison between temperature phase shift φ, and second harmonic amplitude parameter T_2 data is shown. Parameters maps, parameters curves and fatigue limit found, are compared with the value of Standard Statistical Dixon method.

The adopted processing procedure leads to obtain statistical data for assessing fatigue limits, by analyzing the parameter values in the gauge length of the specimen (data matrix 40 × 140 pixels for AISI 316, and 60 × 160 pixels for X4 CR NI 16-4).

The data series considered in the processing were: difference percentiles 98° − 2° for temperature phase shift parameter and max value for second order harmonic amplitude parameter.

The experimental data set considered for phase shift analysis was the difference of 98° and 2° percentiles of phase series : due to large scattering of phase data this procedure sets the results free of outliers points.

The 98° − 2° percentiles difference is applied pixel by pixel of the matrix representing the gauge length of specimen, for all the loading steps.

To eliminate any external influence, e.g. the thickness of matt black coating layer, a residual analysis is made on 98° − 2° percentile data series according to similar procedure proposed by the author [2].

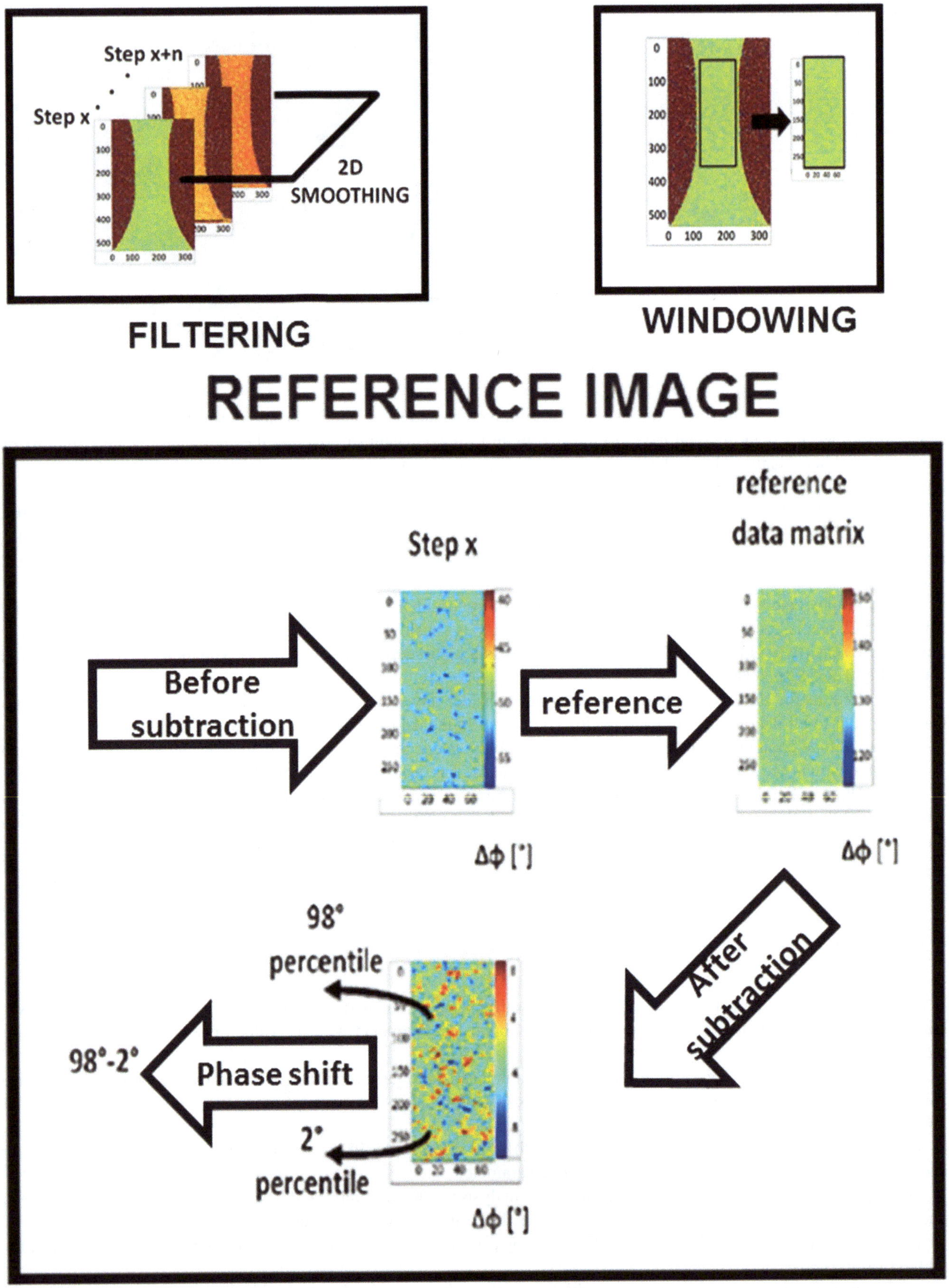

Fig. 1.2 Smoothing procedure for phase data matrix

For each sample, the residual analysis refers to phase data of the beginning three loading step levels in which any damage phenomena is present in the bulk material.

A statistical process control chart ensures fixing a threshold to early assess the occurring of damage process and thus, the fatigue limit. The threshold value is given by the mean value of the residuals data onset the test (first three points) plus three times their standard deviation. The threshold method provides a better evaluation of damage and fatigue processes referring to other methods proposed in literature, leaving the researcher their own choice of the breakpoint of data series for assessing fatigue limit [1].

The same data regression analysis was applied on second harmonic amplitude parameter and the fatigue limit found for each specimen has been evaluated by exploiting the shown procedure with another type of statistical analysis with 3σ control chart. In particular in this case, as well, the analysis consists in interpolating early three data and subtracting both the slope and offset of this straight line to the other data. A threshold value is set up to separate from data series, the value at which damage occurs. The choice of two different thresholds has been supported by results of Dixon reference method. It is worth nothing to highlight that the meaning of single threshold ($\mu + \sigma$ chart for phase shift parameter and $\mu + 3\sigma$ chart for dissipation parameter) is relied to the nature of two parameters : temperature phase shift is related to the damage occurring in the material, a significant increase of the value means that the material is experiencing fatigue phenomena. Dissipation parameter, the so called second harmonic

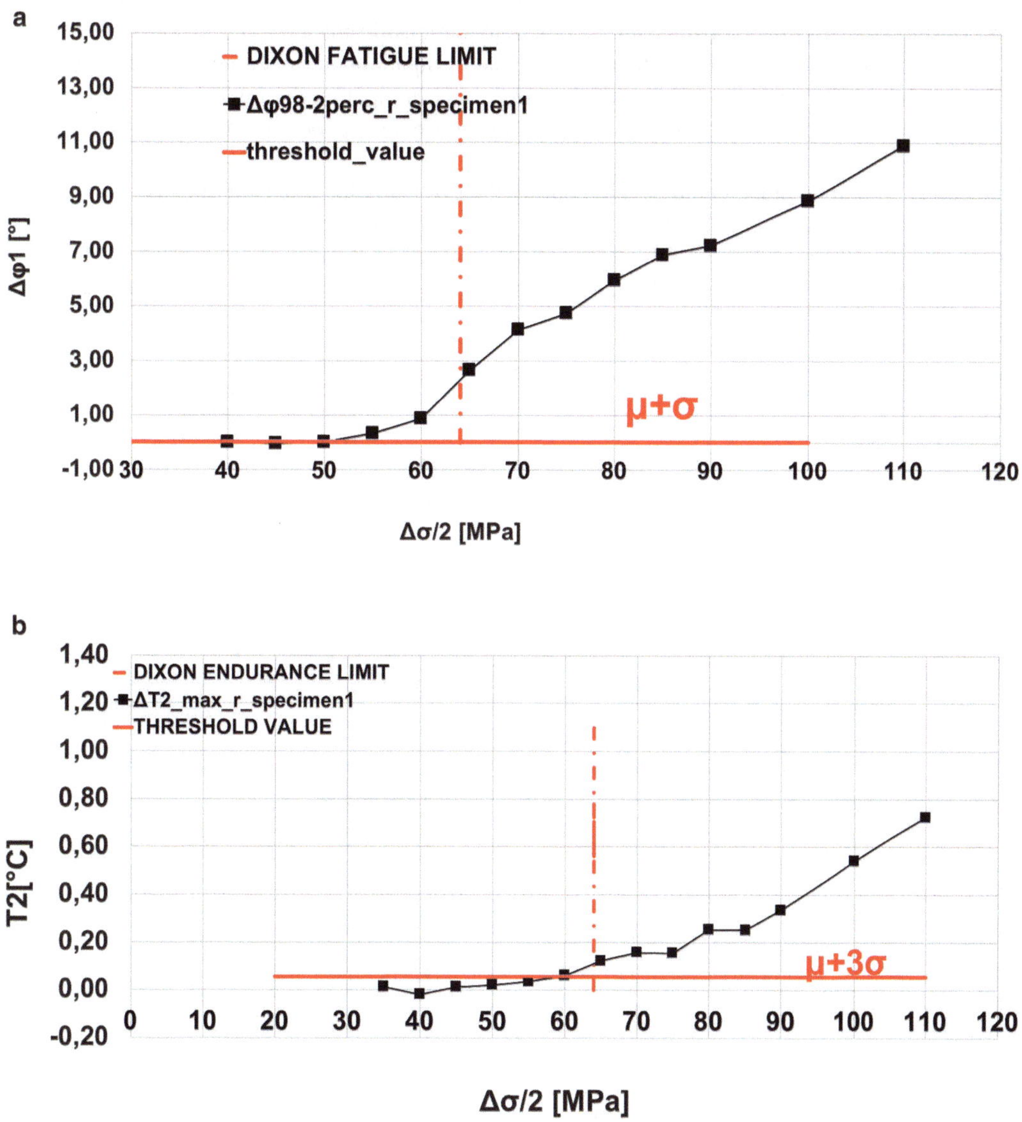

Fig. 1.3 Application of threshold method on AISI 316: phase data (**a**), dissipation data (**b**)

amplitude parameter represents the heat dissipations caused by irreversible phenomena in the material e.g. plasticity. The appearance of a dissipative phenomenon not necessarily implies damage.

Generally the range of magnitude of this parameter is lower than the magnitude of phase shift parameter i.e. 0.001–0.005 °C vs. 1–10°.

As an example, for AISI 316 specimen 1, Fig. 1.3 shows the phase parameter curve (a) and the dissipation parameter curve (b) with applied thresholds.

Figure 1.3 shows also the fatigue limit of 65 MPa (bolt red line) found with Dixon experimental tests carried out on AISI 316. The standard test method consists in run-in/run-out based procedure with infinite threshold life fixed at 2×10^{-7} cycles of loading machine. The load ratio for all the test was fixed at 0.5, and test frequency was 39 Hz.

1.5 Discussion

As drawn in Fig. 1.4 for AISI 316 austenitic material (right column in the pictures) , due to its ductile behaviour, in the phase map it is possible to observe that the damage is spread in the gauge length of specimen. In this case, for austenitic stainless steels the high yield strength allows to a global plastic work affecting homogeneously the whole material (as shown in phase and T2 matrix Figs. 1.4 and 1.5 right columns) except for some clusters appearing at the last cycles of the test phase map Fig. 1.4. Figure 1.5 shows the dissipation parameter ΔT_2 for both martensitic and austenitic steels. Referring to right column, the austenitic behaviour of AISI 316 indicates that no localized dissipations occur in the material, excepting for the final loading step. Due to the ductile beahviour the material seems easily dissipating the imposed high stress gradients without leading to some localized plasticizations. The Figs. 1.4 and 1.5 show also, a good reproducibility for all the curves of AISI 316 material.

For brittle materials, a fatigue crack could develop during the cyclically loading experiments.

By analyzing Figs. 1.4 and 1.5 (left column) representing respectively the temperature phase shift and second harmonic amplitude parameter maps, the monitoring of the fatigue crack growing in the material is performed. It is worth noting the potential of this technique in detecting material failure or damage processes of real and complex shaped component during

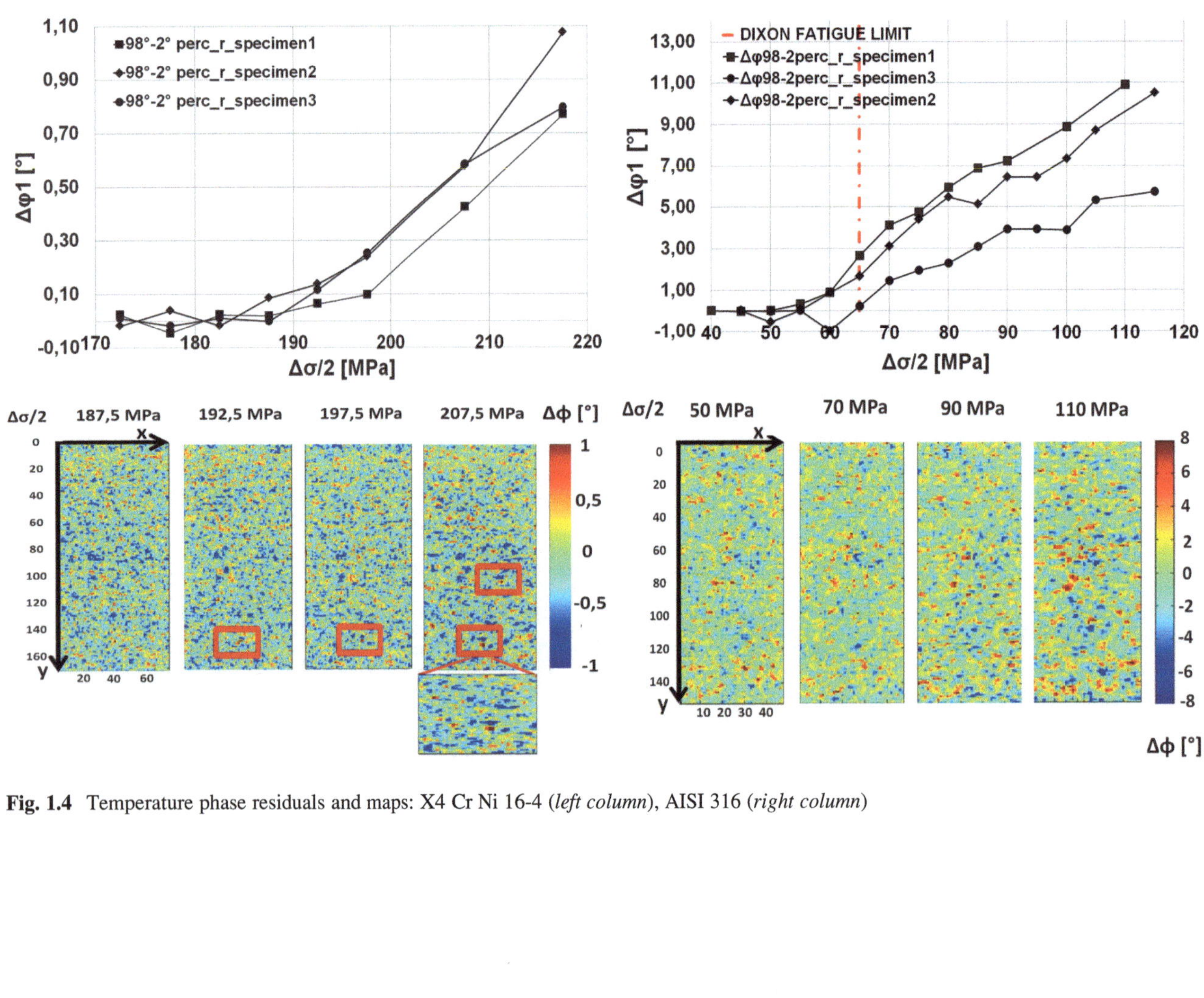

Fig. 1.4 Temperature phase residuals and maps: X4 Cr Ni 16-4 (*left column*), AISI 316 (*right column*)

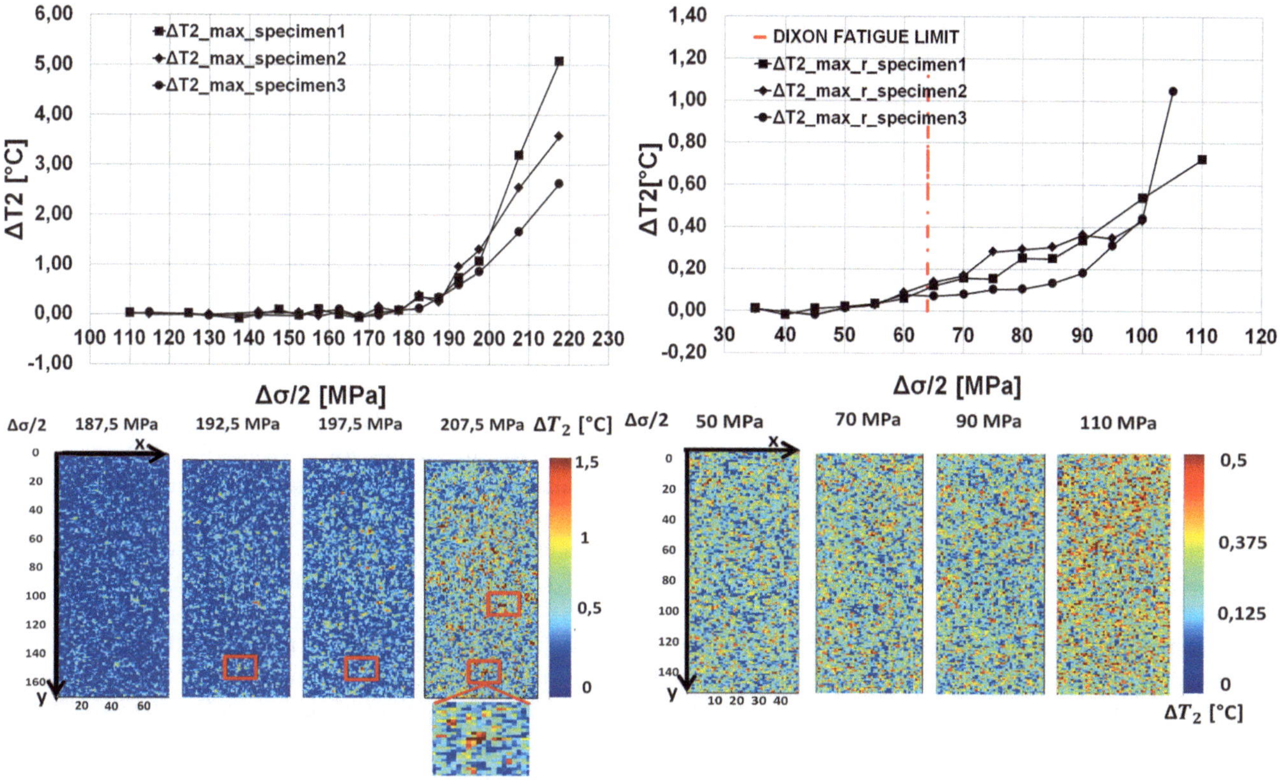

Fig. 1.5 Second order harmonic amplitude parameter T_2 residuals and maps: X4 Cr Ni 16-4 (*left column*) , AISI 316 (*right column*)

Table 1.1 AISI 316 Endurance limit results: different methods

AISI 316	Themperature phase shift parameter (MPa)	2° harmonic amplitude parameter (MPa)
AVERAGE	**60.00**	**60.00**
STD. DEV.	5.00	0.00
	FATIGUE LIMIT 'DIXON' METHOD [MPa]	
	64.58 (3.51 MPa standard deviation)	

Table 1.2 X4 Cr Ni 16-4 Endurance limit results: different methods

X4 Cr Ni 16-4	Themperature phase shift parameter (MPa)	2° harmonic amplitude parameter (MPa)
AVERAGE	**190.83**	**180.83**
STD. DEV	2.89	2.89

maintenance. As disclosed before, the appearance of dissipation phenomena could not be necessarily related to damage. In this case, by confronting Figs. 1.4 and 1.5 (left column) for X4 Cr Ni 16-4 material, a low signal dissipation appears at 187.5 MPa while the temperature phase shift map in Fig. 1.4 shows the appearance of damage at 192.5 MPa referred to the same position. By confronting the phase map (Fig. 1.4) and dissipation map (Fig. 1.5) it is possible to observe the results are in good agreement. At this regard it can be concluded that dissipations appear before the crack. The crack , as noticed in this case, develops at higher stress value in the same position. It is possible to claim that the dissipation parameter fulfil important role in detecting the inception of plastic phenomena in the material, and as confirmed by temperature phase parameter the combined use of these two parameters could help the researcher in predicting fatigue limit and failure localization.

Results regarding fatigue limits are showed in Tables 1.1 and 1.2, for both parameters for both materials.

For AISI 316 materials the results in Table 1.1 are in good agreement with the value found by means of Dixon Standard method, the phase mean values are very close to 64.58 MPa.

Referring to tested martensitic stainless steel (Table 1.2), the average value of phase result is higher than T2 average value. This phenomenon can be explained by considering the intrinsic difference in meaning between phase and second order amplitude components: the first is a manifestation of occurring damage while the second represents the dissipative heat sources due to a localized stress concentration in the material. The obtained ΔT_2 value could be lower than phase mean value because the dissipative heat sources appear before any material failure. Further studies will be focused on the comparison of found values with the reference given by standard test method.

1.6 Conclusions

This work deals with the characterisation of two kind of stainless steels : AISI 316 and X4 Cr Ni 16-4. In particular as knowledge of authors, not so many works concerns the fatigue characterisation of showed martensitic stainless steel.

The fatigue limit characterisation has been made by exploiting potentialities of a thermal methods. Infrared measurements provide a signal which analysis concerns the extrapolation of two parameters related to thermoelastic signal. The experiments were meant to demonstrate that the temperature phase shift and second order harmonic amplitude represent good indicators for damage and dissipative phenomena appearance. Thus , they allow for assessing fatigue limit of both austenitic and martensitic lattice. In this framework, a method to evaluate a threshold for getting the fatigue limit has been showed. The showed method to post-process the data allows to obtain results in good agreement with theoretical assumptions and they are also very close to Dixon fatigue limit evaluation. Recent advances of the threshold method are addressed to easily and automatic implementation of the algorithm and to calibrate technique by using a Standard Test method.

Following, the research carried out the great dependence from the type of crystal network: the austenitic microstructure dissipates energy by affecting all the surrounding lattice planes, while the martensitic microstructure leads to a stress concentration only in localized zones that may become fatigue cracks. As shown by analysis, and as confirmed by results, the phase shift $\Delta\varphi$ parameter is capable to detect the damaged zone and the $\Delta T2$ analysis involves in an assessment of dissipative heat sources not-necessarily close to damage (a dissipative heat sources can be localized inside the material but these one could not cause a failure).

In this paper the fatigue test with infrared thermal methods was meant to show the thermoelastic signal processing provides different parameters describing the overall process involved in the damage behavior of material. The potential of this method is moreover, the reliability to real component monitoring during maintenance.

References

1. Luong, M.P.: Infrared thermographic scanning of fatigue in metals. Nucl. Eng. Des. **158**, 363–376 (1995)
2. De Finis, R., Palumbo, D., Ancona, F., Galietti, U.: Fatigue limit evaluation of various martensitic stainless steels with new robust thermographic data analysis. Int. J. Fatigue **74**, 88–96 (2015)
3. Freire, J.L.F., Waugh, R.C., Fruehmann, R., Dulieu-Barton, J.M.: Using thermoelastic stress analysis to detect damaged and hot spot areas in structural components. J. Mech. Eng. Autom. **5**, 623–634 (2015)
4. La Rosa, G., Risitano, A.: Thermographic methodology for rapid determination of the fatigue limit of materials and mechanical components. Int. J. Fatigue **22**, 65–73 (2000)
5. Boulanger, T., Chrysochoos, A., Mabru, C., Galtier, C.: Calorimetric analysis of dissipative and thermoelastic effects associated with the fatigue behavior of steels. Int. J. Fatigue **26**, 221–229 (2006)
6. Palumbo, D., Galietti, U.: Characterization of steel welded joints by infrared thermographic methods. Quant. Infrared Thermogr. J. **11**(1), 29–42 (2014)
7. Enke, N.F., Sandor, B.I.: Cyclic plasticity analysis by differential infrared thermography. Proceding of the VII International Congress on Experimental Mechanics, pp. 830–835, 7 June 1988
8. Krapez, J.C., Pacou, D., Gardette, G.: Lock-in thermography and fatigue limit of metals. Quantitative Infrared Thermography Conference QUIRT 2000, Reims (France), 18–21 July 2000
9. Galietti, U., Palumbo, D., De Finis, R., Ancona, F.: Fatigue damage evaluation with new thermal methods. 3rd International Workshop on Advanced Infrared Technology and Applications, Tourin (Italy), 11–14 September 2013
10. Wang, W.J., Dulieu-Barton, J.M., Li, Q.: Assessment of non-adiabatic behaviour in thermoelastic stress analysis of small scale components. Exp. Mech. **50**, 449–461 (2010)
11. Stanley, P.: Beginnings and early development of thermoelastic stress analysis. Strain **44**, 285–287 (2008)
12. Wong, A.K., Sparrow, J.G., Dunn, S.A.: On the revised theory of thermoelastic effect. J. Phys. Chem. solid. **49**(4), 395–400 (1988)
13. Galietti, U., Palumbo, D., De Finis, R., Ancona, F.: Fatigue limit evaluation of martensitic steels with thermal methods. Quantitative Infrared Thermography 12th International QIRT Conference, Bordeaux (France), 7–11 July 2014
14. McGuire, M.F.: Martensitic stainless steels. In: Stainless Steels for Design Engineers, pp. 123–135. ASM International, Materials Park (2008)

Chapter 2
Measurement of Mechanical Dissipation in SMAs by Infrared Thermography

Didier Delpueyo, Xavier Balandraud, Michel Grédiac, Sergiu Stanciu, and Nicanor Cimpoesu

Abstract The reversibility of the phase transformation in shape-memory alloys (SMAs) directly governs the mechanical response of these materials. Infrared thermography is used in this study to measure the mechanical dissipation produced by copper-based SMAs under cyclic loading at ambient temperature. Several specimens with different chemical compositions are tested. Mechanical dissipation which is produced by the material is deduced from the temperature change using a 0D version of the heat equation. Results obtained show that the chemical composition as well as the nature of the phase involved (martensite or austenite) influence the mechanical dissipation produced by the specimens during cyclic mechanical loading.

Keywords SMAs • Phase transformation • Infrared thermography • Cyclic loading • Mechanical dissipation • Heat sources

2.1 Introduction

Shape memory alloys (SMAs) have peculiar thermomechanical properties. Superelasticy and memory effect are the most famous ones. The high reversibility of the austenite-martensite (A-M) transformation occurring in these materials is at the origin of their macroscopic properties. The present work proposes to assess the low level of irreversibility of the stress-induced transformation in Cu-Zn-Al SMAs by using infrared (IR) thermography. Specimens featuring different chemical compositions were prepared. The objective is to measure the mechanical (intrinsic) dissipation in specimens subjected to cyclic uniaxial loading at ambient temperature. The paper is organized as follows. The manufacturing procedure to obtain the specimens is first described. The experimental set-up and the data processing are then detailed. Preliminary tests are then presented: monotonic tensile test and natural return to ambient temperature. Results obtained for the cyclic tests are finally presented and discussed.

2.2 Experimental Conditions and Processing

The specimens were copper-based polycrystals. Materials were first elaborated in a standard induction furnace. Parallelepipedic ingots were obtained for each composition. The upper and lower surfaces of the ingots were machined to obtain sheets, nearly 4 mm in thickness, with flat and regular surfaces. The specimens were then heated up to 750 °C and hot rolled. The resulting sheets were finally cut to obtain the specimens for the tests. All specimens were finally tempered by heating up to 750 °C and then water quenched. Metallic samples were mechanically polished in order to remove the coarse oxide layer. Square aluminum tabs were bonded at the ends of the specimens to prevent any slippage within the grips of the testing machine, see Fig. 2.1. Table 2.1 provides the chemical compositions of two specimens discussed here as well as the dimensions of the zone where IR measurements were performed. Chemical compositions were measured using an

D. Delpueyo • M. Grédiac
Université Clermont Auvergne, Université Blaise Pascal, Institut Pascal, BP 10448, 63000 Clermont-Ferrand, France

CNRS, UMR 6602, Institut Pascal, 63178 Aubière, France

X. Balandraud (✉)
CNRS, UMR 6602, Institut Pascal, 63178 Aubière, France

Université Clermont Auvergne, Sigma Clermont, Institut Pascal, BP 10448, 63000 Clermont-Ferrand, France
e-mail: xavier.balandraud@sigma-clermont.fr

S. Stanciu • N. Cimpoesu
University Gh. Asachi of Iasi, Bd. D. Mangeron 61A, 700050 Iasi, Romania

S. Quinn, X. Balandraud (eds.), *Residual Stress, Thermomechanics & Infrared Imaging, Hybrid Techniques and Inverse Problems, Volume 9*, Conference Proceedings of the Society for Experimental Mechanics Series, DOI 10.1007/978-3-319-42255-8_2

Fig. 2.1 Schematic view of the specimen and two reference sheets placed in the jaws of the testing machine

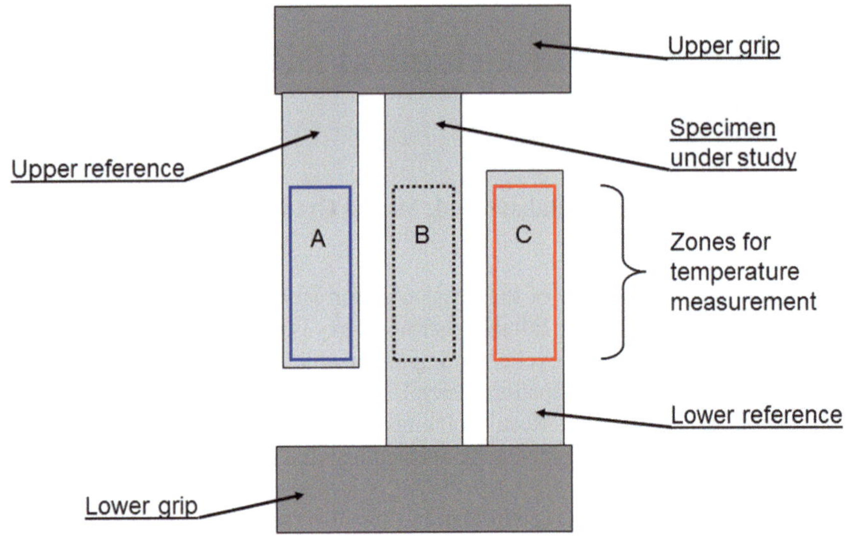

Table 2.1 Specimens under study

Specimen	State	Chemical composition (at. %)	Dimension of the gauge zone
#1	Austenitic	Cu-18.76Zn-15.42Al	$50 \times 11.03 \times 0.68$ mm^3
#2	Martensitic	Cu-15.04Zn-18.07Al	$64 \times 10.48 \times 0.99$ mm^3

EDS-Bruker detector with Xflash—Silicon Drift Detector (SDD) connected at a VegaTescan LMH II SEM equipment. The results were obtained from a 10 kcps signal and a work distance of 15.5 mm (supplier's recommended values). For analysis, Element List mode was used to determine the atomic percentage of Cu, Zn and Al. Compositions in Table 2.1 correspond to average values (measurement repeated twenty times) over on a zone of 3×3 mm^2.

A MTS ±15 kN testing machine was used to apply the mechanical loading at ambient temperature. For each test, two copper reference sheets were also placed in the jaws of the testing machine (see Fig. 2.1). These references enabled us to extract properly the temperature change in the specimen under study (Zone B) due to mechanical loading only. In practice, temperatures in the two references (Zones A and C) were taken into account to remove parasitic effects: change in ambient temperature, change in the temperature of the grips of the testing machine (in particular the loading grip), reflection due to environment, etc. The dimension for the three zones was 50×11 mm^3. The specimens, the two reference sheets, as well as the background, were painted in black to maximize the thermal emissivity of the thermal scene. A CEDIP Jade III-MWIR camera featuring a noise equivalent temperature difference of about 20 mK was used for temperature measurements. Note that the averaging operation over Zones A, B and C enabled us to improve the resolution of the temperature measurement. In practice, the resolution which can be reached by this averaging operation is of some mK.

All tests were performed as follows. After placing the specimen and its two references in the jaws of the testing machine, a waiting time was applied to ensure that thermal equilibrium was reached. A cyclic force-controlled loading was then applied. The loading frequency was 2 Hz. It consisted of blocks of 1200 sinusoidal cycles (i.e. 10 min) with increasing amplitudes by step of 100 MPa, up to specimen failure. The ratio between maximum and minimum forces was equal to 0.1 for all the blocks.

The acquisition frequency of the temperature fields was equal to 0.5 Hz with each recorded image, corresponding to the average of the temperature fields over 2 s captured à 100 Hz. This procedure enabled us to reduce both the size of the recorded files, and to extract the mean trend of the temperature variation.

For the present study, we applied a so-called 0D approach. This approach can be found for instance in [1–3]. It requires the use of thin specimens in such a way that the surface temperature field is representative of the temperatures in the thickness. It also requires that the heat sources produced or absorbed by the material due to stretching are homogeneous in the useful zone of the specimen. This hypothesis is reasonable for specimens featuring a constant section along the useful length, at least if there is no localization. The 0D approach was used in the present study because of very small temperature changes obtained during the cyclic tests, requiring a strong spatial averaging to improve the measurement resolution. Let us define by $\Theta(t)$ the average temperature change in time t over gauge Zone B. The heat source variation was deduced using a 0D version of the heat equation:

$$\frac{d\Theta}{dt} + \frac{\Theta}{\tau} = s \tag{2.1}$$

where s is the heat source divided by the product of the density and specific heat of the material. This quantity, expressed in °C/s or K/s, will be named "heat source" in the following for the sake of simplicity. Parameter τ is a time constant characterizing the heat exchanges of Zone B with its environment.

The heat source produced or absorbed by a material subjected to mechanical loading is composed of a quantity, which is related to thermomechanical couplings (e.g. thermoelastic coupling) as well as a quantity related to mechanical irreversibility. The latter is referred to as *mechanical dissipation* or *intrinsic dissipation*. The calorific energy corresponding to the thermomechanical couplings is null over a thermodynamical cycle. As each recorded thermal data here corresponds to the average of the temperature changes over 4 cycles (2 s × 2 Hz), the processing enabled us to directly extract the mechanical dissipation.

2.3 Preliminary Tests

Before performing the cyclic tests, we performed two preliminary tests for each composition: a monotonic tensile test at ambient temperature, and a natural return to ambient temperature. The former provided the stress-strain response of the material, while the latter enabled us to identify the value of the time constant τ in Eq. (2.1).

Figure 2.2 shows the stress-strain curves at a stress rate of 100 MPa/s for the two specimens. For specimen #1, a linear response is first observed (corresponding to the elasticity of the austenite). From about 200 MPa, an austenite-to-martensite (A → M) transformation occurred. For specimen #2, a martensite reorientation process occurred while loading.

Figure 2.3 shows the variation of the temperature change Θ for both tests. For specimen #1, the temperature change is first negative due to thermoelastic coupling. Then temperature strongly increases due to the latent heat production (exothermic A → M transformation). Mechanical dissipation due to plasticity probably also contributes to the heating in the last part of the test. For specimen #2, temperature changes are smaller because martensite reorientation is not associated to latent heat production. As for specimen #1, first temperature decreases due to thermoelastic coupling which is predominant. Then it increases due to a stronger mechanical dissipation associated to irreversibilities in the martensite reorientation process, and probably also to plasticity in the last part of the test.

Figure 2.4 shows the variation of the temperature change during a natural return to ambient temperature. The time constant τ was estimated to 9.5 s for specimen #1 and to 30 s for specimen #2. The difference can be explained by the difference in thickness of the specimens (see Table 2.1). Indeed, the thicker the specimen, the lower the time constant τ.

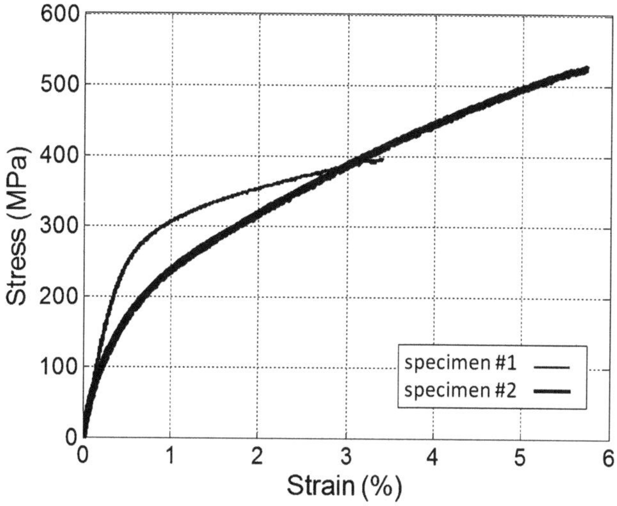

Fig. 2.2 Tensile test at 100 MPa/s up to specimen failure: stress-strain curve

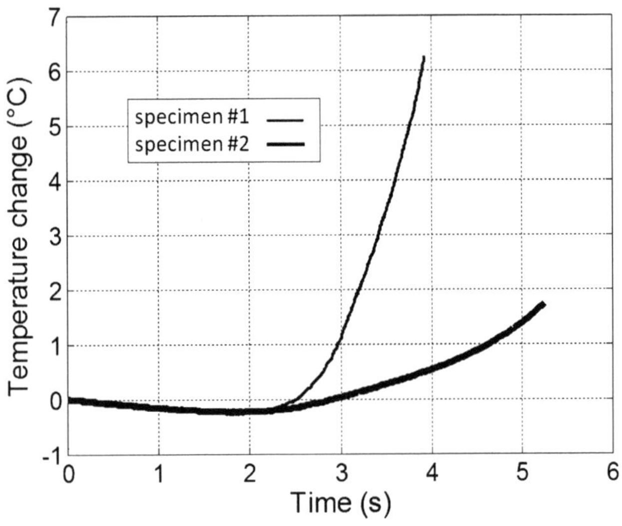

Fig. 2.3 Tensile test at 100 MPa/s up to specimen failure: temperature change *vs.* time

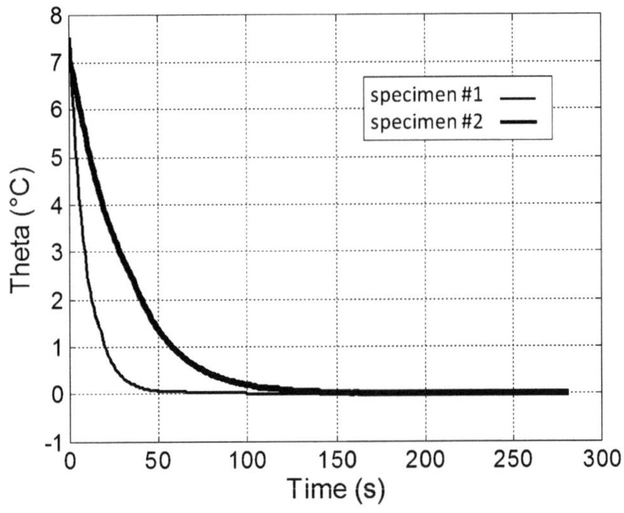

Fig. 2.4 Natural return to ambient temperature: temperature change *vs.* time

2.4 Results

Figure 2.5 shows the maximum and minimum values of the strain during the fatigue test for specimen #1. This specimen failed during the third block. Within each block, the maximum and minimum strains evolved over the mechanical cycles. These strains quickly stabilized in the first two blocks. The variation of these maximum and minimum strains is the consequence of mechanical irreversibilities, whose mechanical dissipation is the calorific signature.

Let us recall that the heat source is a priori composed of three parts: the thermoelastic coupling term, the latent heat source due to phase transformation, if any, and the mechanical dissipation (associated to an irreversibility in the phase transformation or in the martensite reorientation process, and to plasticity). Under cyclic loading, the *stabilized* temperature change due to thermoelastic coupling is equal to zero. Similarly, the *stabilized* temperature change due to a cyclic A↔M transformation, if any, is also equal to zero. The mechanical dissipation is a quantity which is always positive. As a consequence, the *stabilized* temperature change obtained during the experiments is expected to be due to the mechanical dissipation only [3].

Figure 2.6 shows the variation of the temperature change during the first two blocks for specimen #1. At the beginning of each block, temperature immediately decreases as a consequence of the thermoelastic coupling or A → M transformation (let us recall that the mechanical loading started by a stress increase). Temperature increases for about 1 min and stabilizes above the ambient temperature. The order of magnitude of the temperature change in the stabilized regime is 10 mK. It is worth noting that the averaging operations mentioned above enable us to observe a stabilized temperature change of this order of magnitude for about 9 min. Mechanical dissipation is deduced from the temperature change in the stabilized regime by dividing by the time constant τ. Table 2.2 provides the values which are obtained for specimen #1. For each block, three

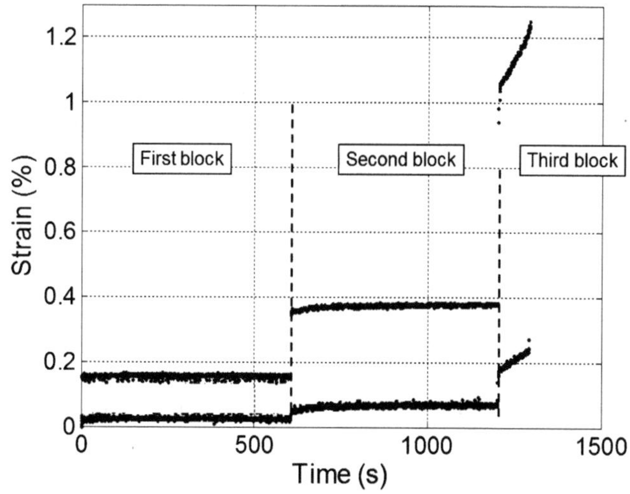

Fig. 2.5 Maximum and minimum values of the strain during the fatigue test for specimen #1

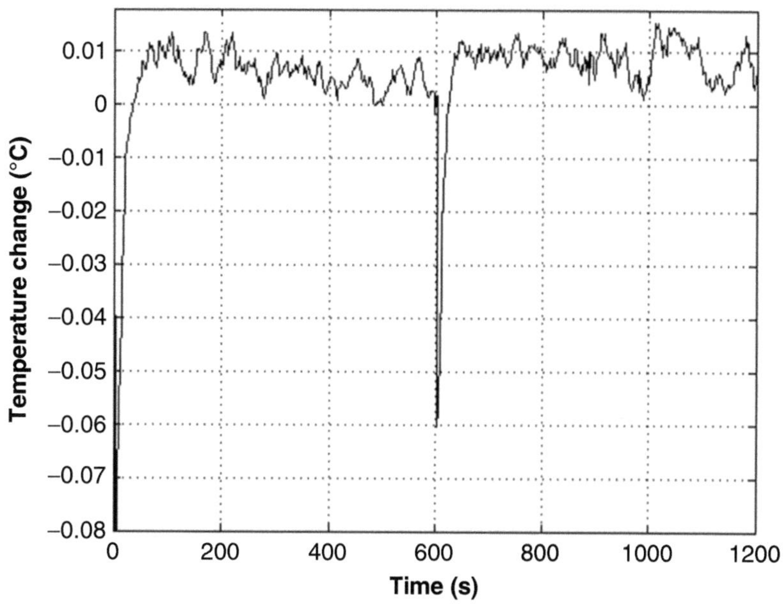

Fig. 2.6 Temperature change during the first two blocks for specimen #1

Table 2.2 Mechanical dissipation during the cyclic test

		Specimen #1
First block	s_{mean} (mean value)	0.67 mK/s
$\sigma_{min} = 10$ MPa	s_{max} (max)	0.91 mK/s
$\sigma_{max} = 100$ MPa	s_{end} (at the end of the block)	0.46 mK/s
Second block	s_{mean}	0.97 mK/s
$\sigma_{min} = 20$ MPa	s_{max}	1.02 mK/s
$\sigma_{max} = 200$ MPa	s_{end}	≈1 mK/s
Third block	s_{mean}	–
$\sigma_{min} = 30$ MPa	s_{max}	–
$\sigma_{max} = 300$ MPa	s_{end}	–

quantities are given: the maximum mechanical dissipation, the mechanical dissipation at the end of the block and a mean value over the 9 min time period. It can be noted that the mechanical dissipation slightly decreases over the cycles. The comparison with the results obtained for the second specimen shows that the chemical composition, as well as the nature of the phase which is involved (martensite or austenite), influence the mechanical dissipation which is produced by the specimen during the cyclic mechanical loading.

2.5 Conclusion

SMAs are characterized by a very high reversibility of the solid-to-solid phase transformation. This study proposes an experimental procedure based on infrared thermography to assess the mechanical dissipation in SMA specimens subject to mechanical cyclic loading. Tests were performed on Cu–Zn–Al specimens in both the austenitic and the martensitic states, each of them featuring a specific thermal response. In particular, different values of the mechanical dissipation were measured, thus revealing different levels of mechanical irreversibility.

References

1. Chrysochoos, A.: Analyse du comportement des matériaux par thermographie infrarouge. In: Berthaud, Y. (ed.) Photomécanique 95, pp. 203–211. Eyrolles, Paris (1995)
2. Samaca Martinez, J.R., Le Cam, J.B., Balandraud, X., Toussaint, E., Caillard, J.: Mechanisms of deformation in crystallizable natural rubber. Part 2: quantitative calorimetric analysis. Polymer **54**, 2727–2736 (2013)
3. Bubulinca, C., Balandraud, X., Grédiac, M., Stanciu, S., Abrudeanu, M.: Characterization of the mechanical dissipation in shape-memory alloys during stress-induced phase transformation. J. Mater. Sci. **49**, 701–709 (2014)

Chapter 3
The Effect of Microstructure on Energy Dissipation in 316L Stainless Steel

P.J. Seelan, J.M. Dulieu-Barton, and F. Pierron

Abstract The link between dissipated energy in a cyclically loaded AISI 316L stainless steel and its microstructural changes is investigated. After a brief introduction to energy dissipation in materials, the experiments devised to investigate if different microstructures, resulting from different heat treatments, has a measurable effect on the dissipated energy are detailed. The data processing procedure and some preliminary results are described. Future work aimed at characterising the dissipative heat source in welds is presented.

Keywords Dissipative heat source • Internal friction • Infrared thermography • Welded structures

3.1 Introduction

Welding is an important joining technique which induces residual stresses in a component. The presence of residual stress in components can be detrimental to their service life as residual stress can interact with various degradation mechanisms like stress corrosion cracking, fatigue and fracture. At present residual stresses are mainly assessed using two categories of measurement techniques: destructive and non-destructive. The destructive (or semi-destructive) techniques such as the hole-drilling method gives accurate residual stress measurements but are not suitable to be used in residual stress measurement of safety critical components. Non-destructive techniques such as X-ray and neutron diffraction have limitations such as being costly and not portable. Therefore, there is scope for a full-field, non-destructive and portable technique like thermography to be applied in residual stress assessment. Researchers have attempted to assess residual stress using Thermoelastic Stress Analysis (TSA) [1, 2] but none of the proposed approaches have been successfully transformed into an industrial technique. Energy dissipation which results in the temperature rise in materials (also known as self-heating) is another effect that accompanies mechanical deformation of material. This phenomena has not yet been applied in residual stress assessment, though many studies have focused on using this as a marker of fatigue damage [3, 4]. It is well known that weld joints consist of at least three different regions with different microstructures: the base metal, the fusion zone, and the heat affected zone. In the context of assessing residual stress, one has to first estimate the intrinsic dissipation of these zones without residual stresses. The work in the present paper examines the effect of different microstructures, such as those developed during welding processes on the energy dissipation in AISI 316L stainless steel (EN1.4404), a material commonly used to produce nuclear power plant components. It is expected that these results will contribute to the understanding of the dissipative heat source around the weld for the purpose of residual stress measurement.

3.2 Energy Dissipation

Energy dissipation in metallic materials subjected to mechanical deformation is a result of the movement of dislocations, among other less important mechanisms. It has been shown that even cyclic loads below the macroscopic yield stress can lead to significant energy dissipation [5]. Dissipation manifests itself as a temperature rise in the material being loaded. However, this temperature rise is not intrinsic to the material behaviour as it's affected by thermal exchanges between the specimen and its environment. Therefore, the heat source has to be calculated as this relates directly to the material behaviour. It can be calculated from the temperature by using the heat diffusion equation:

P.J. Seelan (✉) • J.M. Dulieu-Barton • F. Pierron
Faculty of Engineering and the Environment, University of Southampton Highfield, Southampton SO17 1BJ, UK
e-mail: P.Jayaseelan@soton.ac.uk

S. Quinn, X. Balandraud (eds.), *Residual Stress, Thermomechanics & Infrared Imaging, Hybrid Techniques and Inverse Problems, Volume 9*, Conference Proceedings of the Society for Experimental Mechanics Series, DOI 10.1007/978-3-319-42255-8_3

$$\rho c\frac{\partial\theta}{\partial t} - k\Delta\theta + \rho c\frac{\theta}{\tau} = H_D + H_{th} \tag{3.1}$$

where ρ is the density, c is the specific heat capacity, θ is the temperature variation from the first temperature map, t is time, k is the thermal conductivity, τ is the time constant for linearised convective and radiative exchanges [6], H_D is the dissipative heat source and H_{th} is the thermoelastic heat source.

3.3 Experimental Work

A range of heat treatments were carried out on AISI 316L to produce different microstructures that replicate microstructures found in AISI 316L welds. Strip specimens of dimension 300 mm × 30 mm × 2 mm were uniformly heated at 1150 °C (annealing temperature of AISI 316L) for 2 min and then cooled by the following methods: cooled at room temperature, water quenched and furnace cooled at 100 °C h^{-1}. Another set of identical specimens were tested in the as received condition and used as control specimens. Micrographs of the resulting microstructure in each case and their respective average grain sizes are shown in Fig. 3.1.

The specimens were subjected to uniaxial cyclic load of various stress amplitudes below and just about the yield stress, at a stress ratio, R_σ of 0.1 using an Instron 8800 servo-hydraulic testing machine. The resulting temperature variations were recorded using a Cedip Silver 480M infrared detector. As the temperature variations expected is lower than the Noise Equivalent Temperature Difference (NETD) of the infrared detector, data processing methods developed by Maquin and Pierron [5] have been used as detailed in the section below. Due to limited detector sensitivity it is also important to reduce background radiation which is one of the main sources of noise. The necessary thermal insulation was provided in the form of a chamber (polystyrene covered with Aluminium foil) placed surrounding the specimen (see Fig. 3.2a). The entire path connecting the camera to the opening of the chamber was also protected using the same chamber material. Besides that, a reference specimen [5] made of the same material as the specimen was placed next to the specimen (see Fig. 3.2b). This ensures that the fluctuations in the environmental temperature can be taken into account.

3.4 Data Processing, Detection Threshold, and Preliminary Result

The data processing methods employed to improve the detection threshold include image subtraction, spatial averaging, filtering and smoothing. The first image (before the start of the cyclic loading) was subtracted from the subsequent images (see Fig. 3.3 step 1). The effect of thermal reflections were supressed through this operation. Spatial averaging (see Fig. 3.3 step 2) refers to the calculation of the mean value of all of the pixels contained in the region of interest (specimen and reference specimen) in a frame. The region of interest was averaged horizontally and then vertically. The spatial mean of the reference specimen temperature variation was subtracted from the spatial mean of the specimen temperature variation (see Fig. 3.3 step 3). The resulting overall temperature variation of the specimen consists of monotonic rise superimposed to a cyclic thermoelastic temperature variation. Extraction of the temperature rise from the overall temperature variation was done by implementing a notch filter. This was possible due to the fact that the thermoelastic temperature variation was known to be harmonic and occurs at the same frequency as the loading. The extracted temperature rise was not free from noise and therefore smoothing was performed. Smoothing replaces a data point with the average of the neighbouring data points. A suitable temporal smoothing window size was chosen based on the camera frame rate and loading frequency so as to maintain accuracy of the result (see Fig. 3.3 step 4). The smoothed temperature variation was then used to calculate the dissipative heat source.

The method employed in this work enables the determination of the dissipative heat source in the very first few cycles of fatigue loading. However, the heat source levels involved in this study is very low. As such, the detection threshold of the method or noise floor needs to be determined so that the results can be presented with known confidence level. In this study, the detection threshold or noise floor is the threshold above which the results have a confidence level of 95 %. It was determined by recording the data without applying the cyclic load. The dissipative heat source is not activated in this case and any temperature variation is likely to be noise caused by the detector or thermal exchanges with the surrounding.

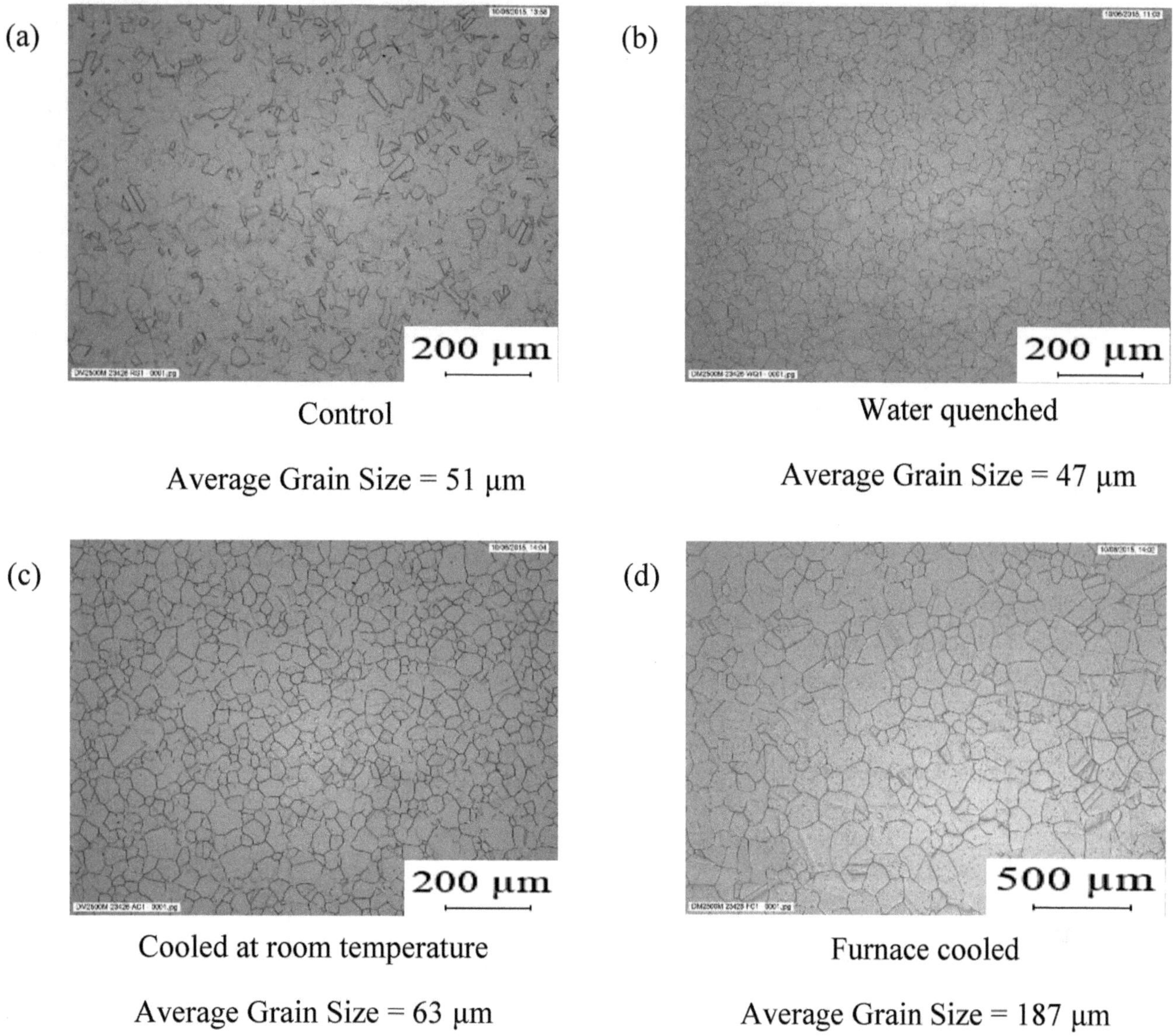

Fig. 3.1 Micrograph of the specimens: (**a**) control (**b**) water quenched (**c**) cooled at room temperature (**d**) furnace cooled [7]

This process was repeated three times and the resulting data was subsequently processed with the same method as detailed above. The mean and the standard deviations of the resulting heat source was calculated using these three sets of data collected. The two standard deviation interval gives the detection threshold above which measurements are significant at a confidence level of 95 %. Since the standard deviation was shown to be constant in time, it was averaged to provide a single value. The heat source detection threshold was found to be 0.566×10^4 Wm^{-3}. When divided by ρc (density and specific heat capacity), a detection threshold is obtained in terms of heating rate which corresponds here to 1.4 mK s^{-1}. This implies that, in a period of one second, the combination of the thermographic measurement and data processing method has a noise equivalent to 1.4 mK. The red lines in Fig. 3.4 show the two standard deviation levels. As the noise heat source computed is random, positive and negative thresholds are shown. The heat source of the as received AISI 316L cyclically loaded at a stress ratio, R_σ of 0.1 with a maximum stress, σ_{max} of yield stress (250 MPa) is also shown in Fig. 3.4. The results are similar to the ones obtained by Connesson et al. in [8] where the initial peak was attributed to microplasticity due to irreversible movement of dislocations and the stabilised source was attributed to viscoelasticity due to reversible movement of dislocations.

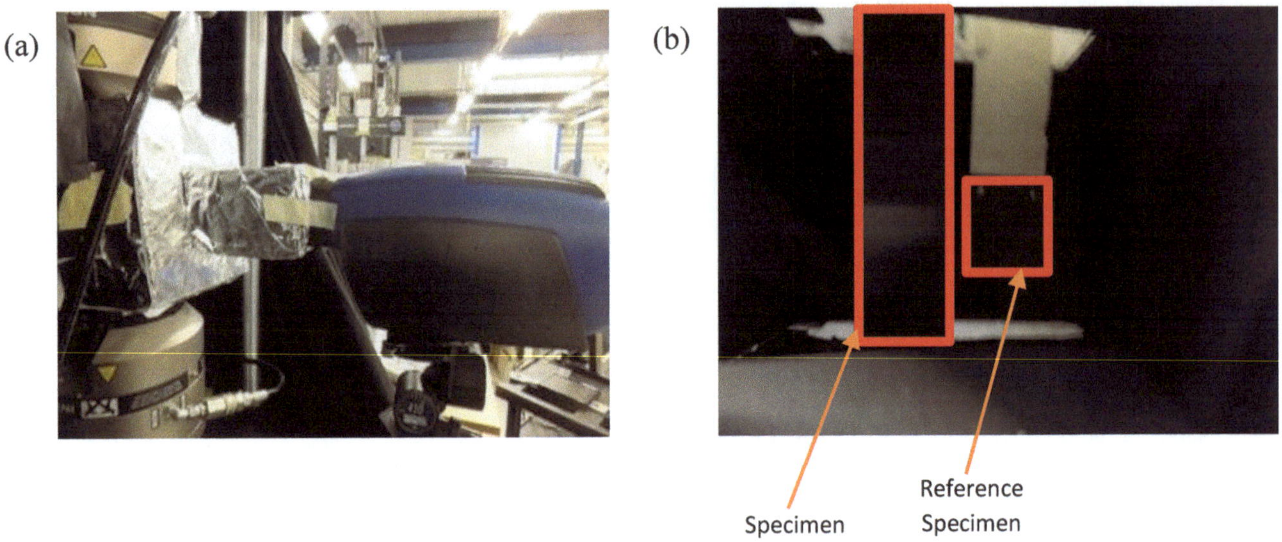

Fig. 3.2 (**a**) Setup of the actual experiment (**b**) specimen and reference specimen

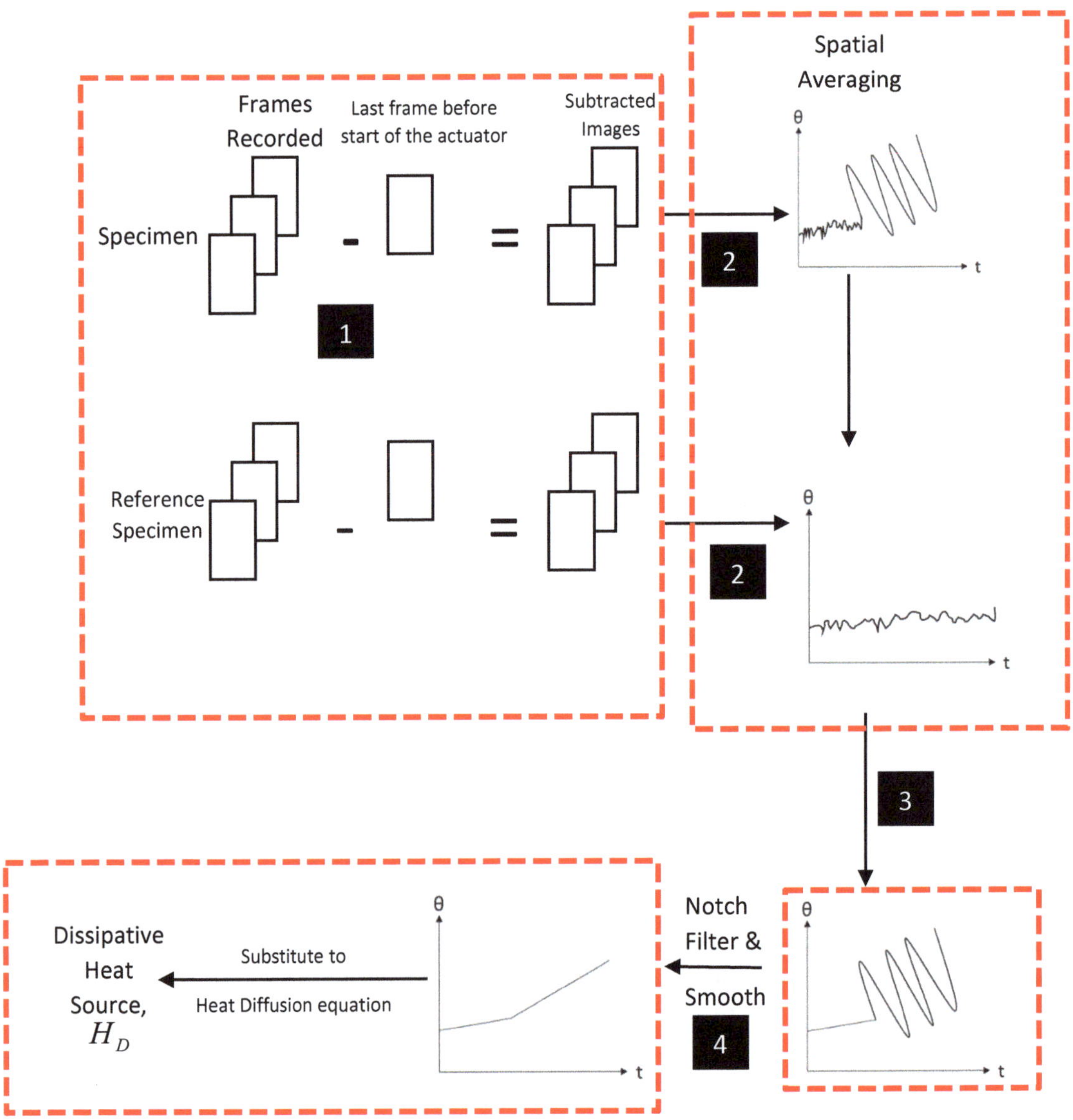

Fig. 3.3 Summary of the data processing procedure

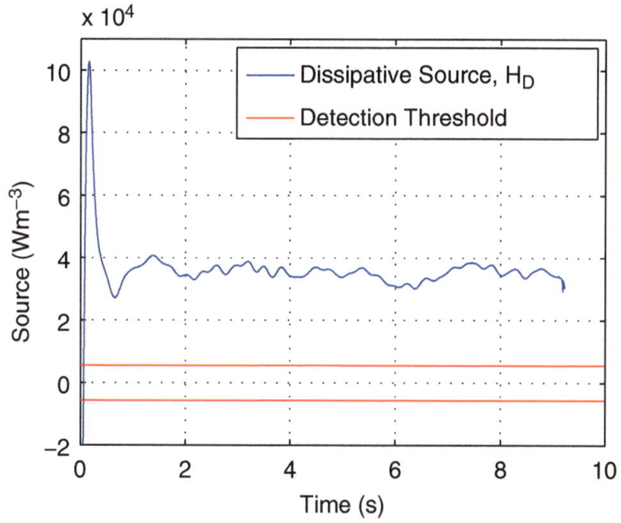

Fig. 3.4 The dissipative heat source and the detection threshold

3.5 Conclusion and Future Work

The current setup is adequate to measure the spatially averaged heat source generated by a macroscopically homogenous stress field. Future work involving the characterisation of the dissipative heat source of a weld will require spatial averaging on smaller areas within the weld, thus increasing the detection threshold. This was tested in [9] and proved to be feasible if heat sources are high enough. This may need loading at higher stress levels.

Acknowledgements The work was supported by a Ph.D. studentship funded by Innovate UK. The authors would also like to thank Elise Chevallier for carrying out the heat treatment procedures and producing the micrographs. Professor Pierron gratefully acknowledges support from the Wolfson Foundation and the Royal Society through a Royal Society Wolfson Research Merit Award.

References

1. Wong, A., Dunn, S., Sparrow, J.: Residual stress measurement by means of the thermoelastic effect. Nature **332**, 613–615 (1988)
2. Robinson, A.F., Dulieu-Barton, J.M., Quinn, S., Burguete, R.L.: The potential for assessing residual stress using thermoelastic stress analysis: a study of cold expanded holes. Exp. Mech. **53**, 299–317 (2012)
3. Luong, M.P.: Fatigue limit evaluation of metals using an infrared thermographic technique. Mech. Mater. **28**, 155–163 (1998)
4. Boulanger, T., Chrysochoos, A., Mabru, C., Galtier, A.: Calorimetric analysis of dissipative and thermoelastic effects associated with the fatigue behavior of steels. Int. J. Fatigue **26**, 221–229 (2004)
5. Maquin, F., Pierron, F.: Heat dissipation measurements in low stress cyclic loading of metallic materials: From internal friction to micro-plasticity. Mech. Mater. **41**, 928–942 (2009)
6. Louche, H., Chrysochoos, A.: Thermal and dissipative effects accompanying Lüders band propagation. Mater. Sci. Eng. A **307**, 15–22 (2001)
7. Chevallier, E.: Private Communication, 2016
8. Connesson, N., Maquin, F., Pierron, F.: Dissipated energy measurements as a marker of microstructural evolution: 316L and DP600. Acta Mater. **59**, 4100–4115 (2011)
9. Connesson, N.: Contribution à l'étude du comportement thermomécanique de matériaux métalliques. Ph.D. Thesis, Troyes, France (2010)

Chapter 4
Large Area Nondestructive Evaluation of a Fatigue Loaded Composite Structure

Joseph N. Zalameda, Eric R. Burke, Michael R. Horne, and Eric I. Madaras

Abstract Large area nondestructive evaluation (NDE) inspections are required for fatigue testing of composite structures to track damage initiation and growth. Of particular interest is the progression of damage leading to ultimate failure to validate damage progression models. In this work, passive thermography and acoustic emission NDE were used to track damage growth up to failure of a composite three-stringer panel. Fourteen acoustic emission sensors were placed on the composite panel. The signals from the array were acquired simultaneously and allowed for acoustic emission location. In addition, real time thermal data of the composite structure were acquired during loading. Details are presented on the mapping of the acoustic emission locations directly onto the thermal imagery to confirm areas of damage growth leading to ultimate failure. This required synchronizing the acoustic emission and thermal data with the applied loading. In addition, processing of the thermal imagery which included contrast enhancement, removal of optical barrel distortion and correction of angular rotation before mapping the acoustic event locations are discussed.

Keywords Large area composite inspection • Thermography • Acoustic emission • Fiber break • Delamination

4.1 Introduction

The purpose of this paper is to describe the methodology of measuring damage onset and growth in a composite structure during fatigue loading. Thermal and acoustic emission nondestructive evaluation (NDE) techniques were used to track the damage in real time. When damage growth was detected, the loading was stopped for a more detailed inspection using non-immersion ultrasound. The non-immersion ultrasound was used to provide a detailed assessment of the damage growth through-the-thickness and thus ultimately provided a better understanding of damage progression modes that can lead to ultimate failure [1]. NDE inspections are necessary to provide structural engineers a tool to incrementally control and document damage growth as a function of fatigue cycles before failure. This allows for the comparison of NDE results to develop and validate progressive damage analysis (PDA) models [2–5]. The ultimate goal is to use the validated PDA models to decrease the time required to certify composite structures and therefore save development costs.

Real time NDE can document the progression of damage in between the ultrasonic measurements and also provide the documentation of ultimate failure mechanisms. The technologies used for real time NDE were thermography and acoustic emission. In the past there have been numerous studies combining thermography with acoustic emission [6–9]. Our technique is different where the acoustic emission data were directly mapped onto the thermal imagery. In addition, image rotation and removal of optical distortions on the thermal images were required to accurately map the acoustic emission data onto the thermal images. Mapping the acoustic emission events onto the thermal images provided confirmation of damage growth at areas of heating (due to fiber breaks, rubbing of disbond areas and matrix cracks) and where ultimate failure will occur. This provided a measurement capability to detect growing damage (location and size) for improved structures testing. Multiple infrared (IR) cameras and multiple acoustic emission sensors were employed for full coverage of the structure during testing.

J.N. Zalameda (✉) • E.R. Burke • E.I. Madaras
NASA Langley Research Center, MS231, Hampton, VA 23681, USA
e-mail: joseph.n.zalameda@nasa.gov

M.R. Horne
National Institute of Aerospace, NASA Langley Research Center, MS231, Hampton, VA 23681, USA

S. Quinn, X. Balandraud (eds.), *Residual Stress, Thermomechanics & Infrared Imaging, Hybrid Techniques and Inverse Problems, Volume 9*, Conference Proceedings of the Society for Experimental Mechanics Series, DOI 10.1007/978-3-319-42255-8_4

4.2 Sample

The hat stiffened composite panel flat side and stringer side with acoustic emission locations (stringer side) are shown in Fig. 4.1a and b respectively. The panel skin is 17 plies with a thickness of 0.32 cm. The hat flange is 11 plies with a thickness of 0.21 cm. Barely visible impact damage (BVID) were induced at four locations (I1, M1, I2, M2) on the panel skin on the hat stiffeners flange or right next to the flange. The BVID was used to simulate existing delamination damage and to serve as potential damage growth sites. Cyclic compressive loads were applied from −20,000 lb up to −150,000 lb at 2 Hz using the setup shown in Fig. 4.2. The sample was painted with a speckle pattern for high speed digital image correlation imaging. The high speed digital image correlation cameras required illumination using red light emitting diode (LED) lights. The LED lights were not a problem for the IR cameras operated in the mid IR waveband. The load was applied from the bottom while the top was held stationary. Passive thermography was used to track the damage on both the flat and stringer sides during loading using two IR cameras. Shown in Fig. 4.1b are the locations of the acoustic emission sensors in relation to the BVID. The acoustic emission sensors were used for real time detection of acoustic events indicating damage growth.

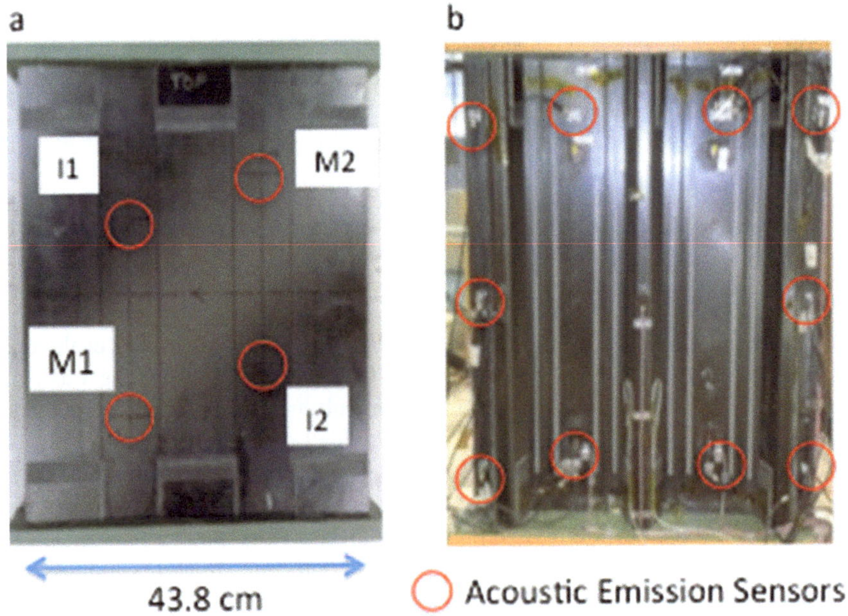

Fig. 4.1 Three stringer composite panel tested with impact and acoustic emission sensor locations. (**a**) Flat side, (**b**) Stringer side

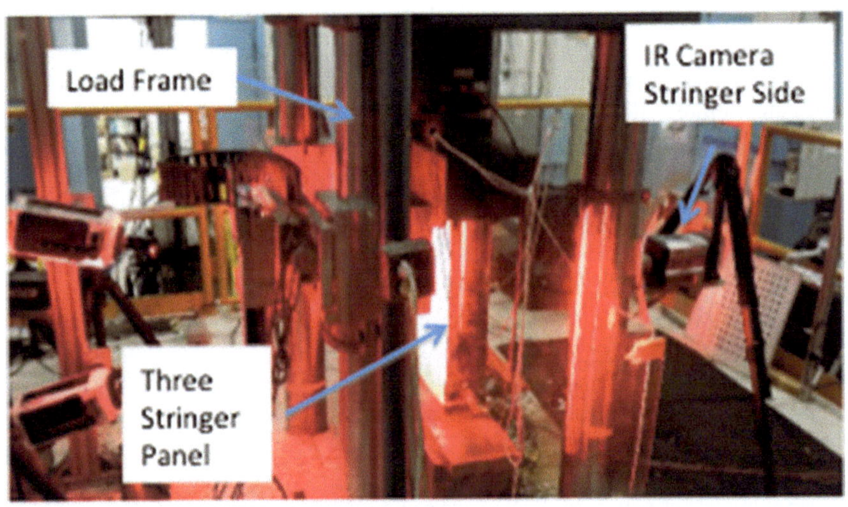

Fig. 4.2 Fatigue loading setup for three-stringer panel

4.3 Inspection Systems

The thermal measurement setups are shown in Fig. 4.3. The basic system consists of two IR cameras operating in the 3–5 μm (midwave) IR band and an image data acquisition computer. The IR cameras were both configured with 25 mm germanium optics. The focal plane array sizes for the cameras were 640 × 512 and 1024 × 1204. The passive inspection captured the thermal variations during the fatigue loading. The setup required a Plexiglas® shield to contain fragments if ultimate failure occurred. The IR cameras were located behind the shield with the lens positioned in front of a viewing hole. The added benefit of the shield is it filters out spurious IR background sources. The cameras were synchronized using the load signal from the test machine. The sinusoidal load signal was offset corrected and the zero crossover points were used to produce a TTL trigger signal. The trigger signal was converted to a series of pulses using a signal generator. The IR cameras were operated at a frequency range of 16–90 Hz depending on acquisition time duration. Higher frequencies were used to capture the damage progression as the panel approached ultimate failure. The thermography data were processed using an image differencing technique [5]. Ultrasonic inspections were implemented at intervals determined by the indication of damage growth from thermography and acoustic emission. The ultrasonic inspection technique utilizes a broadband 10 MHz transducer (1.27 cm diameter with 5.08 cm focus) contained in a captive water column with a flexible membrane tip design [1]. Water is misted onto the surface for ultrasonic coupling. The probe is scanned over the entire panel on the flat side. A picture of the ultrasonic probe is shown in Fig. 4.4.

Fig. 4.3 Three-stringer panel setup for passive thermography inspection of both flat and stringer sides

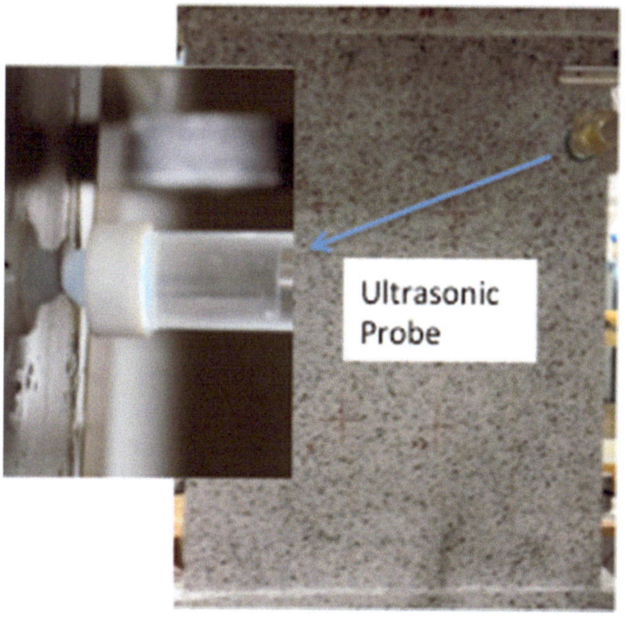

Fig. 4.4 Close up of ultrasonic probe

Fourteen acoustic emission sensors were bonded to the composite panel on the stringer side (shown in Fig. 4.1b). They were connected to a data acquisition computer via Digital Wave™ PA0 preamp/line drivers and a Digital Wave™ FM1 signal conditioning 16 channel amplifier for capture of acoustic data. The acoustic emission computer also captured the load signal for synchronization. Acoustic emission systems collect structure-borne sound, typically not audible, in the ultrasonic frequency band (approx. 50–500 kHz), generated by dynamic displacements such as damage initiation and growth at picometer scales [10]. Pre-cursors to failure are monitored in real-time. The total signal energy, *SE*, of an acoustic signal is given as:

$$SE = \sum_{i=1}^{n} V_i^2 \, \Delta t \qquad (4.1)$$

where V_i is the signal voltage, i is the time reference point, n is the number of time points in the signal, Δt is the sampling time per point. Trends of the signal energy over time, or load, are used to identify and track damage development (matrix cracks, fiber breaks, delamination formation). Signals from the distributed array of sensors are acquired simultaneously when any one sensor threshold triggers. This allows calculating the source or event location [10, 11].

4.4 Measurement Results

Examples of processed thermal images of the flat side are shown in Fig. 4.5. The processing technique used to improve defect contrast involves a delayed image subtraction [5, 12]. A moving buffer is required in the acquisition software in order for this processing technique to be implemented in real time. The delayed subtraction is given as:

$$\mathit{Difference\ Image} = \mathit{Acquired\ Image}[i] - \mathit{Previous\ Image}[i - k/2] \qquad (4.2)$$

where $k =$ number of images per cycle and i is the current acquired image number. Dividing k by 2 insures the maximum and minimum temperature images are subtracted within a cycle. This provides a series of images with optimal defect contrast varying with damage depth. Improved defect signal to noise can be obtained by averaging the difference images

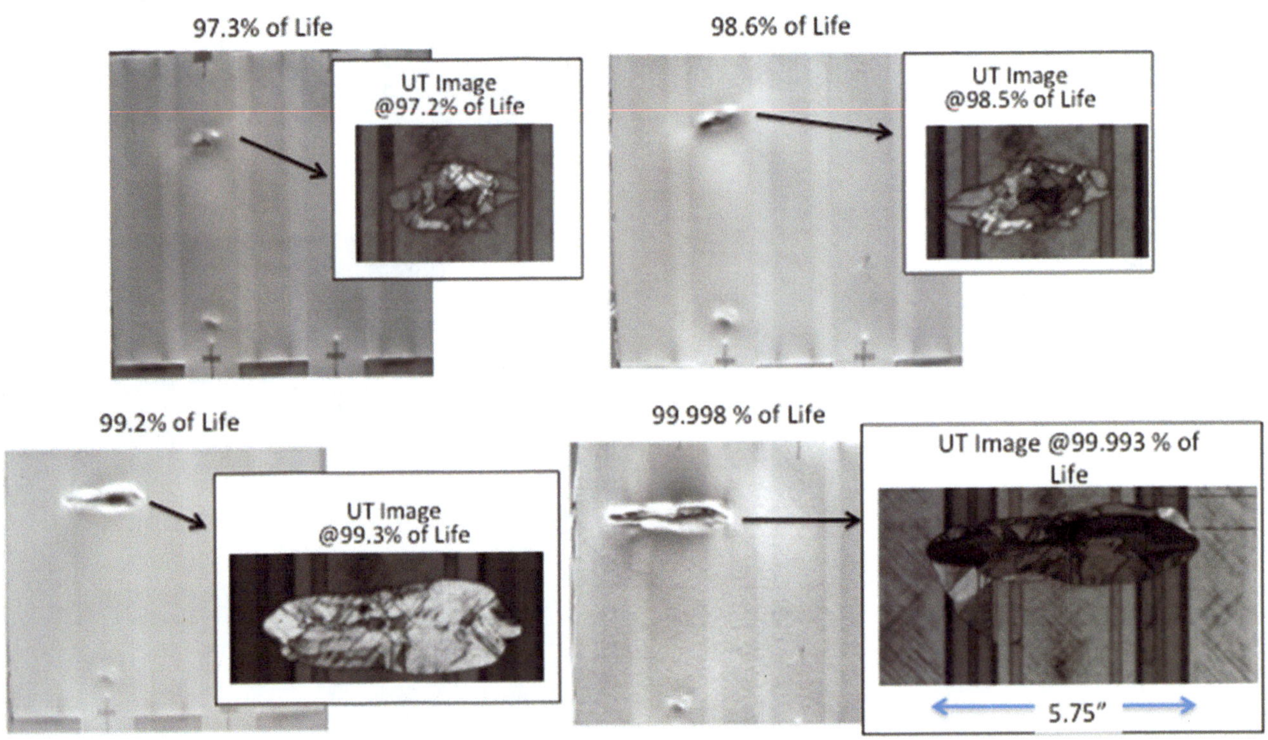

Fig. 4.5 Comparison of processed thermography images to ultrasonic inspection results

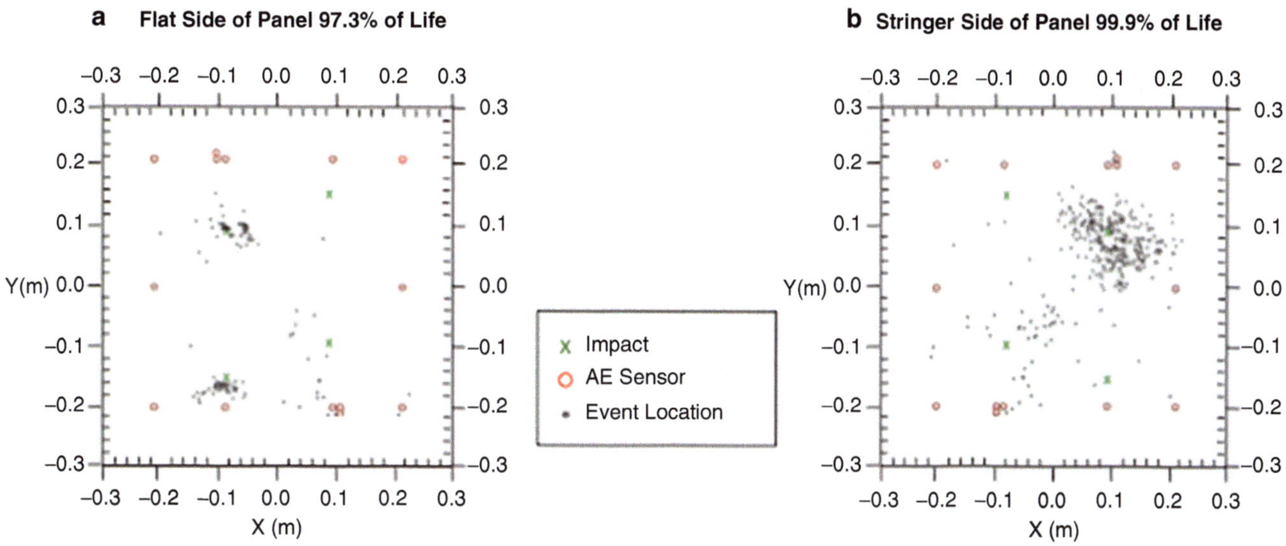

Fig. 4.6 Three-stringer panel acoustic emission event locations

over a number of cycles if desired. Also shown in Fig. 4.5 are the respective ultrasonic images obtained at intervals determined by the indication of damage growth from thermography and acoustic emission. The acoustic emission events are shown in Fig. 4.6a and b where the acoustic emission events are shown for different fatigue tests. The accumulated acoustic events for Fig. 4.6a were for approximately 50 s and for Fig. 4.6b were for 16.5 s. As the structure nears failure the accumulated acoustic events increases as a function of time as shown in Fig. 4.6a and b.

4.5 Mapping Acoustic Emission Events onto Thermal Imagery

An image processing routine was developed to map the acoustic emission data onto the thermal imagery. The procedure is shown in Fig. 4.7 along with some example thermal images. This method required improving image contrast, removing angular rotation, and correcting for optical barrel distortion. In addition, the resolution per pixel was determined by measuring known distances on the composite structure. The acoustic emission data were then mapped onto thermal imagery, revealing the cluster of acoustic emission event locations around the thermal signatures of interest. The removal of angular rotation is performed using an affine transformation [13]. The barrel distortion is corrected using an image forward transformation given by the equation below [14].

$$r_{undistorted} = r_{distorted}\left(1 - d\,(r_{distorted})^{2}\right) \tag{4.3}$$

Where $r_{undistorted}$ is the corrected Cartesian coordinate vector distance from the center of the image and $r_{distorted}$ is the distorted

Cartesian vector distance from the center of the image. The parameter d was set to 0.2 to remove the distortion. The pixel resolution was calculated to be 8.2 pixels per cm. A comparison of the contrast enhanced, rotation corrected and barrel distortion thermal images are shown in Fig. 4.7b and c. The image in Fig. 4.7c is used to map the acoustic emission data.

The mapped acoustic emission data as a function of remaining life are shown in Figs. 4.8 and 4.9 for the flat side and stringer side respectively. The impact locations (X) are also shown. Both the thermal data and the acoustic emission data were synchronized to the load and therefore the timing of the acoustic emission events are matched to the thermal image temporally. For a given location, the mapped indicator (star graphic) size and shading are based on the event energy given by:

$$Indicator = Round\left[\,(event\ energy)^{1/4} + 10\right] \tag{4.4}$$

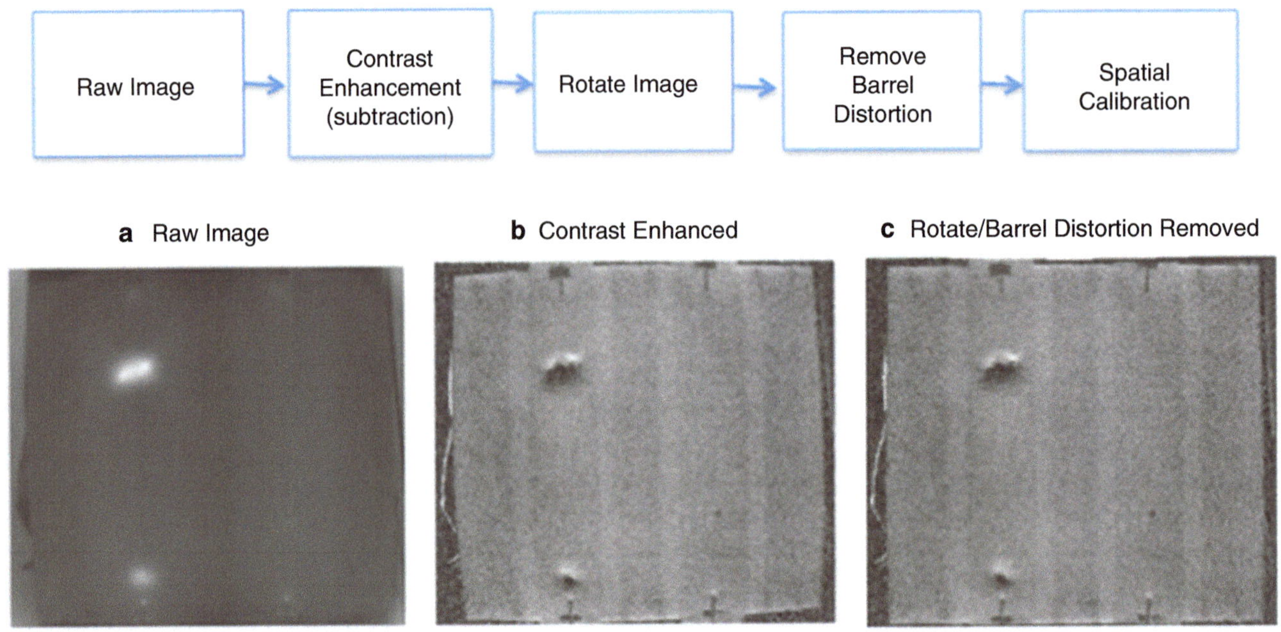

Fig. 4.7 Procedure to process thermal imagery for mapping of acoustic emission data

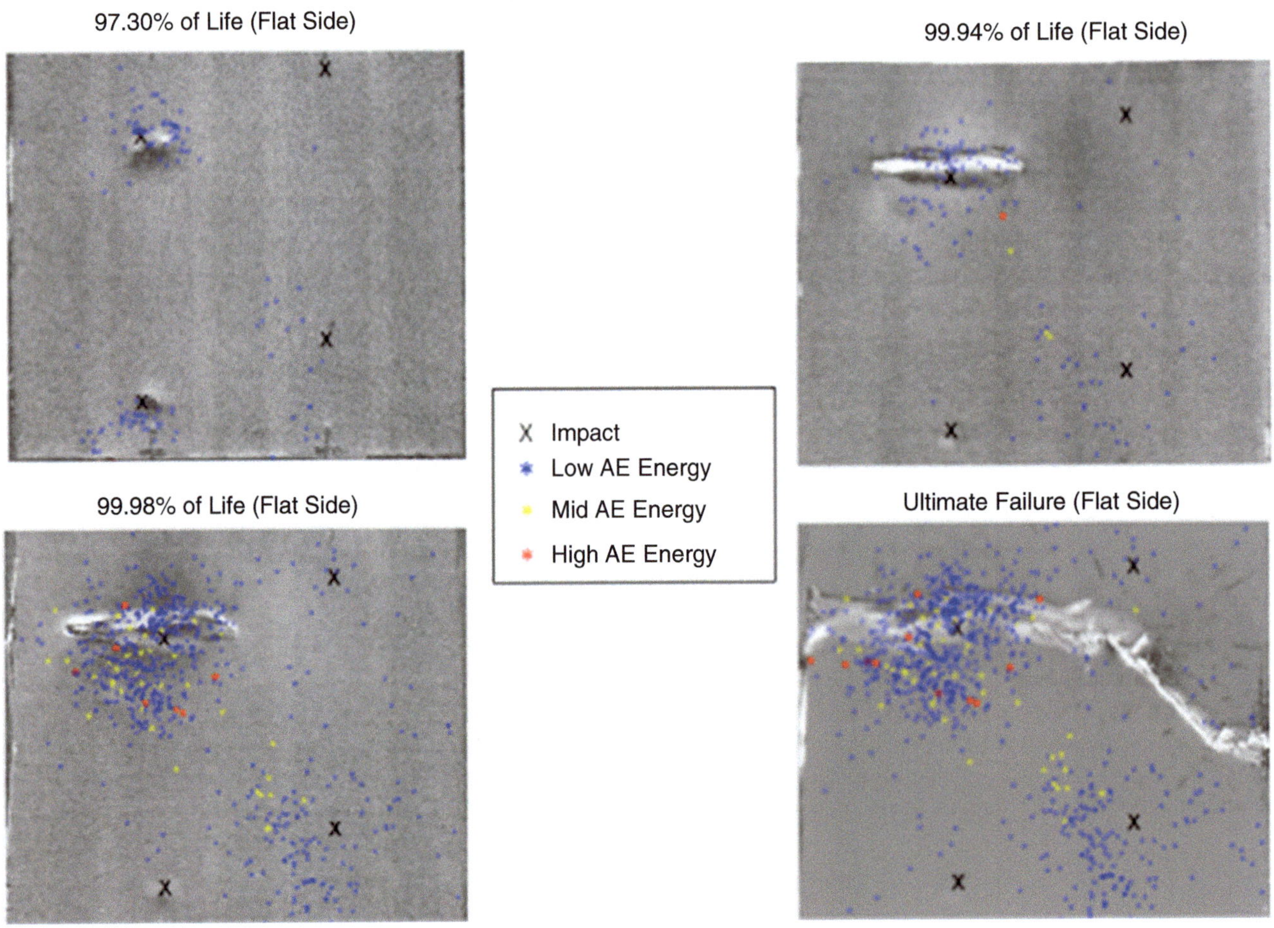

Fig. 4.8 Mapping of acoustic emission data onto processed thermal imagery (Flat Side)

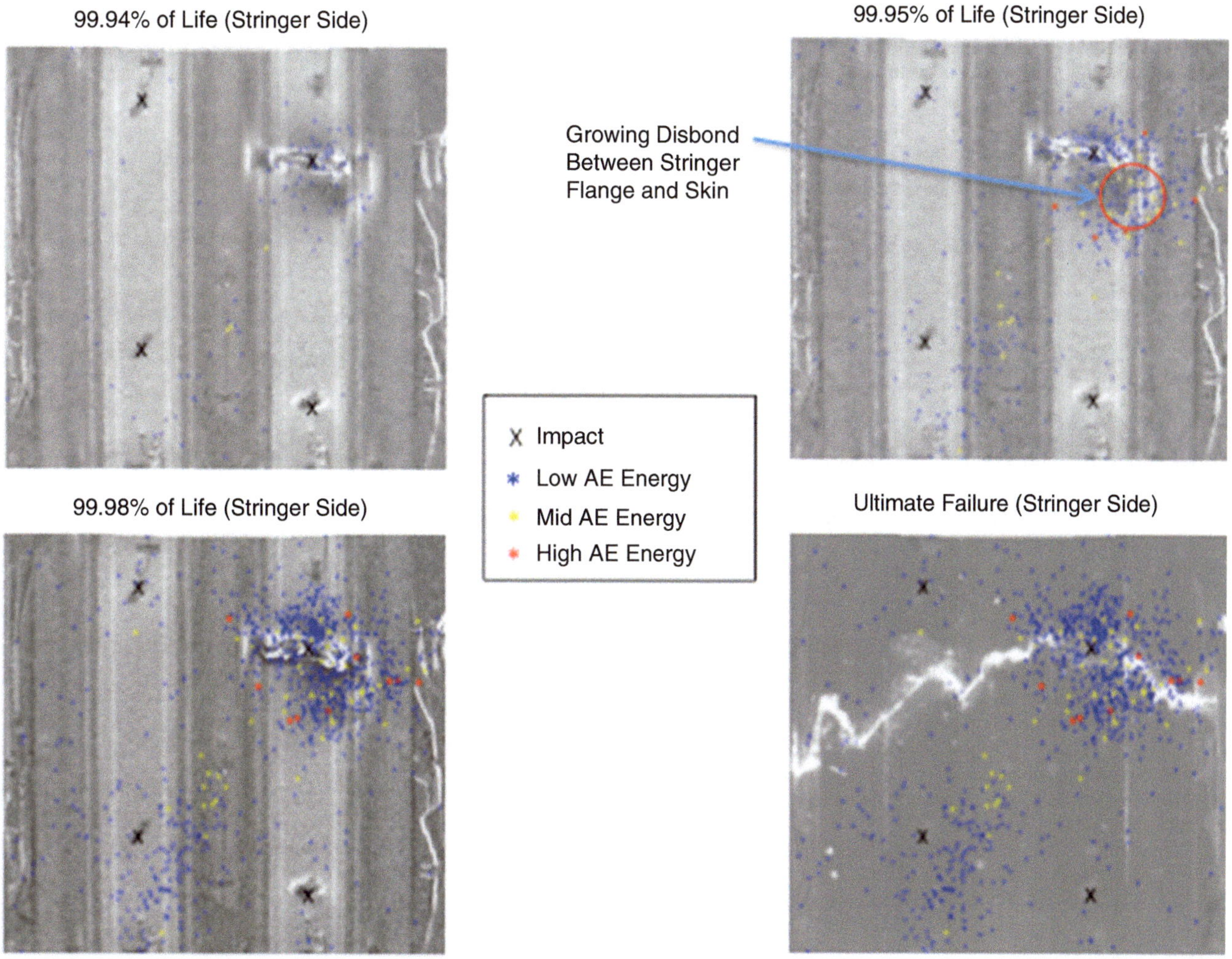

Fig. 4.9 Mapping of acoustic emission data onto processed thermal imagery (Stringer Side)

where the event indicator values of less than 13 (with a minimum size of 10) were given a low energy (opacity of 0.5), indicator values between 13 and 15 were designated as a mid energy event (opacity of 0.7), and greater than 15 was a high energy event with (opacity of 1.0). As the structure nears failure, the accumulated acoustic events and indicator values increase as shown in Figs. 4.8 and 4.9. Ultimate failure occurred at the location where the red high energy indicators and mostly yellow indicators are clustered. It is important to note the location of the acoustic emission events are prone to errors due to the progression of damage (fiber breaks, delaminations, matrix cracking) as the structure nears failure. These defects can affect the propagation of the acoustic emission signal in terms velocity and waveform mode conversions resulting in processing errors; however, overall the clustered locations are comparable to the thermal indications of the growing damage areas. In particular for the left image (thermal image from the stringer side) in Fig. 4.9, some of the clustered points tend to accumulate below the delamination on the right side. This clustering was due to a growing disbond between the stringer flange and the skin below the delamination. This disbond between the stringer flange and skin was determined, from the thermal data, to have caused ultimate failure.

4.6 Conclusions

By combining thermography with acoustic emission a large area, real time NDE inspection technique was developed. This technique allows for the determination of damage growth areas and failure location of a loaded composite structure to be accurately determined. A particular area of heating, in addition to confirmation of clustered acoustic emission events, were a good indicator of damage growth and where ultimate failure occurred. An image processing procedure has been presented to allow for mapping of the acoustic emission events onto the thermal images. This technique has shown potential for real time inspection of large area aerospace structures during load testing.

Acknowledgments The authors would like to acknowledge Mr. Wade Jackson of NASA Langley for ultrasonic inspections, sample preparation, and mounting. Also the authors would like to acknowledge Mr. William Johnston of Lockheed Martin for testing support and Mr. Kenneth McLarney, NASA Internship, Fellowship, and Scholarships (NIFS) student for data processing.

References

1. Johnston, P.H., Wright, C.W., Zalameda, J.N., Seebo, J.P.: Ultrasonic monitoring of ply crack and delamination formation in composite tube under torsion load. In: Ultrasonic Symposium (IUS), IEEE, pp. 595–598, (2010)
2. O'Brien, T.K.: Development of a delamination fatigue methodology for composite rotorcraft structure. In: NASA Aviation Safety Technical Conference Denver, CO, October 2008
3. Krueger, R.: An approach to assess delamination propagation simulation capabilities in commercial finite element codes. NASA/TM-2008-215123 (2008)
4. Bisagni, C., Dávila, C.G., Rose, C., Zalameda, J.N.: Experimental evaluation of damage progression in postbuckled single stiffener composite specimens. In: American Society for Composites 29th Technical Conference Proceedings, US-Japan 16, ASTM D30, February 2014
5. Zalameda, J.N., Burke, E.R., Parker, R.F., Seebo, J.P., Wright, C.W., Bly, J.B.: Thermography inspection for early detection of composite damage in structures during fatigue loading. In: Burleigh, D., Stockton, G.R. (eds) Thermosense XXXIV, Proceedings of SPIE, vol. 835403, April 2012
6. La Rosa, G., Clienti, T., Savio, F. Lo.: Fatigue analysis by acoustic emission and thermographic techniques. In: Procedia Engineering, XVII International Colloquium on Mechanical Fatigue of Metals, (ICMFM17), vol. 74, pp. 261–268 (2014)
7. Kordatos, E., Dassios, K., Aggelis, D., Matikas, T.: Rapid evaluation of the fatigue limit in composites using infrared lock-in thermography and acoustic emission. Mech. Res. Commun. **54**, 14–20 (2013)
8. Munoz, V., Vales, B., Perrin, M., Pastor, M.L., Welemane, H., Cantarel, A., Karama, M.: Damage detection in CFRP by coupling acoustic emission and infrared thermography. Compos. B Eng. **85**, 68–75 (2015)
9. Ringermacher, H.I., et al.: System and method for locating failure events in samples under load. United States Patent No. 7,516,663 B2, 14 April 2009
10. Horne, M.R.: Rayleigh wave acoustic emission during crack propagation in steel. Ph.D. Dissertation, Virginia Polytechnic Institute and State University (2003)
11. Ziola, S.M., Gorman, M.R.: Source location in thin plates using cross-correlation. J. Acoust. Soc. Am. **90**(5), 2551–2556 (1991)
12. Winfree, W.P., Zalameda, J.N., Howell, P.A.: Improved flaw detection and characterization with difference thermography. Proc. SPIE **8013**, 80130U (2011)
13. Weisstein, E.W.: Affine transformation. From MathWorld—A Wolfram Web Resource. http://mathworld.wolfram.com/AffineTransformation.html. Accessed 5 Oct 2015
14. Gribbon, K.T., Johnston, C.T., Bailey, D.G.: A real-time FPGA implementation of a barrel distortion correction algorithm with bilinear interpolation. In: Image and Vision Computing New Zealand, pp. 408–413 (2003)

Chapter 5
Sensitivity Analysis of Hybrid Thermoelastic Techniques

W.A. Samad and J.M. Considine

Abstract Stress functions have been used as a complementary tool to support experimental techniques, such as thermoelastic stress analysis (TSA) and digital image correlation (DIC), in an effort to evaluate the complete and separate full-field stresses of loaded structures. The need for such coupling between experimental data and stress functions is due to the fact that experimental techniques offer discrete information of stresses or displacements, e.g. isopachic stresses in the case of TSA, as well as unreliable data near edges. For TSA, additional information is needed to separate stresses, as it is often necessary for fatigue analysis and a general better understanding of structural integrity. This separation is often accomplished by using an *Airy* stress function, which stems from compatibility and equilibrium conditions, and is frequently represented in the form of an indefinite series of coefficients. To date, only *ad hoc* estimates for the number of coefficients necessary for accurate representation of a loaded structure are used, with the estimates being influenced by quality of experimental data, experimental noise, and complexity of loading and boundary conditions. Information presented here attempts to systematize the selection of the *Airy* stress function's indefinite series coefficients relative to experimental thermographic data.

Keywords TSA • Hybrid • Sensitivity Analysis • *Airy* stress function

Nomenclatures

TSA	Thermoelastic stress analysis
PSA	Photoelastic stress analysis
DIC	Digital image correlation
m	Number of TSA data points
k	Number of Airy coefficients
S^*	Raw TSA image proportional to $\sigma_1 + \sigma_2$
K	TSA calibration factor
S	Calibrated TSA image
R	Disk radius
t	Disk thickness
x & y	Cartesian coordinates
r & θ	Polar coordinates
RMS	Root mean square

5.1 Introduction to Hybrid-TSA

Hybrid-experimental approaches, where full-field experimental data is coupled with an analytical expression allows for the determination of fundamentally important separate stresses or constitutive parameters, such as stress concentration factors and stiffness constants. When full-field experimental techniques such as thermoelastic stress analysis (TSA), moiré,

W.A. Samad (✉)
Rochester Institute of Technology—Dubai Campus, Dubai, United Arab Emirates
e-mail: wascad@rit.edu

J.M. Considine
Forest Products Laboratory, USDA, Madison, WI, USA

S. Quinn, X. Balandraud (eds.), *Residual Stress, Thermomechanics & Infrared Imaging, Hybrid Techniques and Inverse Problems, Volume 9*, Conference Proceedings of the Society for Experimental Mechanics Series, DOI 10.1007/978-3-319-42255-8_5

photoelastic stress analysis (PSA) , digital image correlation (DIC), etc., are utilized, raw results are discrete, with their resolution being a function of experimental setup and equipment (sensor resolution, lens used, focal distance, etc.), and data near discontinuities are unreliable due to different edge effects. More importantly, raw data offer one piece of information, e.g. combined stresses for TSA and PSA, and displacements for DIC.

With the focus here being on thermoelectricity, raw data from a TSA test are proportional to the summation of the first two principal stresses when addressing a two-dimensional plane stress loading scenario as shown in Eq. (5.1) below:

$$S^* = K\Delta S = K\Delta(\sigma_1 + \sigma_2) \tag{5.1}$$

Where S^* is the recorded TSA signal and K is the TSA calibration factor related to the relevant physical properties of the material of interest, surface condition and TSA system parameters [1–3], and S is the calibrated TSA image. With most design problems in engineering fatigue and failure analysis, the separation of the above TSA coupled stresses becomes essential and has been the topic of interest for many research efforts.

The first determination of individual thermoelastic stresses was presented in [4] by taking into account the boundary conditions and the expected form of the stress distribution. More work was done in the following decade with the focus on performing such separation on structures with various discontinuities, structure with a circular hole [5], beneath concentrated loads [6], elliptical cutout [7], and irregularly shaped discontinuities [8]. In some instances, stress separation was accomplished without combining the experimental data with the analytical expression of Eq. (5.1) where TSA was instead combined with PSA [9–11] and with the results remaining discrete as opposed to having a complete continuous separate stress field. More recently, hybrid-TSA was applied to obtain full-field stress information on a complicated contact stress problem as is the case for a pinned joint [12].

5.2 Airy Stress Function

Individual stress components can be determined by combining TSA thermal data of Eq. (5.1) with relevant analytical information. Many such schemes process the recorded thermal information with equations based on an *Airy* stress function. For plane-stress isotropy [13], the *Airy* stress function satisfying conditions of equilibrium and compatibility is given in Eq. (5.2) below:

$$\begin{aligned}\phi &= a_0 + b_0\ln r + c_0 r^2 + d_0 r^2 \ln r + (A_0 + B_0 \ln r + C_0 r^2 + D_0 r^2 \ln r)\theta + \left(a_1 r + b_1 r\ln r + \frac{c_1}{r} + d_1 r^3\right)\sin\theta + \\ &\left(a_1' r + b_1' r\ln r + \frac{c_1'}{r} + d_1' r^3\right)\cos\theta + (A_1 r + B_1 r\ln r)\theta\sin\theta + (A_1' r + B_1' r\ln r)\theta\cos\theta + \\ &\sum\nolimits_{n=2}^{\infty}\left(a_n r^n + b_n r^{n+2} + \ln r + c_n r^{-n} + d_n r^{2-n}\right)\sin n\theta + \sum\nolimits_{n=2}^{\infty}\left(a_n' r^n + b_n' r^{n+2} + c_n' r^{-n} + d_n' r^{2-n}\right)\cos n\theta\end{aligned} \tag{5.2}$$

Where S in Eq. (5.1) can be expressed in terms of the *Airy* stress function coefficients since,

$$\sigma_1 + \sigma_2 = \sigma_r + \sigma_\theta = \frac{1}{r}\frac{\partial\phi}{\partial r} + \frac{1}{r^2}\frac{\partial^2\phi}{\partial\theta^2} + \frac{\partial^2\phi}{\partial r^2} \tag{5.3}$$

To date, little work has been done in the area of determining the effects of: the number of *Airy* coefficients, the number of TSA data points, and the level of noise on the overall accuracy of Hybrid-TSA.

5.3 Numerical Experiment

The sensitivity approach performed here is applied to the case of a diametrically-loaded isotropic circular disk, Fig. 5.1, with the TSA expression of $\sigma_1 + \sigma_2$ simulated using the exact analytical solution given in [14] as shown in Eq. (5.4):

Fig. 5.1 Diametrically loaded disk

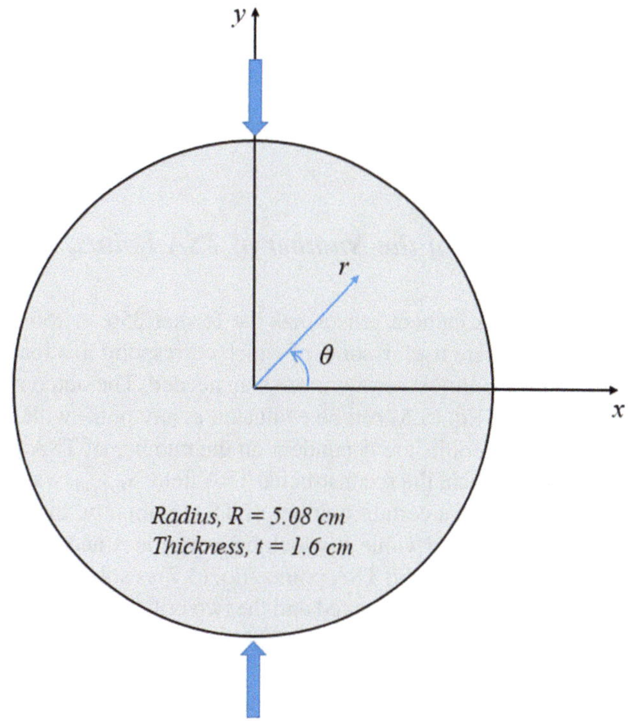

$$S_{TSA} = \sigma_1 + \sigma_2 = \sigma_x + \sigma_y = \frac{-2P}{\pi t}\left[\frac{(R-y)^3 + (R-y)x^2}{\left(x^2 + (R-y)^2\right)^2} + \frac{(R+y)^3 + (R+y)x^2}{\left(x^2 + (R+y)^2\right)^2} - \frac{1}{R}\right] = S_{exact} \tag{5.4}$$

The analytical expression, Eq. (5.4), was used to more easily simulate experimental parameters such as number of data points and noise levels. Due to symmetry, only the first quadrant was examined, with the center and Cartesian and polar coordinates displayed in Fig. 5.1. The *Airy* expression of *S* in Eq. (5.3) was reduced to Eq. (5.5) below for the symmetry and location of the origin in the diametrically-loaded disk at hand:

$$S_{Airy} = 4c_0 + \sum\nolimits_{n=2,4,\ldots}^{\infty}\left[4b_n'(n+1)r^n \cos n\theta\right] \tag{5.5}$$

The solution to the hybrid-TSA involves solving an over-determined linear-least squares problem where S_{Airy} is equated to S_{TSA}, in Eq. (5.6)

$$\{S_{Airy}\}_{m \times 1} = [A]_{m \times k}\{\text{coef}\}_{k \times 1} = \{S_{TSA}\}_{m \times 1} \tag{5.6}$$

where: m is the number of TSA data points used, k is the number of coefficients, and $[A]$ is the *Airy* matrix composed of m 1st stress invariant expressions of the form of Eq. (5.5) in terms of r and θ associated with the m TSA data points used. Upon solving for the coefficients in Eq. (5.6), those are then substituted back into Eq. (5.5) such that the TSA stress field can be reconstructed as shown in Eq. (5.7); where the reconstructed image is referred to as S_{hybrid} and is the result of this hybrid-TSA method.

$$\{S_{hybrid}\}_{m \times 1} = [A]_{m \times k}\left\{[A]_{m \times k} \backslash \{S_{TSA}\}_{m \times 1}\right\} \tag{5.7}$$

The sensitivity study presented in this work looks into the effects of m, k as well as noise on the obtained hybrid-TSA solution, S_{hybrid}. The two criteria used to evaluate a certain combination of m and k with or without noise are (a) the RMS of the S_{hybrid} when compared to the original S_{TSA} and (b) the maximum error in any one point in S_{hybrid}. These two criteria will be referred to throughout this manuscript as criteria A and criteria B respectively.

5.3.1 *Effect of the Number of TSA Points,* m

A typical TSA camera sensor has the format 256 × 256 which yields 65,536 pixels or data points. While not all of the 65,536 pixels are useful, some of which correspond to a loading fixture, a grip, etc., the number of available useful pixels on a structure are almost always more than needed. The outcome of a hybrid-TSA approach results in infinity many points as the expression of Eq. (5.5) can be evaluated at any point with the *Airy* coefficients determined. However, the validity of those infinite many points are dependent on the number of TSA points, m, used while solving the linear-least squares problem of Eq. (5.7). As such, the reconstructed TSA field, S_{hybrid}, was evaluated based on criteria A and B for different values of m. For each scenario of a certain number of TSA points, m, the number of coefficients, k, was varied such that the best value of k was determined; k value at which both criteria A and B were minimized (minimum error in the reconstructed, S_{hybrid}). For instance, for m = 1000 TSA points, Eq. (5.7) is solved for while varying k from two coefficients to 50 coefficients. At each value of k, S_{hybrid} is commuted and the two criteria of RMS and maximum error are evaluated and plotted as seen in Fig. 5.2a and b. For the case where m = 1000, the appropriate value of k is determined to be 24 coefficients.

Table 5.1 summarizes the effect of the number of TSA points on hybrid-TSA revealing some interesting observations. First, the RMS and maximum percent error increase with the number of TSA points, m, used. Second, the value k for each of the different scenarios changes and is not a constant. For instance the minimum RMS and error for $m = 6362$ points was at achieved at $k = 22$ coefficients when only $k = 24$ coefficients where necessary to arrive at the best solution for $m = 1038$ points.

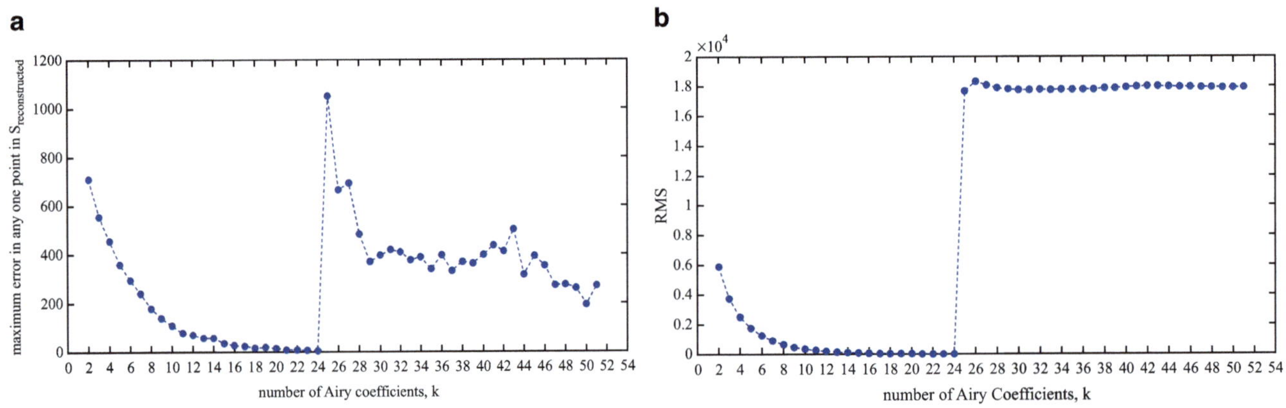

Fig. 5.2 (**a**) Maximum error VS k (**b**) RMS VS k for $m = 1000$ points

Table 5.1 Effect of m number of necessary coefficients k

m	k (necessary)	**Maximum % error** (criteria A)	**RMS** (criteria B)
267	25	0.5	1.589
1038	24	4.8	7.3
2849	23	6.5	16.3
6362	22	8.2	30.7
11,279	22	10.4	42.1
25,261	21	11.2	75.4
31,165	21	11.7	85.5
34,929	21	12.2	91.1
44,837	21	12.3	103.1

Third, the lesser the TSA data points, m, used, the more number of Airy coefficients, k, are needed to arrive at the best result for each scenario. So in conclusion, better results are achieved using fewer points but while at the same time necessitating more *Airy* coefficients. Note that in all scenarios, all the m points are equally spaced and lying between 0.25 mm and 45.72 mm.

5.3.2 *Number of Airy Coefficients,* k

In this section, the number of TSA points used is set to $m = 1000$, while a closer look at the number of coefficients, k, is considered. Figure 5.2a and b shown earlier are the plots of the RMS versus k and the maximum error versus k respectively. Notice that for this value of m, the ideal value of k was observed to be 24 coefficients as it yielded the lowest RMS and error. Both RMS and maximum error keep decreasing with increasing k till $k = 24$. After which, the plots of the two criteria show a sudden increase and fail to arrive at another meaningful value of k. That sudden increase seen in Fig. 5.2 a and b is due to the *Airy* matrix $[A]_{1,000 \times 25}$ becoming rank deficient for $k = 25$. Upon solving for $\{\text{coef}\}_{25 \times 1}$ by performing linear-least-squares of Eq. (5.7), the first coefficient c_0 in Eq. (5.5) goes to zero as tabulated in Table 5.1 along with all the other coefficient values. The column on the far right of Table 5.2 shows the values of the coefficients for $k = 25$ coefficients but while fixing c_0 to the value $c_{0\ @\ k=24} = -140$. Results (RMS and maximum error) do improve upon setting c_0 to its $c_{0\ @\ k=24}$ value, however no improvement is seen over $k = 24$ and thus rendering it unnecessary. Another note worth mentioning, the *jump* seen in Fig. 5.2 a and b are not unique to the transition from k_{24} to k_{25} since a very similar *jump* is seen at other values k_x to k_{x+1} for different values of m. For instance for m = 40,000 TSA points, the sudden deterioration in S_{hybrid} is observed between k_{21} to k_{22} (Table 5.2).

Table 5.2 Airy coefficient values at the "*jump*" observed in Fig. 5.3a and b

$m = 1000$			
	$k = 24$	$k = 25$	$k = 25$ with c_0 fixed
c_0	−140.051	0	−140 (fixed)
b'$_2$	−23.3396	−23.3396	−23.34162996
b'$_4$	−3.49998	−3.51148	−3.574389852
b'$_6$	−0.62509	−0.62509	−0.625354758
b'$_8$	−0.12151	−0.12529	−0.126586946
b'$_{10}$	−0.02485	−0.02485	−0.024877378
b'$_{12}$	−0.00526	−0.00606	−0.004816275
b'$_{14}$	−0.00114	−0.00114	−0.001142183
b'$_{16}$	−0.00025	−0.0003	−0.000247195
b'$_{18}$	−5.61E−05	−5.61E−05	−5.63E−05
b'$_{20}$	−1.28E−05	−1.83E−05	−1.47E−05
b'$_{22}$	−2.92E−06	−2.92E−06	−2.90E−06
b'$_{24}$	−6.74E−07	−1.24E−06	−5.80E−07
b'$_{26}$	−1.58E−07	−1.58E−07	−1.54E−07
b'$_{28}$	−3.73E−08	−1.30E−07	−2.52E−08
b'$_{30}$	−8.73E−09	−8.73E−09	−8.38E−09
b'$_{32}$	−2.09E−09	−6.17E−09	−2.34E−09
b'$_{34}$	−4.77E−10	−4.77E−10	−4.64E−10
b'$_{36}$	−1.17E−10	−5.30E−10	−8.30E−10
b'$_{38}$	−2.84E−11	−2.84E−11	−2.48E−11
b'$_{40}$	−7.12E−12	1.06E−11	−3.09E−11
b'$_{42}$	−1.76E−12	−1.76E−12	−1.36E−12
b'$_{44}$	−4.26E−13	2.44E−12	−3.14E−12
b'$_{46}$	−9.85E−14	−9.85E−14	−8.20E−14
b'$_{48}$	N/A	7.46E−14	1.38E−13
max error	**4.9 %**	**1051 %**	**347 %**
RMS	**7.2**	**1767**	**286**

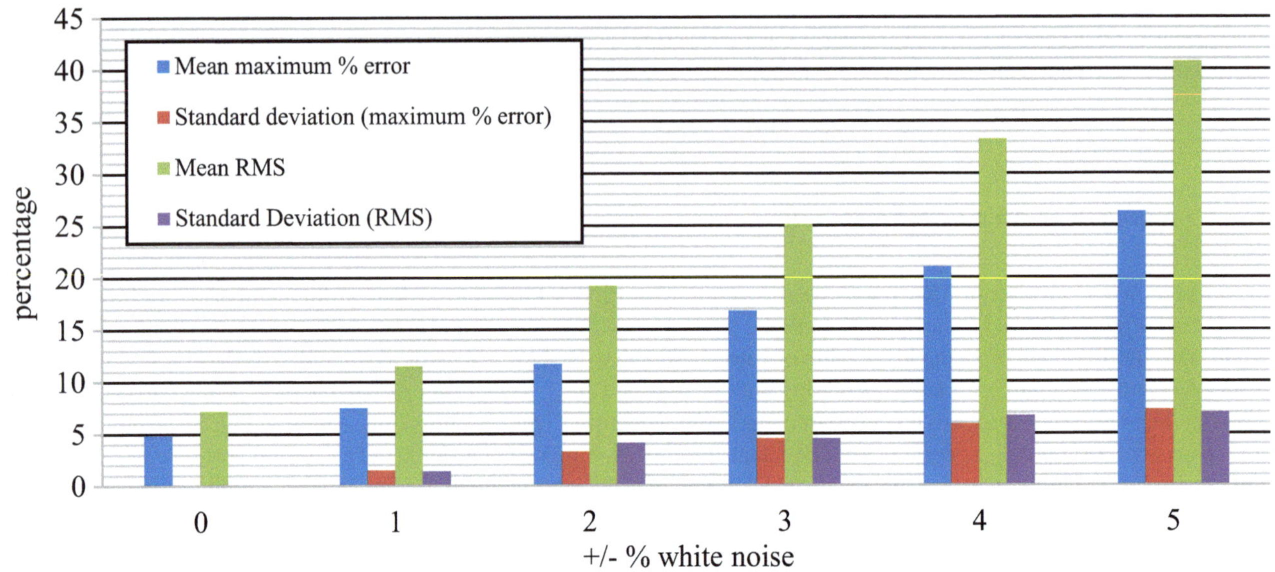

Fig. 5.3 Deterioration of reconstructed thermal image upon introducing different noise levels

5.3.3 *Effect of Noise*

Thermoelastic stress analysis is typically accompanied by white noise which has a negative effect on the determined *Airy* coefficients upon solving the over-determined problem $[A]_{m \times k} \backslash \{S_{TSA}\}_{m \times 1}$. In this section, *m* is to 1000 points and *k* is set to 24 coefficients, and different white noise levels are introduced to the analytical soliton for *S* shown in Eq. (5.4). The white noise was generated at every point as $[S_{noise} = S_{exact} \pm rand \times S_{average}]$ where *rand* is the MATALB operator that generates a random number from zero to one and $S_{average}$ is the average value of $\sigma_1 + \sigma_2$ over the first quadrant of the diametrically-loaded disk. Similar to the sections before, the linear-least squares problem is solved, and S_{hybrid} is reconstructed. Figure 5.3 shows a bar chart of the mean values of the maximum percent error and RMS along with their corresponding standard deviations over 100 trials for $m = 1000$ and $k = 24$ at different noise levels. Both max percent error and RMS increase by ~12 % upon increasing the white noise level by 1 %.

Note that the simulations of the 100 trials in Fig. 5.3 are all using $m = 1000$ points and $k = 24$ coefficients, which in a way assumes that the appropriate number of coefficients is independent of the noise levels and is only a function of the number of TSA points used. This independency of the appropriate number of coefficients, k, on noise levels is verified in Fig. 5.4 where it is seen that the drastic increase in the RMS value is always between k_{24} and k_{25}, for all white noise levels (0 to ±8 %). While not shown here, a similar trend is observed for the maximum error VS *k* plot at the different noise levels.

Now that the appropriate number of *Airy* coefficients is better understood, a closer look at the sensitivity of the *Airy* coefficients themselves can be analyzed with *k* again being set to 24 coefficients and *m* to 1000 TSA points. Figure 5.5 shows the standard deviation in the 24 *Airy* coefficients for different noise levels (±1 % to ±5 %). It is clear that the higher order coefficients carry higher variations and uncertainty, and in a way are the main contributors for the increases in the errors in the reconstructed S_{hybrid}. While it seems appropriate to eliminate some of those higher order coefficients (e.g. b_{44} and b_{46}), the simulations carried out earlier and summarized in Fig. 5.4 prove otherwise. While not shown here, eliminating *specific* higher order coefficients (e.g. b_{36}) as opposed to the highest order coefficient does not yield in an improved S_{hybrid}.

5.4 Conclusion and Future Work

Four main conclusions can be derived from the work presented here:

1. an increase in the number of TSA data points, *m*, does not yield a better solution
2. the lesser the number of TSA data points, *m*, the more *Airy* coefficients, *k* are needed to arrive at the best solution

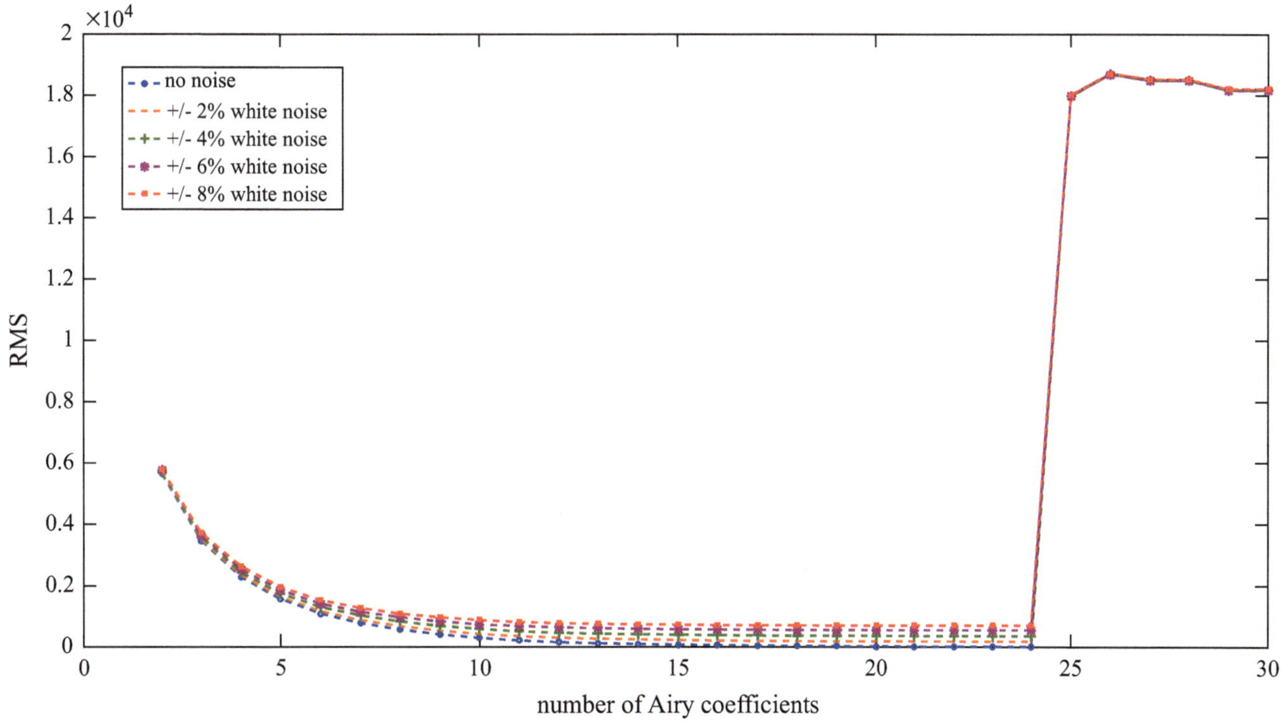

Fig. 5.4 RMS VS k for $m = 1000$ points at different noise levels

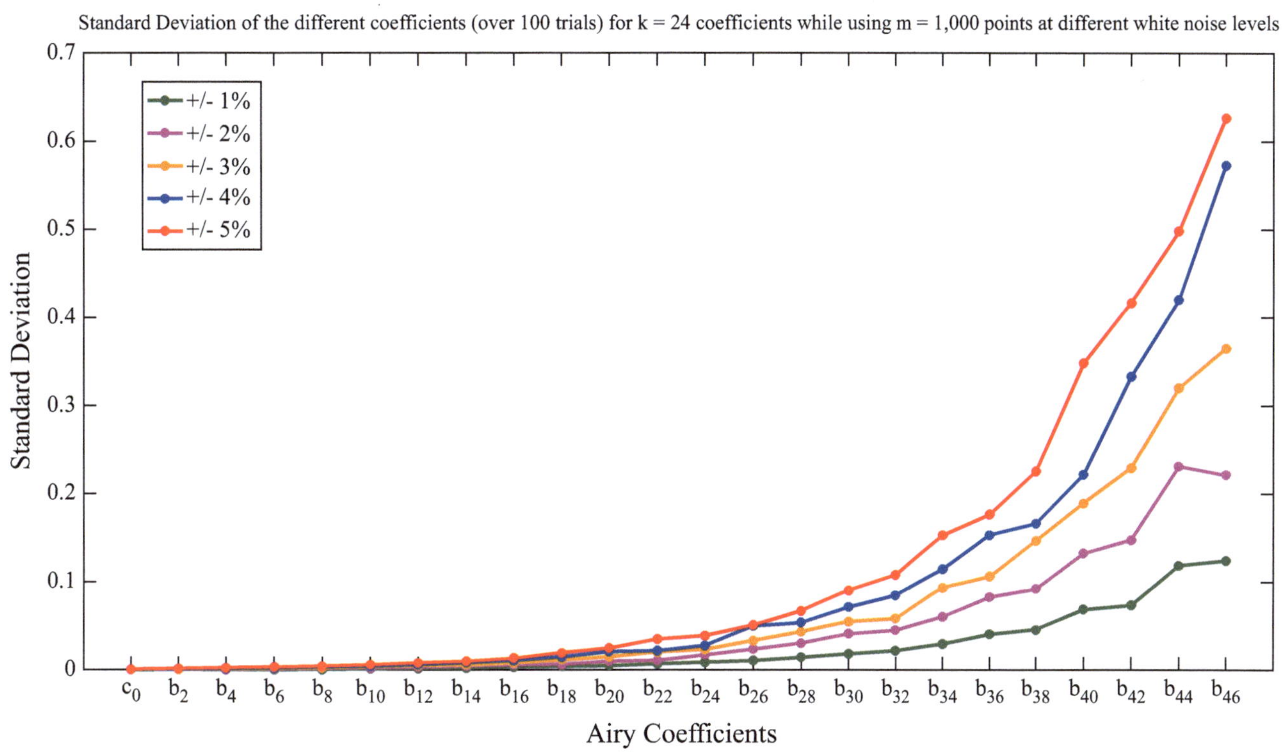

Fig. 5.5 Standard deviation in the determined coefficients for $m = 1000$, $k = 24$, and at different noise levels over 100 trials

3. fixing one of the *Airy* coefficients to a pre-known value does need yield a better result
4. the hybrid approach is fairly sensitive to white noise levels
5. different noise levels do not have an effect on the appropriate number of *Airy* coefficients, k, with the latter being strictly dependent on m
6. higher order coefficients have larger uncertainty than lower order ones

Some items worth looking into in future work:

1. the effect of varying point spacing within a single hybrid-TSA approach (e.g. increasing the TSA point density in some specific locations)
2. the effects of smoothing noisy data on the reconstructed S_{hybrid}
3. the effect of the geometry (complexity of the TSA S field)

References

1. Stanley, P., Barton-Dulieu, J.M.: Development and applications of thermoelastic stress analysis. J. Strain Anal. Eng. Des. **33**(2), 93–104 (1998)
2. Stanley, P., Chan, W.: Assessment and deleopment of the thermoelasic technique for engineering application: four years of progress. In: SPIE 0731, Stress Analysis by Thermoelasic Techniques in SPEI 731, London (1987)
3. Greene, R., Patterson, E., Rowlans, R.E.: Thermoelastic stress analysis. In: Handbook of Experimental Solid Mechanics. Springer, New York (2008)
4. Ryall, A.W.T.G.: Determining stress components from the thermoelastic data: a theoretical study. Mech. Mater. **7**(3), 205–214 (1988)
5. Lin, S.J., Matthys, D.R., Rowlands, R.E.: Separating stresses thermoelastically in a central circularly perforated plate using an airy stress function. Strain **45**(6), 516–526 (2009)
6. Lin, S.J., Quinn, S., Matthys, D.R., New, A.M., Kincaid, I.M., Boyce, B.R.K.A.A., Rowlands, R.E.: Thermoelastic determination of individual stresses in vicinity of a near-edge hole beneath a concentrated load. J. Exp. Mech. **51**(6), 797–814 (2010)
7. Khaja, A.A., Rowlands, R.E.: Experimentally determined stresses associated with elliptical holes using polar coordinates. Strain **49**(2), 116–124 (2013)
8. Samad, W.A., Rowlands, R.E.: Full-field thermoelastic stress analysis of a finite structure containing an irregularly-shaped hole. J. Exp. Mech. **54**(3), 457–469 (2014)
9. Sakagami, T., Kubo, S., Jujinami, Y., Kojima, Y.: Experimental stress separation technique using thermoelasticity and photoelasticity and its application to fracture mechanics. JSME Int. J. Ser. A **47**(3), 298–304 (2004)
10. Barone, S., Patterson, E.A.: Full-field separation of principal stresses by combined thermo- and photoelasticity. J. Exp. Mech. **36**(4), 318–324 (1996)
11. Greene, R.J., Patterson, E.A.: An integrated approach to the separation of principal surface stresses using combined thermo-photo-elasticity. J. Exp. Mech. **46**(1), 19–29 (2006)
12. Samad, W.A., Khaja, A.A., Kaliyanda, A:R., Rowlands, R.E.: Hybrid thermoelastic stress analysis of a pinned joint. J. Exp. Mech. **54**(4), 515–525 (2014)
13. Little, R.W.: Elasticity. Dover Publications, New York (1998)
14. Frocht, M.M.: Photoelasticity. Wiley, New York (1948)

Chapter 6
Determining Stress Intensity Factors Using Hybrid Thermoelastic Analysis

R.B. Vieira, G.L.G. Gonzáles, and J.L.F. Freire

Abstract This paper presents and discusses a technique suited for the determination of mode I Stress Intensity Factors (SIF) of fatigue-initiated and propagated cracks at the keyhole of polycarbonate specimens. A hybrid approach combined Thermoelastic Stress Analysis (TSA) results with Linear Elastic Fracture Mechanics solutions using Westergaard's stress function to describe the stress field near the crack tip. The TSA results used an experimental approach that does not require an infrared camera with lock-in capability. The experiments used a micro-bolometer camera A655sc from FLIR Inc. and a data processing software DeltaTherm2 from StressPhotonics Inc. Two distinct data fitting methods are presented. The first method measures the crack length, which makes the problem become linear, allowing for a simple Least Squares Method (LSM) approach. The second method, highlighting the true power of TSA as a fatigue analysis technique, uses the crack tip position as an adjustable parameter, making the problem non-linear and solvable by a complex numerical algorithm known as the Downhill Simplex Method (Nelder-Mead). The paper describes automated methodologies for making good initial estimates for the position of the crack, required by the non-linear approach, as well as for selecting data points to be fitted, both based on the loss of linearity of the TSA data due to non-adiabatic conditions.

Keywords Thermoelastic Stress Analysis • Micro-bolometer detector • Fatigue • Stress Intensity Factor • Polycarbonate

6.1 Introduction

Thermoelastic Stress Analysis (TSA) is a non-contact experimental technique that measures full-field stress maps for the surface of a component. It is based on the thermoelastic effect [1], which states that a cyclic loaded body experiences small temperature variations (ΔT), and that, if adiabatic conditions are satisfied, this variation is proportional to the first stress invariant.

Practical applications of TSA use infrared thermography to measure the very small pixel-by-pixel temperature variations. With these variations having been measured, Eq. (6.1) describes how the stress invariant is determined.

$$\Delta(\sigma_1 + \sigma_2) = A\,S \tag{6.1}$$

where $\Delta\sigma_1$ and $\Delta\sigma_2$ are the principal stress ranges acting on the observed point located at the surface, A is the thermoelastic calibration coefficient, and S is the magnitude of the TSA signal.

Combining the experimental data for the first stress invariant range, obtained through TSA, with an analytical solution for the crack tip stress-field, such as Westergaard stress function, the stress intensity factor (SIF) can be estimated.

Infrared detectors with enough sensitivity to measure such small temperature variations were very expensive in the past, but with recent advancements in micro-bolometer technology, the cost is going down quickly. The camera used to gather the experimental data presented in this paper is the A655sc from FLIR Inc., which has an uncooled micro-bolometric focal plane array detector and, as stated in [2], costs much, much less than the cooled quantum detectors that other authors have used.

DeltaTherm2, a new software program from Stress Photonics Inc., was used to convert the temperature data into TSA data. The typical TSA analysis involves a lock-in process, which correlates the ΔT data with a reference signal, usually obtained from the function generator that feeds the cyclic actuator. DeltaTherm2 automates this process and takes the reference signal directly from the ΔT measurements, making the whole TSA test much simpler to set up.

The SIF measurement algorithm was tested using numerically generated data that emulates the experimental TSA data for the stress field around a crack tip, and then was used to measure the SIF of a cracked SAE Keyhole specimen.

R.B. Vieira (✉) • G.L.G. Gonzáles • J.L.F. Freire
Pontifical Catholic University of Rio de Janeiro, Rua Marquês de São Vicente, 225, Gávea, Rio de Janeiro, RJ 22451-900, Brazil
e-mail: renato.bicharav@hotmail.com

S. Quinn, X. Balandraud (eds.), *Residual Stress, Thermomechanics & Infrared Imaging, Hybrid Techniques and Inverse Problems, Volume 9*, Conference Proceedings of the Society for Experimental Mechanics Series, DOI 10.1007/978-3-319-42255-8_6

Many authors have used TSA to measure SIFs, and they have also used varied numerical and experimental approaches. Stanley and Chan [3] used the first two terms of Westergaard's stress function to fit the thermoelastic data around cracks in mode I and II. Stanley and Dulieu-Smith [4] based their approach on the cardioid format of the isopachic fringes around a crack tip. Lesniak *et al.* [5] used Williams' stress field equations to fit the data using the least squares method (LSM). Lin *et al.* [6] approached the problem using the J-integral formulation in the data fitting process. Tomlinson *et al.* [7] developed a technique based on Muskhelishvili's stress field equations and a multipoint-over-deterministic method (MPODM). Díaz *et al.* [8] and [9] proposed a methodology for finding the crack tip position by looking at thermoelastic data, and then fitting the data using a non-linear approach in order to measure SIFs of propagating cracks. Marsavina *et al.* [10] used TSA to investigate crack closure during fatigue. Tomlinson and Olden [11] presented a broad review on the use of TSA in fracture mechanics and fatigue cracks.

The present work attempts to follow in these steps and develop an automated technique that can be used with a inexpensive infrared camera in combination with DeltaTherm2 software, which does not require a lock-in procedure.

6.2 Description of the SIF Calculation Algorithm

Figure 6.1 shows a flowchart of the SIF calculation algorithm. The following sections describe each step.

6.3 Algorithm Inputs

The algorithm inputs are the outputs derived from the TSA analysis. The TSA data is usually expressed as a complex variable that has a magnitude and a phase angle. The magnitude is proportional to the temperature variation and thus, through Eq. (6.1), proportional to the variation of the first stress invariant. The phase angle carries the signal information, that is, if a stress should be treated as compressive or tensile.

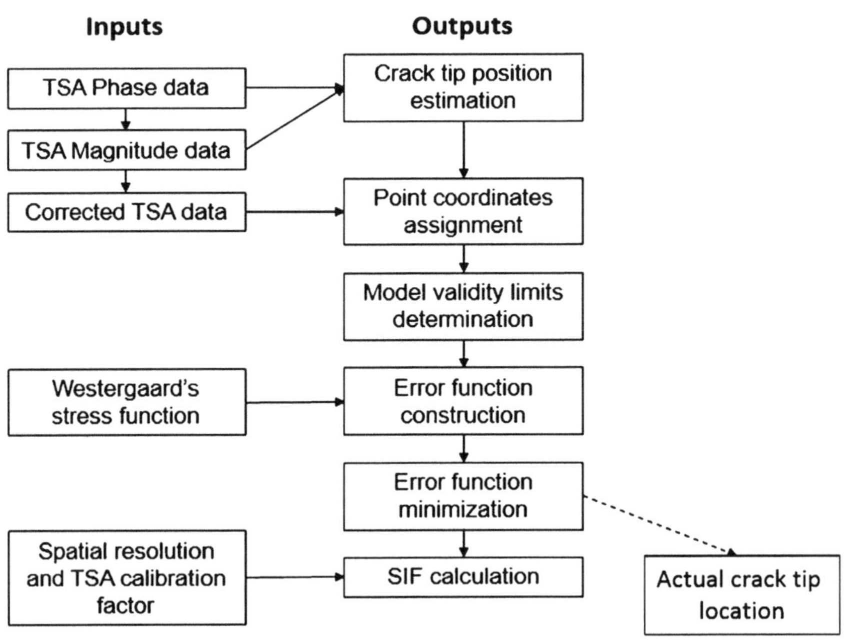

Fig. 6.1 Flowchart of the SIF calculation algorithm

In order to generate the phase data, the TSA software uses a reference signal and determines its phase difference from the temperature measurements. This is called the lock-in process. The most usual way to do this is to use an input from the cyclic loading machine as the reference; but as stated above, the software used in the present work does it differently. DeltaTherm2 uses the temperature of a selected area of the specimen's surface as the reference.

Subsequently, with the TSA magnitude and phase data, it is possible to generate corrected TSA data,

$$S = \begin{cases} Mag \text{ } if \text{ } |phase| < 45^o \\ -Mag \text{ } if \text{ } |phase| > 45^o \end{cases} \tag{6.2}$$

where S is the corrected TSA data, Mag is the magnitude data value in "camera units", and |phase| is the absolute value of the phase angle, in degrees.

Another input required is the mathematical model that is being used. In this case, a Westergaard's stress function (We) was used to describe the stress field around the crack tip [12]:

$$We(x, y) = \sigma_1 + \sigma_2 = 2[Re(Z) + Re(Y)] \tag{6.3}$$

with

$$\begin{aligned} Z &= \sum_{n=0}^{N} \left[A_n \, z^{n-0.5}\right] \\ Y &= \sum_{m=0}^{M} \left[B_m \, z^m\right] \end{aligned} \tag{6.4}$$

where N and M determine the number of coefficients to be used, A_n and B_m are the data fitting coefficients, and z is the complex variable of the point coordinates:

$$z = (x - x_o) + i(y - y_o) \tag{6.5}$$

Where x and y are the horizontal and vertical coordinates of a point (placing the origin at the estimated crack tip location), and $i = \sqrt{-1}$ and x_o and y_o are the horizontal and vertical errors of the estimated crack tip location. These errors are solved only when applying the non-linear method; for the linear method they are set to 0.

The final required inputs are the spatial resolution (SR) and the TSA calibration factor (A). The spatial resolution is simply determined via a known dimension in the specimen,

$$SR = W/p \tag{6.6}$$

where SR is the spatial resolution in mm/pixel, W is a known dimension of the specimen in mm (i.e. width), and p is the number of pixels the dimension measures on the thermographic image.

There are many possible TSA calibration processes [13], but this work uses what may be the simplest one, which consists of using a specimen, machined from the same sheet of polycarbonate, with an analytically determined stress. Then, using Eq. (6.1) to correlate the calculated change of the stress invariant ($\Delta\sigma_1 + \Delta\sigma_2$) and the measured TSA signal (S), A is determined for the camera-material pair.

6.4 Crack Tip Position Estimation and Point Coordinates Assignment

This work uses a specially improved, automated version of the methodology proposed in [8] to estimate the position of the crack tip.

First, the image needs to be positioned so that the crack is horizontal. Then, using the TSA magnitude map, the vertical position of the crack tip (Y_o') is estimated by plotting y vs. $1/S^2$ curves for each vertical line in the image (perpendicular to the crack). The resulting graphs, for the lines ahead of the crack tip, have a clear minimum value. The mean value of y where this minimum value occurs for all the vertical lines is equal to Y_o'. Figure 6.2a shows the magnitude map around a crack, while Fig. 6.2b shows a typical y vs. $1/S^2$ plot.

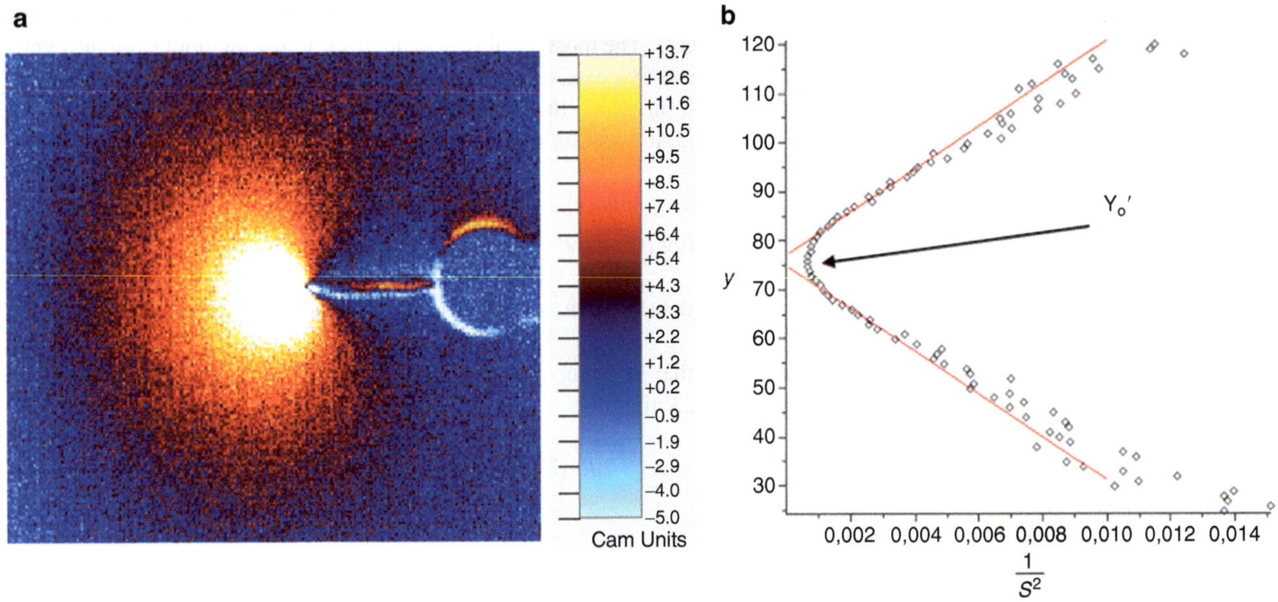

Fig. 6.2 (**a**) Magnitude map around a crack and (**b**) Typical y vs. $1/S^2$ plot

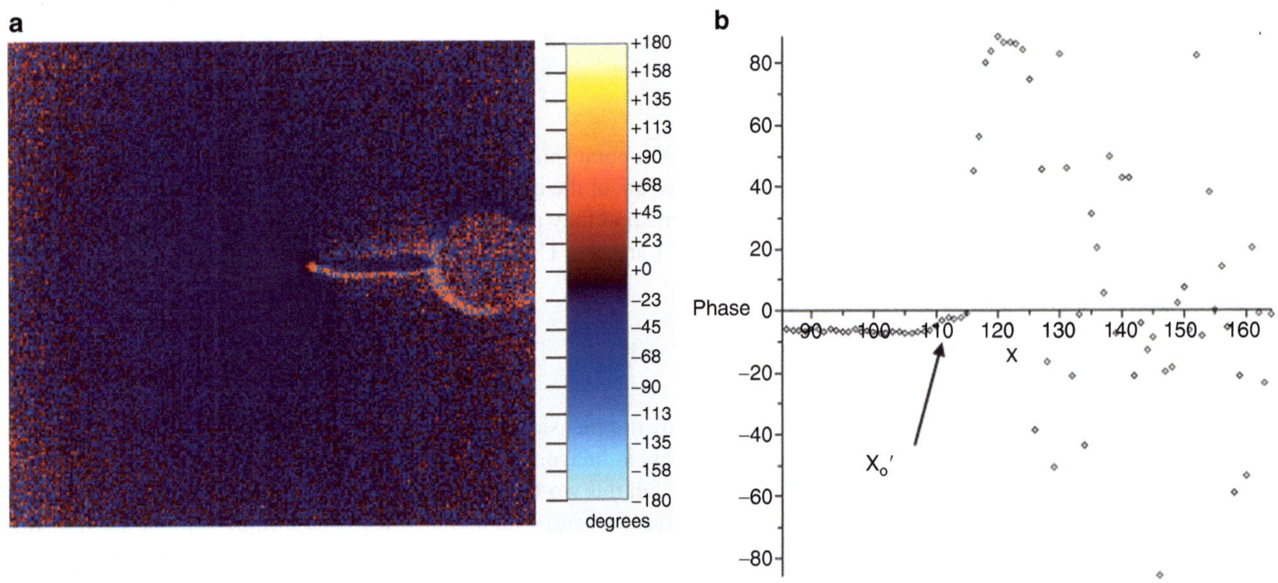

Fig. 6.3 (**a**) Phase map around a crack and (**b**) Typical phase vs. x plot

The horizontal position of the crack tip (X_o') is estimated using what is called the phase shift. Due to the plasticity near the crack tip, the adiabatic conditions required by TSA are lost. This can be seen as an abrupt change in phase angle directly in front of the crack tip. In the horizontal phase vs. x curve at Y_o', the position where this abrupt change occurs is equal to X_o'. Figure 6.3a shows the phase map around a crack and Fig. 6.3b shows a typical phase vs. x plot.

Having determined the values of X_o' and Y_o', the coordinates assignment step is simple: each point of the image gets assigned to an x and a y value, which are the horizontal and vertical distances from the estimated crack tip, respectively.

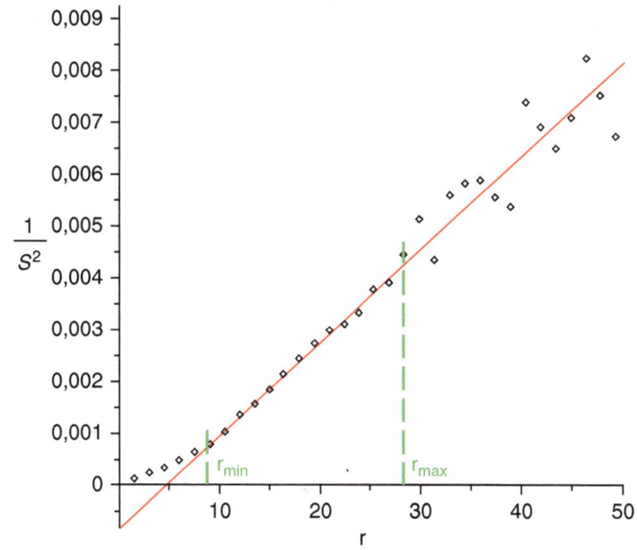

Fig. 6.4 Typical $1/S^2$ vs. r plot

6.5 Model Validity Limits Verification

The points used for the data fitting process must be in accordance with the mathematical model. In [9], the authors concluded that the model predicts a linear behavior of $1/S^2$ with respect to the distance from the crack tip. They proposed that only the points of Fig. 6.2b that presented a linear behavior should be used in the data fitting process.

The methodology used in the present work uses the same argument, but instead of using the vertical line plots, it uses plots of $1/S^2$ vs. r, where r is the radial distance from the estimated crack tip. With r and θ being the polar coordinates centered at X_o' and Y_o', and $\theta = 0$ being the horizontal line ahead of the crack, $1/S^2$ vs. r curves are plotted for each 1° increment of θ between -120° and $+120^o$. Figure 6.4 shows an example of one of these plots.

The loss of linearity near the crack tip is explained by non-adiabatic conditions that interfere with the TSA results. For the region far away from the crack tip, there are two possible explanations, or a combination of both. The higher noise in TSA data due to smaller stresses and consequent smaller temperature variation, and the complexity of the actual stress field far from the crack, which requires not only the SIF term, but also more terms of the stress function to be predicted accurately.

For each of the $1/S^2$ vs. r plots, r_{min}, r_{max} values are determined using two margins of tolerance of non-linearity that are manually adjusted to fit the data. The authors assume that the r_{min} margin should be related to the loading level and that the r_{max} margin could be related to the overall noise levels of the data. No attempts were made to automate the optimization process of these margins.

The values R_{min} (and R_{max}), below (and above) which no data points are collected, are defined as the mean value of all the r_{min} (and r_{max}).

6.6 Error Function Construction and Minimization

After determining which data points can be fitted by the model, an error function (e) is constructed:

$$e = \sum\nolimits_{Data\ points} [S - We(x, y)]^2 \tag{6.7}$$

where e is the error function to be minimized, S is the TSA signal of the point, and We(x,y) is defined by Eq. (6.3).

For the linear approach, the minimization process uses an LSM algorithm and for the non-linear approach it uses a Downhill Simplex method (Nelder-Mead) [14].

6.7 SIF Calculation and Real Crack Tip Location

After having calculated all the coefficients, the SIF can be calculated as

$$\Delta K_I = A_0\, A\, \sqrt{2\pi\, SR} \tag{6.8}$$

where ΔK_I is the mode I SIF range in MPa√m, A_0 is the first coefficient as described in Eq. (6.4), A is the TSA calibration factor, and SR is the spatial resolution.

In the non-linear approach, the values of x_o and y_o, as seen in Eq. (6.5), are determined, as well as the real location of the crack tip:

$$\begin{aligned} X_o &= X'_o - x_o \\ Y_o &= Y'_o - y_o \end{aligned} \tag{6.9}$$

where X_o and Y_o give the calculated actual position of the crack tip.

6.8 Description of the Equipment

As stated above, the camera used in the experiments described below was the FLIR A655sc. It has a focal plane array detector composed of 640 × 480 uncooled micro-bolometers and can record full-frame data at up to 50 Hz. It has a spatial resolution of 17 μm and a sensitivity of 30 mK.

The camera is pointed at the specimen, which is mounted in a pneumatic machine capable of applying cyclic loads at a frequency of 5 Hz. A load cell (5000 N), connected in series with the specimen and the pneumatic muscle, sends the load signal to a computer program that controls the servo-valve. The load magnitude is controlled with a valve that regulates the air pressure entering the muscle. Figure 6.5 shows a photo of the system.

As stated previously, DeltaTherm2 does not require a reference signal from the load cell, so the camera and loading systems are completely separate.

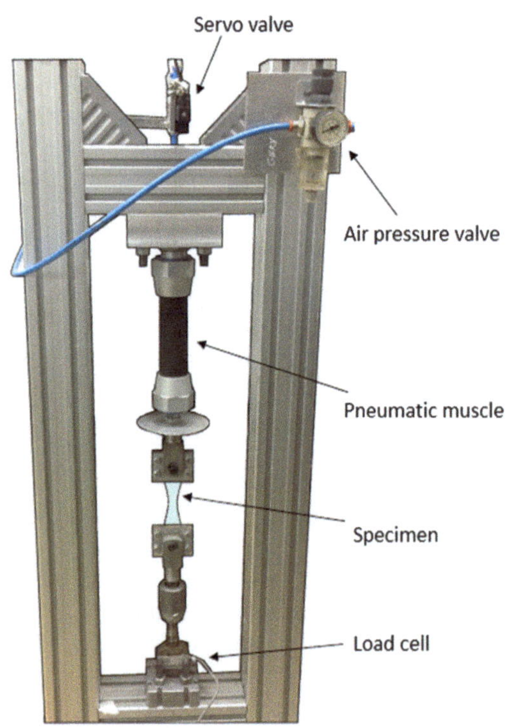

Fig. 6.5 Test system setup

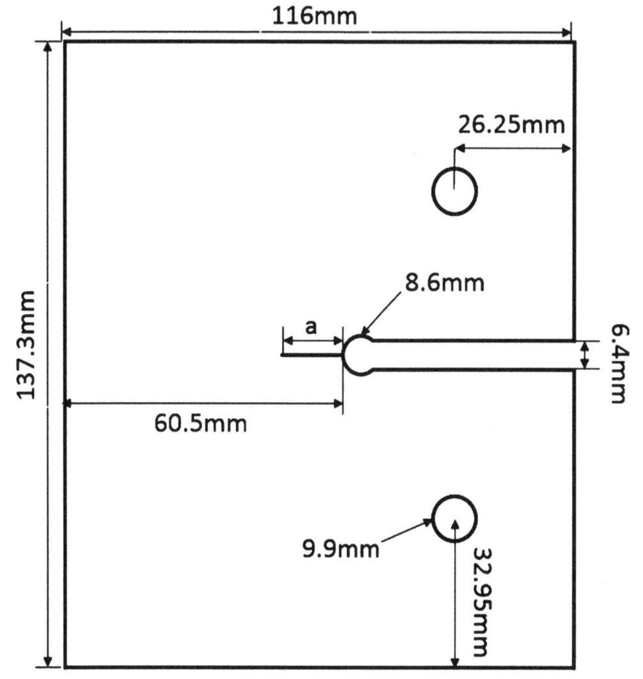

Fig. 6.6 Keyhole specimen dimensions

6.9 Specimen

The specimen used in all the experiments was a SAE Keyhole specimen, which was developed to study both fatigue crack initiation and propagation. The dimensions of the specimen, machined from a 3.9 mm thick PALSUN^{TM} polycarbonate (annealed) plate from PALRAM, are shown in Fig. 6.6.

6.10 Results

First, the algorithm is tested using an emulation of TSA data. Figure 6.7a shows the numerically generated data and Fig. 6.7b shows the fitted model for the crack tip estimation at (−3, 0) relative to the actual position.

For the linear approach, where the crack tip location is not determined, the result of ΔK_I was heavily dependent on the crack tip horizontal position estimation (X_o'), presenting errors of 10 % for just 1 pixel of inaccuracy in the estimation. For the vertical direction, the dependence was much smaller, a 1 pixel inaccuracy in Y_o' resulted in an error of 0.4 % for ΔK_I.

The non-linear approach, on the other hand, gave on-point results for ΔK_I even for initial crack-tip estimation coordinate (X_o' and Y_o') errors of 20+ pixels in the horizontal and vertical directions.

After having validated the methodology, the algorithm was put to the test on several crack lengths (a) for the specimen shown in Figs. 6.2, 6.3 and 6.6. Table 6.1 shows the results for both the linear and non-linear approaches, as well as for Digital Image Correlation (DIC), as described in [15], and finite elements method (FEM) for the ΔK_I [16]. Figure 6.8 shows the $\Delta K_I/\Delta P$ vs. a plot.

The analyses were made using $N = M = 3$, as described in Eq. (6.4), and 800 data points uniformly spaced between R_{min} and R_{max} limiting circles.

It is clear that, for bigger cracks, where the stresses around the crack tip are higher and the experimental data is less noisy, both TSA Non-linear and DIC techniques result in values of ΔK_I closer to the expected value, calculated by FEM. The TSA Linear approach tends to give higher results and increasing errors, probably because with longer cracks, the non-adiabatic zone becomes larger and the crack-tip position estimative error gets bigger.

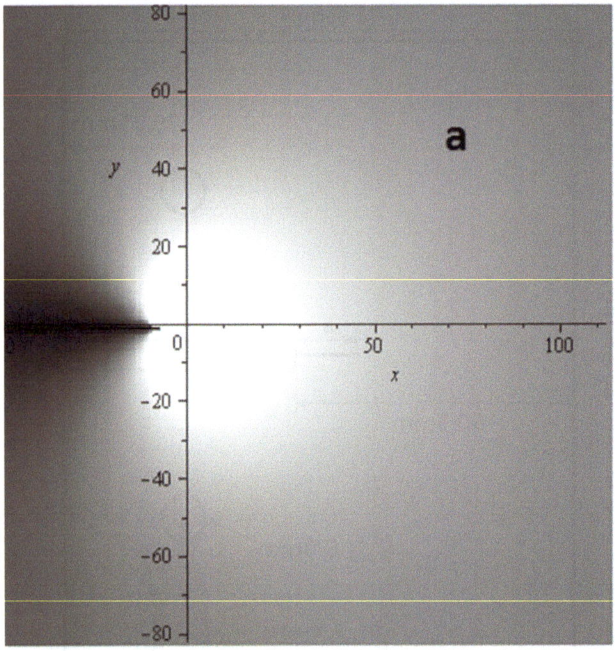

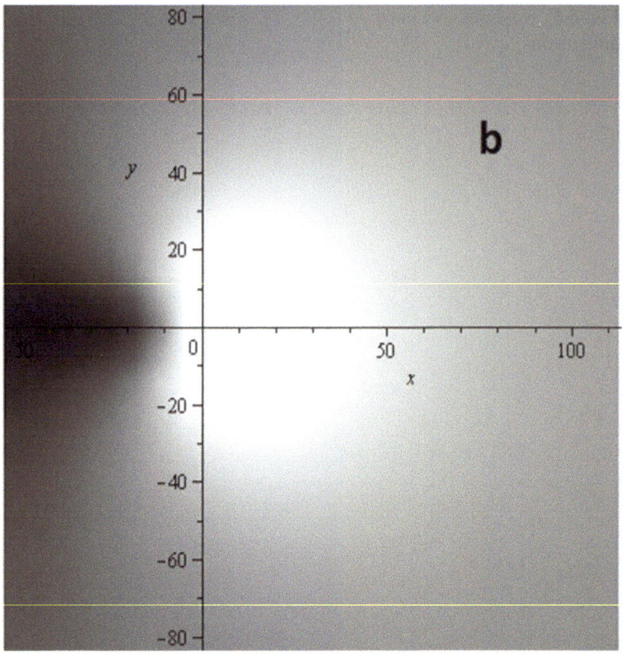

Fig. 6.7 (**a**) Numerically generated TSA data and (**b**) Calculated model for a simulated crack tip located at (−3,−1) pixels

Table 6.1 SIF Results for TSA (Linear and non-Linear), DIC and FEM

a (mm)	ΔP (N)	$\Delta K_I/\Delta P$ TSA Linear (10^{-3} MPa/N √m)	$\Delta K_I/\Delta P$ TSA Non-linear (10^{-3} MPa/N √m)	$\Delta K_I/\Delta P$ DIC (10^{-3} MPa/N √m)	$\Delta K_I/\Delta P$ FEM (10^{-3} MPa/N √m)
4.7	194	5.98	5.46	5.93	4.99
7.4	196	6.63	6.22	6.71	5.55
10	175	7.02	6.51	7.30	6.10
11.9	161	7.76	7.14	7.57	6.66
14.1	153	9.15	8.04	8.81	7.59
15.9	150	9.80	8.80	9.25	8.64
18.4	145	11.72	10.76	10.96	10.67
20.7	137	14.38	13.43	13.40	13.28

Fig. 6.8 $\Delta K_I/\Delta P$ vs. a plot for all experimental techniques and FEM

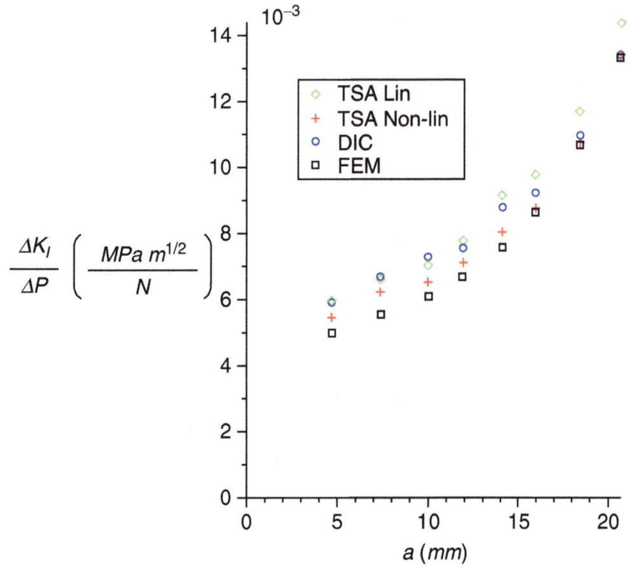

6.11 Conclusions

This work has outlined two methodologies—linear and non-linear—for determining SIF ranges using TSA with a relatively inexpensive, non-cooled micro-bolometric infrared camera coupled with a TSA software program that does not require a lock-in process. The results obtained for both are satisfactory when compared to DIC and FEM results.

The non-linear approach has some clear advantages and presents consistently better results than the linear solution, using FEA as the reference value. The main advantage of the linear approach is simplicity, which makes it faster to run, although no effort was made in the direction of optimizing the algorithms for run speed.

The process of SIF calculation using TSA was automated. The only manual input required at the current stage is the margins of tolerance for non-linearity verification in the model validity limits determination process. If the optimization of these margins were automatic, the algorithm could potentially be used to measure SIF ranges in real-time during crack propagation, although some run-speed improvements would be necessary.

References

1. Dulieu-Barton, J.M.: Introduction to thermoelastic stress analysis. Strain **35**(2), 35–39 (1999)
2. Rajik, N., Rowlands, D.: Thermoelastic stress analysis with a compact low-cost microbolometer system. Quant. Infrared Thermogr. J. **10**(2), 135–158 (2013)
3. Stanley, P., Chan, W.K.: The determination of stress intensity factors and crack tip velocities from thermoelastic infra-red emissions. In: Proceedings of International Conference of Fatigue of Engineering Materials and Structures, c262, IMechE, Sheffield, UK, pp. 105–114 (1986)
4. Stanley, P., Dulieu-Smith, J.M.: Progress in the thermoelastic evaluation of mixed mode stress intensity factors. In: Proceedings of the SEM Spring Conference on Experimental Mechanics, Dearborn, pp. 617–626 (1993)
5. Lesniak, J.R., Bazile, D.J., Boyce, B.R., Zickel, M.J., Cramer, K.E., Welch, C.S.: Stress intensity measurement via infrared focal plane array. Non-Traditional Methods of Sensing Stress, Strain, and Damage in Materials and Structures. ASTM STP 1318, Philadelphia (1997)
6. Lin, S.T., Feng, Z., Rowlands, R.E.: Thermoelastic determination of stress intensity factors in orthotropic composites using the J-integral. Eng. Fract. Mech. **56**, 579–592 (1997)
7. Tomlinson, R.A., Nurse, A.D., Patterson, E.A.: On determining stress intensity factors for mixed mode cracks from thermoelastic data. Fatigue Fract. Eng. Mater. Struct. **20**, 217–226 (1997)
8. Díaz, F.A., Yates, J.R., Tomlinson, R.A., Patterson, E.A.: Some observation on the application of thermoelasticity to fatigue cracks. In: Proceedings of SEM Conference, Milwaukee, USA (2002)
9. Díaz, F.A., Patterson, E.A., Tomlinson, R.A., Yates, J.R.: Measuring stress intensity factors during fatigue crack growth using thermoelasticity. Fatigue Fract. Eng. Mater. Struct. **27**, 571–583 (2004)
10. Marsavina, L., Tomlinson, R.A., Patterson, E.A., Yates, J.R.: Investigation of crack closure by using thermoelastic stress analysis. In: Proceedings of the 16th European Conference of Fracture, Alexandropoulos, Greece (2006)
11. Tomlinson, R.A., Olden, E.J.: Thermoelasticity for the analysis of crack tip stress fields—a review. Strain **35**(2), 49–55 (1999)
12. Shukla, A., Dally, J.W.: Experimental Solid Mechanics. College Enterprises (2010)
13. Dulieu-Smith, J.M.: Alternative calibration techniques for quantitative thermoelastic stress analysis. Strain **31**, 9–16 (1995)
14. Nelder, J.A., Mead, R.: A simplex method for function minimization. Comput. J. **7**, 308–313 (1965)
15. Gonzáles, G.L.G., Díaz, J.G., Gonzáles, J.A.O., Castro, J.T.P., Freire, J.L.F.: Approaches for determining stress intensity factors of fatigue propagating cracks using DIC. In: Proceedings of the SEM Annual Conference on Experimental Mechanics, Orlando, 6–9 June 2016
16. Vieira, R.B.: Thermography applied to the study of fatigue in polycarbonate. MSc thesis, Departamento de Engenharia Mecânica, Pontifícia Universidade Católica do Rio de Janeiro, PUC-Rio, (2016)

Chapter 7
Stress Analysis of a Finite Orthotropic Plate Containing an Elliptical Hole from Recorded Temperature Data

A. Alshaya, X. Shuai, and R. Rowlands

Abstract Individual stresses in an elliptically perforated graphite/epoxy laminated composite are determined from recorded load-induced thermal (TSA) information. Equilibrium and compatibility conditions are satisfied using complex-variable formulation, conformal mapping and analytic continuation. Processing the measured thermal data with a stress function simultaneous smooths the recorded information and evaluates the individual stress components, including on the edge of the hole. Experimental results agree with those from FEM and force equilibrium.

Keywords Composites • Elliptical hole • Stresses • Thermoelasticity • Complex variables

7.1 Introduction

Structures made of composite materials frequently contain holes or notches which produce stress concentrations. Pure analytical or theoretical stress analyses are available for only simple situations involving infinite geometries, whereas many practical problems involve complicated, finite shapes. Like numerical (FEM, FDM) approaches, analytical/theoretical analyses also depend on knowing the external loading. The latter is often unknown in practice. It is therefore advantageous to be able to stress analyze experimentally members made of orthotropic composite materials. This paper demonstrates the ability to determine the stresses in an elliptically perforated orthotropic composite member by TSA. The approach satisfies equilibrium and compatibility using a complex-variable formulation. Experimental reliability is demonstrated by FEM and force equilibrium.

By cyclically loading a structure, the stresses at a location are related to the associated stress-induced temperature changes. Under orthotropy, the recorded signal, S^*, is proportional to the change in the linear combination of the normal stresses, σ_1 and σ_2, in the directions of material symmetry, *i.e.*,

$$S^* = \Delta(K_1\sigma_1 + K_2\sigma_2). \tag{7.1}$$

K_1 and K_2, are traditionally determined experimentally. Recorded TSA data at, and adjacent to, an edge are typically unreliable and raw thermoelastic information in composites is inherently noisy. The present technique overcomes these challenges by avoiding the use of recorded data on and near edges and processing the measured interior data with a stress function employing complex variables, mapping and analytic continuation. The resulting TSA-determined stresses are available on and in the neighborhood of the edge of the hole without knowing the distant geometry or boundary conditions. Reference [1] appears to be the only previous TSA use of a complex variable stress function to stress analyze an orthotropic composite.

A. Alshaya • R. Rowlands (✉)
University of Wisconsin-Madison, Madison, WI, USA
e-mail: rowlands@engr.wisc.edu

X. Shuai
Beijing Aeronautical Science & Technology Research Institute, Beijing, China

S. Quinn, X. Balandraud (eds.), *Residual Stress, Thermomechanics & Infrared Imaging, Hybrid Techniques and Inverse Problems, Volume 9*, Conference Proceedings of the Society for Experimental Mechanics Series, DOI 10.1007/978-3-319-42255-8_7

7.2 Experimental Details

The developed hybrid-TSA approach is utilized to stress analyze a finite tensile $[0_{13}/90_5/0_{13}]$ graphite/epoxy orthotropic plate ($E_{11} = 100.9$ GPa, $E_{22} = 25$ G, $G_{12} = 2.9$ GPA, $\nu_{12} = 0.15$, [2]) containing a central elliptical hole, Fig. 7.1.

The plate was initially very lightly polished with 400 grit sand paper and then sprayed with Krylon Ultra-Flat black paint to provide an enhanced and uniform emissivity. The painted plate was sinusoidally load in tension at a mean load of 7.12 kN, maximum load of 10.67 kN and a minimum load of 3.56 kN at 20 Hz. The TSA data were recorded with Delta Therm model DT1410 system having a sensor array of 256 horizontal by 256 vertical pixels (Stress Photonics, Madison, WI, USA). TSA images were captured and averaged over 2-min durations, and then exported to Excel while converting each pixel into a data point, i.e., 256 by 256 matrix. Pixel size is 0.35 mm. The measured TSA data were averaged about the horizontal and vertical axes of symmetry, Fig. 7.2. The thermoelastic coefficients $K_1 = 12.38$ U/psi and $K_2 = 101.25$ U/psi were determined experimentally.

Due to unreliability, the recorded data along and near the edge of the hole were not used. Only the thermal information within the region covered by $1.1a$ (20.96 mm) and $1.1b$ (10.48 mm) to $1.85a$ (35.24 mm) and $1.85b$ (17.63 mm) was considered. This region, denoted as R^*, contains 2558 values of S^*. Like most experimental data, the recorded S^* values include some noise which necessitates collecting more measured input values than the number of unknown Airy coefficients, *i.e.*, the number of equations, P, will exceed the number of real coefficients, $2N$. The resulting overdetermined system of equations with which to evaluate the Airy coefficients was solved using least-squares.

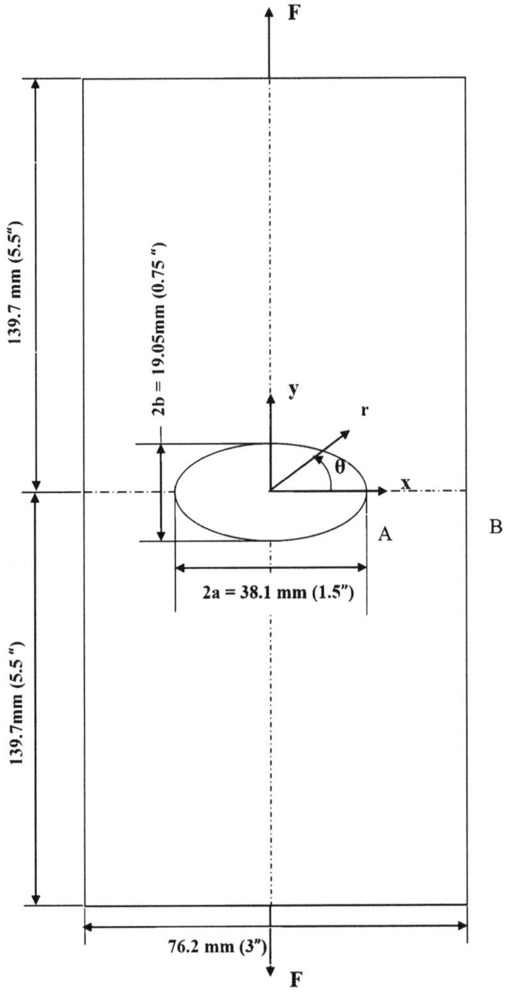

Fig. 7.1 Loaded finite Gr/E $[0_{13}/90_5/0_{13}]$ composite plate with central elliptical hole

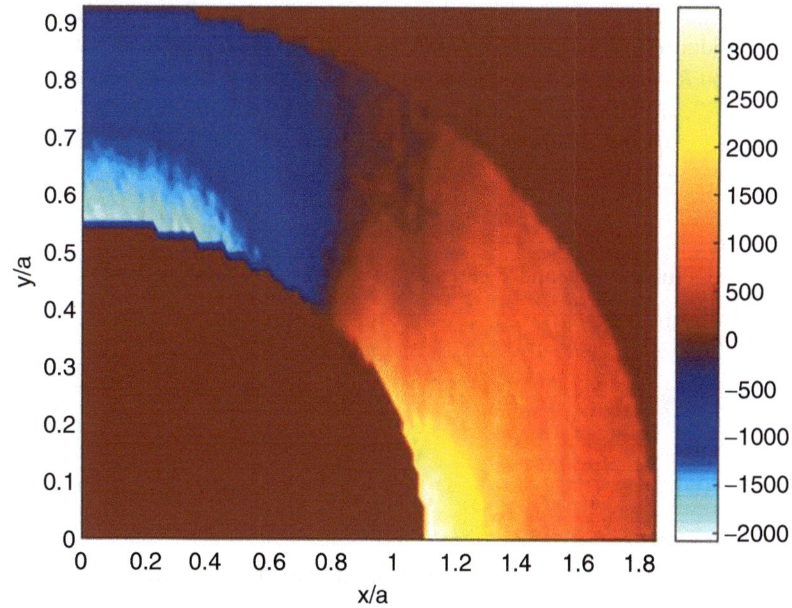

Fig. 7.2 Thermoelastic data, S^*, as averaged throughout the four quadrants (load range = 7.12 kN)

7.3 Relevant Equations

For plane problems having rectilinear orthotropy and no body forces, the Airy stress function, F, can be expressed as a summation of two arbitrary analytical functions, $F_1(z_1)$ and $F_2(z_2)$, in terms of the complex variables, z_1 and z_2, as [3]

$$F = 2Re[F_1(z_1) + F_2(z_2)] \tag{7.2}$$

such that $z_j = x + \mu_j y$ for $j = 1, 2$. The complex material properties μ_1 and μ_2 are distinct roots of the following characteristic equation

$$\mu^4 + \left(\frac{E_{11}}{G_{12}} - 2\nu_{12}\right)\mu^2 + \frac{E_{11}}{E_{22}} = 0 \tag{7.3}$$

The 1- and 2-orientations are the directions of laminate material symmetry, and are in the vertical and horizontal directions, respectively, Fig. 7.1. The roots of Eq. (7.3) are complex, i.e., $\mu_1 = \alpha + i\beta$, $\mu_2 = \gamma + i\delta$, $\mu_3 = \bar{\mu}_1$, and $\mu_4 = \bar{\mu}_2$. Stresses in rectangular coordinates (x, y) of the physical $z(= x + iy)$ plane can be expressed in terms of the stress function, F. By introducing the new stress functions

$$\Phi(z_1) = \frac{dF_1(z_1)}{dz_1}, \quad \text{and} \quad \Psi(z_2) = \frac{dF_2(z_2)}{dz_2} \tag{7.4}$$

the stresses may be written as

$$\sigma_x = 2Re\left[\mu_1^2\Phi'(z_1) + \mu_2^2\Psi'(z_2)\right] \tag{7.5}$$

$$\sigma_y = 2Re[\Phi'(z_1) + \Psi'(z_2)] \tag{7.6}$$

$$\tau_{xy} = -2Re[\mu_1\Phi'(z_1) + \mu_2\Psi'(z_2)] \tag{7.7}$$

For a region of a component adjacent to a traction free-edge, Φ and Ψ can be related to each other by the conformal mapping and analytic continuation techniques. This enables expressing the stresses in terms of the single stress function, Φ. Moreover, Φ will be represented by a truncated power-series expansion whose unknown complex coefficients can all be evaluated experimentally. Knowing Φ and Ψ, the individual stresses are available from Eqs. (7.5) through (7.7). For a significantly large region, it may be necessary to satisfy other boundary conditions at discrete locations. Conformal mapping is introduced to simplify the problem by mapping the region R_z of a complicated physical $z = x + iy$ plane of a structure into a region R_ζ of a simpler shape in the $\zeta = \xi + i\eta$ plane, the latter being a half plane or unit circle. For a mapping function, $z = \omega(\zeta)$, which maps R_ζ of the simpler plane into R_z of the more complicated physical plane, auxiliary planes and their induced mapping functions are defined in terms of $\zeta_j = \xi + \mu_j\eta$ so by substituting $z_j = x + \mu_j y$ into equation

$$z_j = \omega_j(\zeta_j), \quad j = 1,2 \tag{7.8}$$

the stress functions Φ and Ψ can now be expressed as analytic functions of ζ_1 and ζ_2, respectively, such that their derivatives become

$$\Phi'(z_1) = \Phi'(\zeta_1)\frac{d\zeta_1}{dz_1} = \frac{\Phi'(\zeta_1)}{\omega_1'(\zeta_1)}. \tag{7.9}$$

$$\Psi'(z_2) = \frac{\Psi'(\zeta_2)}{\omega_2'(\zeta_2)} \tag{7.10}$$

Using the concept of analytic continuation, the stress functions for a region R_z adjacent to a traction-free boundary of an orthotropic material are related by [4, 5]

$$\Psi(\zeta_2) = B\overline{\Phi(\overline{\zeta}_2)} + C\Phi(\zeta_2) \tag{7.11}$$

where B and C are complex material properties and which depend only on μ_i. Eq. (7.11) enables the elastic state of the structure to be expressed in terms of a single stress function, $\Phi(\zeta_1)$, which can be represented by a truncated Taylor series or Laurent series. Mapping the boundary of the hole in the physical z-plane to the real axis of the ζ-plane uses a Taylor series expansions whereas mappings the boundary of the physical plane to the unit circle in the ζ-plane employs a Laurent series. The finite Taylor series expansion of the stress function is [6]

$$\Phi(\zeta_1) = \sum_{j=0}^{N} A_j(\zeta_1 - \zeta_0)^j \tag{7.12}$$

whereas the stress function expressed as Laurent series expansion is

$$\Phi(\zeta_1) = \sum_{\substack{j=-N,\ \ldots \\ j\neq 0}}^{N} A_j\zeta_1^j \tag{7.13}$$

$A_j = a_j + ib_j$ are the unknown complex coefficients, ζ_0 is some point on the traction-free boundary and the summation in Laurent expansion involves only odd values of j. Substituting Eqs. (7.12) or (7.13) into Eq. (7.11) yields

$$\Psi(\zeta_2) = \sum_{j=0}^{N} \left(\overline{A}_j B + A_j C\right)(\zeta_2 - \zeta_0)^j \tag{7.14}$$

when using a Taylor series while for a Laurent expansion it is

$$\Psi(\zeta_2) = \sum_{\substack{j=-N,\ -N+2,\ \ldots \\ j\neq 0}}^{N} \left(\overline{A}_j B\zeta_2^{-j} + A_j C\zeta_2^j\right) \tag{7.15}$$

$\bar{A}_j$ is the complex conjugate of A_j. At least for a finite simply-connected region R_ζ, $\Phi(\zeta_1)$ is a single-valued analytic function. Upon combining Eqs. (7.4) through (7.12) and (7.14), one obtains the following expressions for the stress when using a Taylor series,

$$\sigma_{xx} = 2\sum_{j=1}^{N} Re\left\{\left[\frac{j\mu_1^2}{\omega_1'(\zeta_1)}(\zeta_1-\zeta_0)^{j-1} + \frac{j\mu_2^2 C}{\omega_2'(\zeta_2)}(\zeta_2-\zeta_0)^{j-1}\right]A_j + \left[\frac{j\mu_2^2 B}{\omega_2'(\zeta_2)}(\zeta_2-\zeta_0)^{j-1}\right]\bar{A}_j\right\} \tag{7.16}$$

$$\sigma_{yy} = 2\sum_{j=1}^{N} Re\left\{\left[\frac{j}{\omega_1'(\zeta_1)}(\zeta_1-\zeta_0)^{j-1} + \frac{jC}{\omega_2'(\zeta_2)}(\zeta_2-\zeta_0)^{j-1}\right]A_j + \left[\frac{jB}{\omega_2'(\zeta_2)}(\zeta_2-\zeta_0)^{j-1}\right]\bar{A}_j\right\} \tag{7.17}$$

$$\sigma_{xy} = -2\sum_{j=1}^{N} Re\left\{\left[\frac{j\mu_1}{\omega_1'(\zeta_1)}(\zeta_1-\zeta_0)^{j-1} + \frac{j\mu_2 C}{\omega_2'(\zeta_2)}(\zeta_2-\zeta_0)^{j-1}\right]A_j + \left[\frac{j\mu_2 B}{\omega_2'(\zeta_2)}(\zeta_2-\zeta_0)^{j-1}\right]\bar{A}_j\right\} \tag{7.18}$$

An analogous set of expressions can be obtained for the stresses by combining Eqs. (7.4) through (7.11), (7.13) and (7.15) when using a Laurent series. The only unknowns in these expressions for the stresses are the complex coefficients, A_j. The latter can be determined from measured thermoelastic data. Choosing the y-axis parallel to the strongest, stiff orientation of the composite, Fig. 7.1, *i.e.*, 1-direction of an orthotropic composite material, and introducing the Taylor series according to Eqs. (7.12), (7.16) and (7.17), the thermoelastic data S^* can be expressed as

$$\begin{aligned} S^* &= K_1\sigma_{yy} + K_2\sigma_{xx} \\ &= 2\sum_{j=1}^{N} Re\left\{\left[\frac{j(K_1+K_2\mu_1^2)}{\omega_1'(\zeta_1)}(\zeta_1-\zeta_0)^{j-1} + \frac{j(K_1+K_2\mu_2^2)C}{\omega_2'(\zeta_2)}(\zeta_2-\zeta_0)^{j-1}\right]A_j \right. \\ &\quad \left. + \left[\frac{j(K_1+K_2\mu_2^2)B}{\omega_2'(\zeta_2)}(\zeta_2-\zeta_0)^{j-1}\right]\bar{A}_j\right\} \end{aligned} \tag{7.19}$$

Upon introducing the Laurent series according to Eq. (7.13), the thermoelastic data S^* becomes [7]

$$\begin{aligned} S^* &= K_1\sigma_{yy} + K_2\sigma_{xx} \\ &= 2\sum_{\substack{j=-N,\,-N+2,\,\ldots \\ j\neq 0}}^{N} Re\left\{\left[\frac{j(K_1+K_2\mu_1^2)}{\omega_1'(\zeta_1)}\zeta_1^{\,j-1} + \frac{j(K_1+K_2\mu_2^2)C}{\omega_2'(\zeta_2)}\zeta_2^{\,j-1}\right]A_j \right. \\ &\quad \left. - \left[\frac{j(K_1+K_2\mu_2^2)B}{\omega_2'(\zeta_2)}\zeta_2^{\,-j-1}\right]\bar{A}_j\right\} \end{aligned} \tag{7.20}$$

The thermoelastic data, S^*, at P different locations are chosen to be inside the region R^*. Eqs. (7.19) and (7.20) each forms a system of simultaneous linear equations, $[M]_{P\times 2N}\{c\}_{2N\times 1} = \{S^*\}_{P\times 1}$, where matrix $[M]$ consists of analytical expression of S^*, vector $\{c\}_{2N\times 1} = \{a_1, b_1, a_2, b_2, \ldots, a_N, b_N\}$ has $2N$ unknown real coefficients (a_j and b_j) and vector $\{S^*\}_{P\times 1}$ has P equations such that $P \gg 2N$. The best values of the coefficients, A_j, in a least-squares numerical sense, can be determined from measured values of S^*. The variables $\zeta_j = \xi + \mu_j\eta$, in Eqs. (7.19) and (7.20) are related to the physical locations $z = x + iy$ through the inverse mapping function $z_j = \omega_j(\zeta_j)$. The individual stresses are then known throughout the region R_z, including on the traction-free edge Γ from Eqs. (7.16) through (7.18) if using the Taylor expansion or from the corresponding set of equations when using the Laurent expansion.

The present objective is to apply the approach to a region R_z adjacent to the traction-free boundary of the hole in Fig. 7.1 by mapping between the region R_ζ in the mapped plane and region R_z in physical plane, where Γ_ζ, a section of the real axis in the ζ-plane, goes to the physical traction-free boundary, Γ. For a region adjacent to the traction-free elliptical

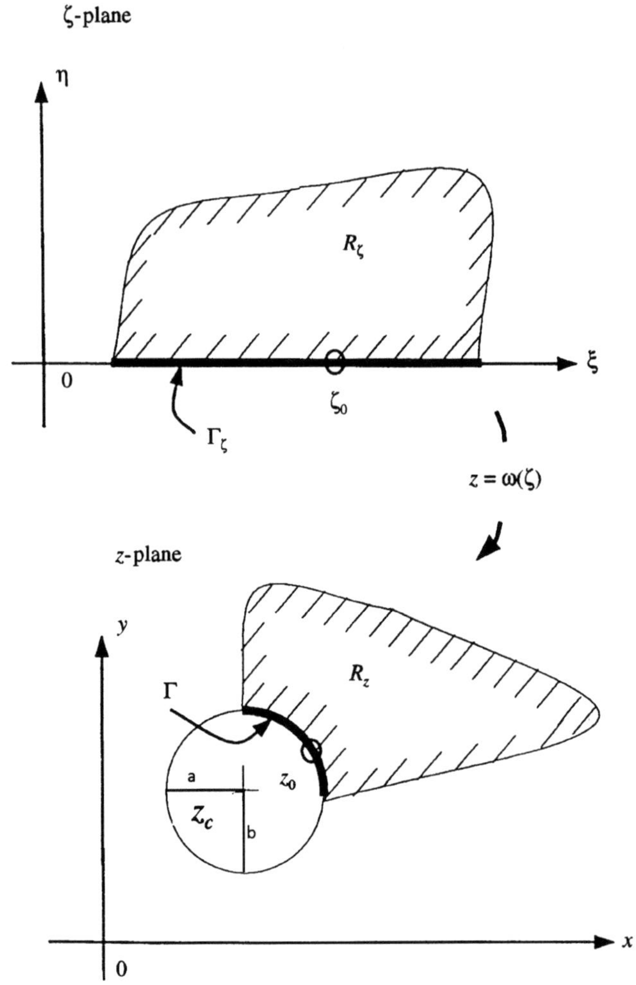

Fig. 7.3 Conformal mapping for elliptical boundary

edge in Figs. 7.1 and 7.2, the following function maps the region R_ζ of the ζ-plane into region R_z of the z-physical plane where Γ_ζ is a section of the real axis of the ζ-plane (Taylor series formulation), Fig. 7.3. Quantities a and b are the lengths of the

$$z = \omega(\zeta) = \frac{a+b}{2}\left(\frac{i-\zeta}{i+\zeta}\right) + \frac{a-b}{2}\left(\frac{i+\zeta}{i-\zeta}\right) + z_c \tag{7.21}$$

major (parallel to the to the horizontal x-direction) and minor (parallel to the vertical y-direction) axes of the ellipse, respectively, and z_c is the center of the ellipse, Figs. 7.1 and 7.3. For convenience, the origin of the coordinate system is chosen at the center of the ellipse, *i.e.*, $z_c = 0$. Substituting Eq. (7.20) into $z_j = x + \mu_j y$ produces

$$z_j = \omega_j(\zeta_j) = \frac{a - ib\mu_j}{2}\left(\frac{i-\zeta_j}{i+\zeta_j}\right) + \frac{a+ib\mu_j}{2}\left(\frac{i+\zeta_j}{i-\zeta_j}\right), \quad j = 1, 2 \tag{7.22}$$

The derivatives of this mapping function are

$$\omega_j'(\zeta_j) = -i\frac{a-ib\mu_j}{(i+\zeta_j)^2} + i\frac{a+ib\mu_j}{(i-\zeta_j)^2}, \quad j = 1, 2 \tag{7.23}$$

and the inverse of the induced mapping functions are

$$\zeta_j = \omega_j^{-1}(z_j) = i\frac{a - ib\mu_j - \left(z_j \pm \sqrt{z_j^2 - a^2 - b^2\mu_j^2}\right)}{a - ib\mu_j + \left(z_j \pm \sqrt{z_j^2 - a^2 - b^2\mu_j^2}\right)}, \quad j = 1,2 \tag{7.24}$$

The branch of the square root in Eq. (7.23) is chosen such that $Im\,\zeta_j \leq 0$ for $j = 1, 2$. For a region adjacent to the elliptical hole, the following function using $z_j = x + \mu_j y$, where $x = a\cos\theta$ and $y = -b\sin\theta$,

$$z_j = \omega_j(\zeta_j) = \frac{a - ib\mu_j}{2}\frac{1}{\zeta_j} + \frac{a + ib\mu_j}{2}\zeta_j, \quad j = 1,2 \tag{7.25}$$

maps the region of a unit circle, R_ζ, in the ζ-plane into the region R_z in the z-physical plane (Laurent formulation). The derivatives of this mapping functions are

$$\omega_j'(\zeta_j) = \frac{a + ib\mu_j}{2} - \frac{a - ib\mu_j}{2}\frac{1}{\zeta_j^2}, \quad j = 1,2 \tag{7.26}$$

and the inverse of the induced mapping functions are

$$\zeta_j = \omega_j^{-1}(z_j) = \frac{z_j \pm \sqrt{z_j^2 - a^2 - \mu_j^2 b^2}}{a + ib\mu_j}, \; j = 1,2 \tag{7.27}$$

The branch of the square root in Eq. (7.26) is chosen such that $|\zeta_j| < 1$ for $j = 1, 2$.

7.4 Finite Element Analysis

For comparison with the TSA results, a finite element analysis (FEA) using ANSYS prediction was prepared of the plate of Fig. 7.1 and the elastic properties. Due to the symmetry, only the upper right quarter of the plate was modeled with symmetrical boundary conditions applied at the bottom and left edges. Based on the static equivalent, a far-field stress of

$$\sigma_0 = \frac{F}{A} = \frac{1600 \text{ lb}}{0.208 \text{ in.} \times 3 \text{ in.}} = 2564 \text{ psi} = 17.68 \text{ MPa}$$

was applied numerically to the top edge of the plate. Plane 82 Isoparametric elements with eight nodes were employed. Very small elements were used in the neighborhood of the elliptical hole to obtain reliable data. A convergence test was applied until the change in results of the maximum stress between two successive meshing was less than 2 %. The FE model utilizes 67,500 elements and 68,101 nodes.

7.5 Results

The complex coefficients, $A_j = a_j + ib_j$, were evaluated by two approaches: the first approach uses Eqs. (7.21) through (7.23) and the Taylor representation of the stress function of Eq. (7.14) to map the physical plane to a half-plane in ζ-plane; the second approach uses Eq. (7.15), maps the physical plane to the unit circle in the ζ-plane and the corresponding stress expressions for the Laurent formulation. The complex coefficients A_j were evaluated individually from Eqs. (7.19) or (7.20) employing the recorded thermoelastic data inside the region, R^*, utilizing the Taylor or Laurent representations for the stress function, respectively. The individual stresses throughout region R_z, including on the edge Γ where no thermoelastic input data were employed, were evaluated using Eqs. (7.16) through (7.18) for the Taylor expansion or corresponding expressions [7] for the Laurent expansion. The number of complex Airy coefficients, N, to retain in the stress functions was selected by evaluating the difference between the magnitude of experimentally-based thermoelastic data and those predicted by the present hybrid method using the root mean square (RMS) method, as well as by comparing the reconstructed and experimentally-based, S^* [7]. These evaluations demonstrate that both the first (Taylor expansion) and second (Laurent expansion) approaches require N = 4 complex coefficients (8 real coefficients) [7]. Having determined how many Airy coefficients to retain, their values were determined by least squares. Knowing the magnitudes of the Airy coefficients, the stresses are available from Eqs. (7.16) through (7.18) for the Tayler representation or corresponding expressions when using the Laurent formulation. The tangential stress, σ_{tt}, and the polar components of stress, normalized with respect to $\sigma_0 = 17.68$ MPa, are plotted on the edge of the hole in Fig. 7.4. Not surprising, the magnitudes of σ_{tt}/σ_0 and $\sigma_{\theta\theta}/\sigma_0$ are very similar to each other, and σ_{rr}/σ_0 and $\sigma_{r\theta}/\sigma_0$ are small, on the edges of the hole. Contour plots of normalized Cartesian components of stress from the hybrid technique using the first approach (Taylor series) and ANSYS are plotted in Fig. 7.5. These TSA-based results agree with the FEM predictions.

TSA reliability was further assessed by checking load equilibrium. This was done by numerically integrating the TSA-determined vertical stress, σ_{yy}, along the line $y = 0$ (line AB in Fig. 7.1),

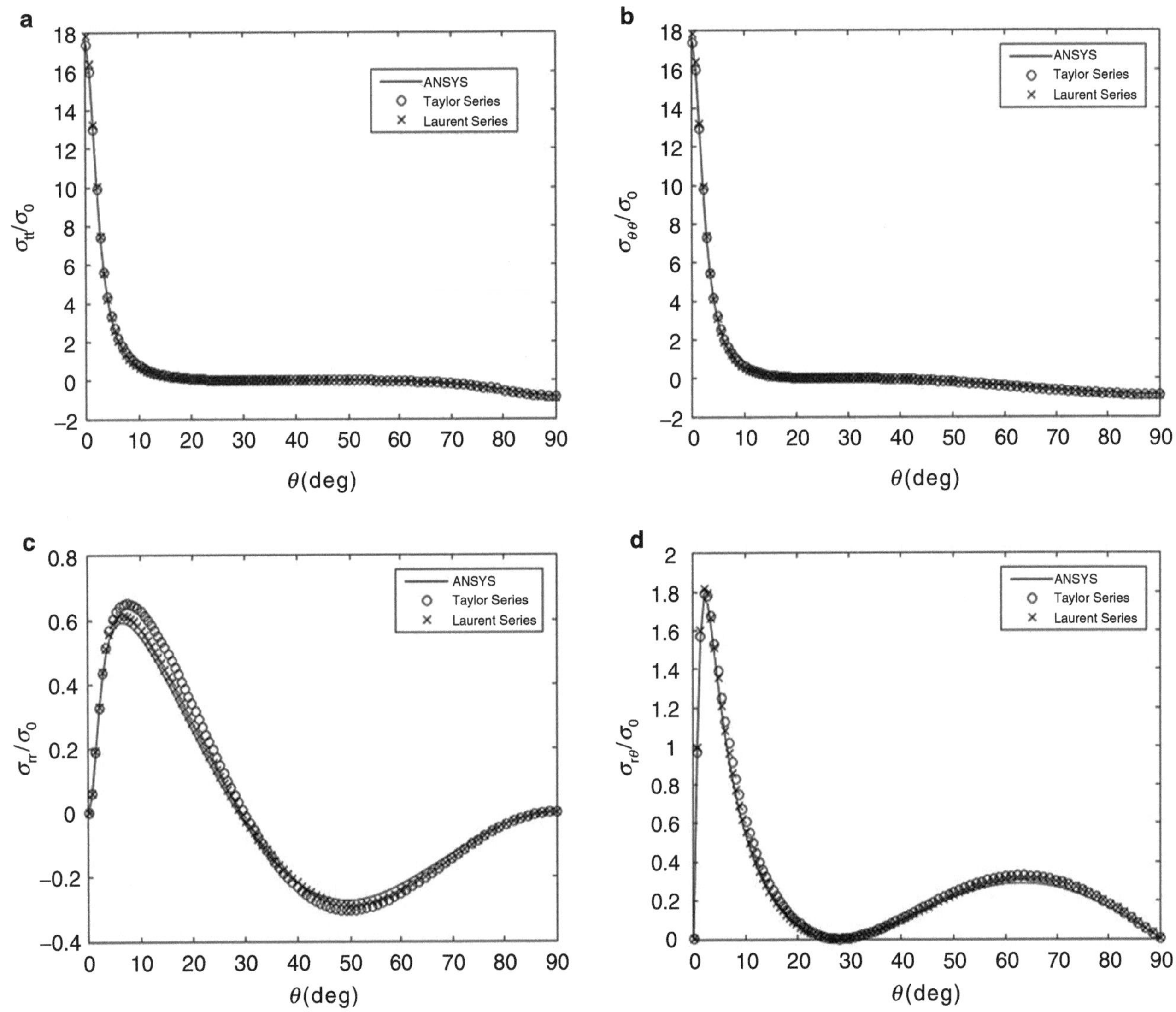

Fig. 7.4 Plot of (**a**) σ_{tt}/σ_0, (**b**) $\sigma_{\theta\theta}/\sigma_0$, (**c**) σ_{rr}/σ_0 and (**d**) $\sigma_{r\theta}/\sigma_0$ along edge of hole from ANSYS and TSA (Taylor and Laurent series)

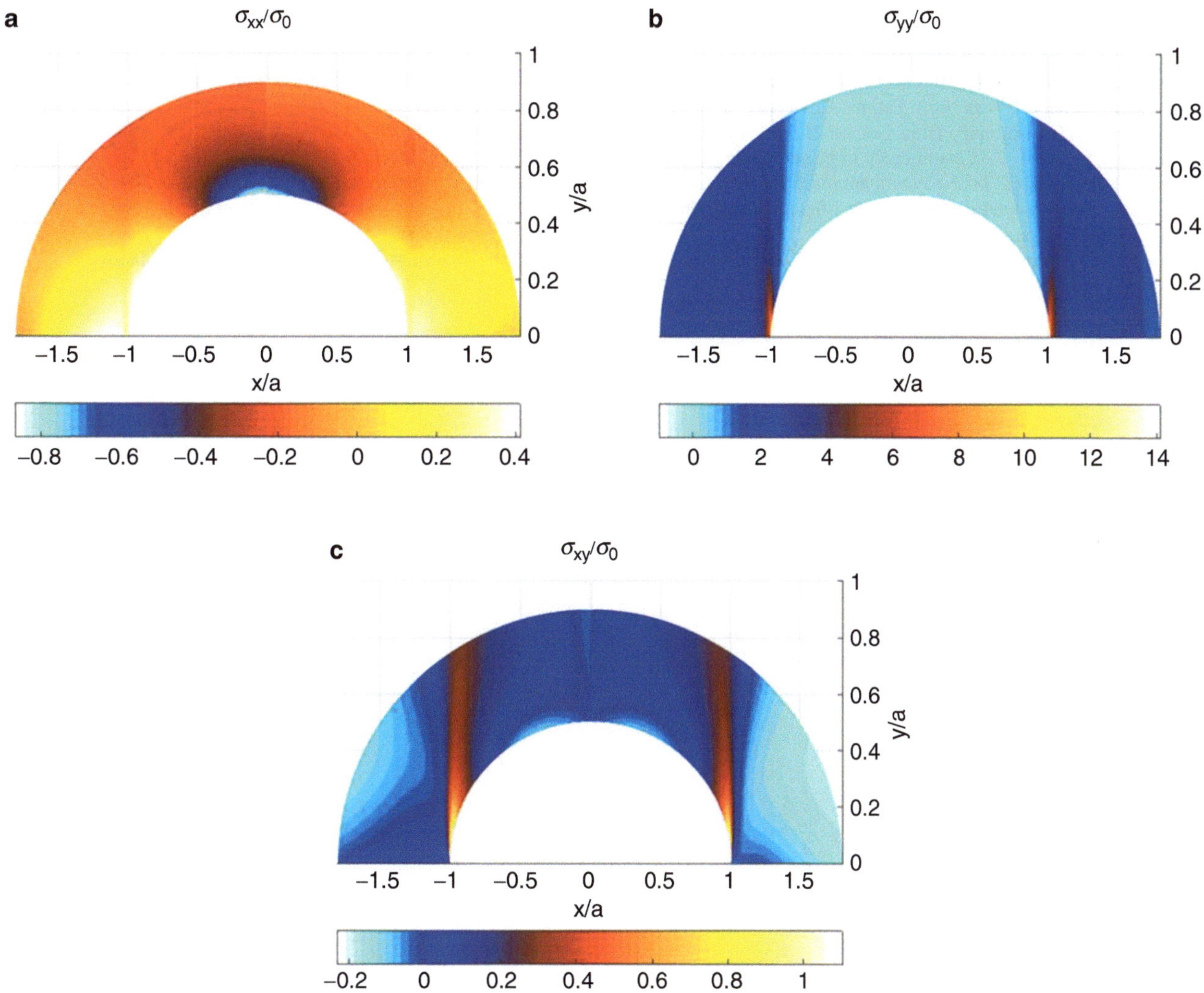

Fig. 7.5 Contour plot of (**a**) σ_{xx}/σ_0, (**b**) σ_{yy}/σ_0, and (**c**) σ_{xy}/σ_0 throughout the region adjacent to hole by FEA (*right*) and (*right*) TSA using the first approach (Taylor series)

$$F = \int \sigma_{yy} dA = 2\int_a^{W/2} \sigma_{yy} t dx$$

where t and W are the plate thickness and width, respectively. Results based on the Taylor and Laurent representations are 7.2 and 7.4 kN, respectively, both of which agree well with the physically applied load of 7.12 kN.

7.6 Summary, Discussion and Conclusions

A hybrid method which processes the load-induced thermal (TSA) signals information with a stress function in complex variables, together with conformal mapping and analytic continuation concepts, provides the individual stresses on and in the neighborhood of an elliptical hole in a finite orthotropic composite plate. Both Laurent and Taylor series representations of the stress function are utilized with equal satisfaction. Unlike pure theoretical or numerical methods, knowledge of the external boundary conditions is unnecessary. TSA results agree with those from FEM and force equilibrium.

Acknowledgement The first author was funded by a graduate scholarship from the University of Kuwait.

References

1. Lin, S.T., Rowlands, R.E.: Thermoelastic stress analysis of orthotropic composites. Exp. Mech. **35**(3), 257–265 (1995)
2. Private Communication with Narin Fatima, University of Wisconsin-Madison, Madison, WI
3. Lekhniskii, S.G.: Anisotropic Plates. Gordon & Breach Scientific Publishers, New York (1956)
4. Bowie, O.L., Freese, C.E.: Central crack in plane orthotropic rectangular sheet. Int. J. Fract. Mech. **8**(1), 49–57 (1972)
5. Gerhardt, T.D.: A hybrid finite element approach for stress analysis of notched anisotropic materials. J. Appl. Mech. **15**(4), 804–810 (1984)
6. Khaja. A.A.: Experimentally determined full-field stress, strain and displacement analysis of perforated finite members. PhD Thesis, University of Wisconsin-Madison, Madison (2012)
7. Alshaya, A.: Experimental, analytical and numerical analyses of orthotropic materials. PhD Thesis, University of Wisconsin-Madison, Madison (2016)

Chapter 8
Using TSA to Identify Regions Having Developed Plastic Strain during Welding

Geoffrey P. Howell, Janice M. Dulieu-Barton, and Mithila Achintha

Abstract Residual stress can be related to plastic strain experienced by a component; therefore the measurement of plastic strain presents the potential for residual stress to be investigated. Previous work has shown that Thermoelastic Stress Analysis (TSA) can identify regions that have undergone plastic strain. To do this, it was necessary to manufacture a test specimen of identical geometry containing zero plastic strain known as a reference specimen. Identifying the regions that have undergone plastic strain is then a simple matter of subtracting two data sets. This approach assumes there is always a reference specimen available, which in an industrial context is not the case. To make the work applicable to in-service components it is necessary to create a simulated reference specimen. The paper presents a means of establishing the simulated TSA reference specimen using Finite Element (FE) modelling through building a linear, elastic model and then including effects of nonlinear plasticity. To validate the idea a welded mock-up is used that contains a known level of plastic strain alongside a strain free specimen of identical material and geometry to the mock-up that has been manufactured to shape using water jet cutting. The work in the paper describes the outcome of using both the experimental and simulated reference specimens and considers the effect of varying material properties.

Keywords TSA • Plastic strain • FEA • NDE

8.1 Introduction

Residual stresses can form during the manufacture, installation, and working life of an engineering component. It is important to understand the residual stress distribution within a component because their interaction with the stresses from in-service loading can cause premature failure of the component [1], which is important for industry both from an economic and a safety point of view. There are a variety of ways to measure residual stresses within a component, the most common of which are hole drilling, and X-ray diffraction techniques. Both of these provide measurements at a single point and many different tests need to be completed before a distribution map can be generated for the surface of the component, which is time consuming. It is also very uncommon for the analysis to be performed on-site as the equipment required for either technique is often limited to laboratory use. It has been proposed that Thermoelastic Stress Analysis (TSA) [2], when combined with Finite Element (FE) modelling, provides an alternative approach that allows for full field, on-site analysis to be undertaken.

TSA is a stress analysis technique which utilises infra-red (IR) thermography to measure the thermal emissions from the surface of a specimen when stress is applied. The temperature change, ΔT, can be related to the change in the first stress invariant, $\Delta(\sigma_1 + \sigma_2 + \sigma_3)$, i.e. the sum of the principal stresses. Since the IR detector only measures surface emissions, these can only be related to the in-plane stresses at the surface of the component so only the first two component of the change in stress are used in the thermoelastic equation (i.e. $\Delta(\sigma_1 + \sigma_2)$). This relationship is quantified in the following:

$$\Delta T = -KT\Delta(\sigma_1 + \sigma_2) \tag{8.1}$$

where T is surface temperature, and K is the thermoelastic constant. $K = \frac{\alpha}{\rho C_P}$ and α, ρ, C_P are the coefficient of linear thermal expansion, density, and specific heat at constant temperature respectively.

Equation (8.1) applies for linear, elastic, isotropic, homogenous materials when the temperature change has taken place under adiabatic conditions. To achieve adiabatic conditions the component needs to be loaded cyclically.

G.P. Howell (✉) • J.M. Dulieu-Barton • M. Achintha
University of Southampton, University Road, Highfield, Southampton SO17 1BJ, UK
e-mail: G.Howell@soton.ac.uk

S. Quinn, X. Balandraud (eds.), *Residual Stress, Thermomechanics & Infrared Imaging, Hybrid Techniques and Inverse Problems, Volume 9*, Conference Proceedings of the Society for Experimental Mechanics Series, DOI 10.1007/978-3-319-42255-8_8

Residual stress cannot be directly measured using TSA, however residual stress can be related to plastic strain as a change in plastic strain causes a change in the thermoelastic constant, *K* [3]. The result is a change in the thermoelastic response. So by comparing the thermoelastic response from a component under investigation to that of an identical component without plastic strain (i.e. a reference component) the regions containing plastic strain can be identified. For specifically manufactured specimens it is relatively simple to create a reference specimen at the same time that can be used to provide the reference dataset the comparison. In-service industrial components are unlikely to have plastic strain free reference specimens, so FE modelling is used to create a synthetic reference dataset. The model, initially elastic with elastic–plastic properties being the next iteration, has a single step stress applied so the resultant surface stress sum $\Delta(\sigma_1 + \sigma_2)$ can be found. The stresses are converted into a thermoelastic response using Eq. (8.1) and the experimental dataset can be compared to the reference dataset to create a resultant dataset or bitmap. This bitmap contains information in the form of thermoelastic response, so a conversion into plastic strain needs to be made. This will be performed using a calibration coefficient that is unique for each material, in this case 316 L stainless steel.

8.2 Methodology

The specimen under investigation has been designed to have a symmetrical heat affected zone at the centre without suffering any distortion of the geometry of the specimen. This was achieved by having weld beads laid down simultaneously on opposite faces of a plate (15 × 270 × 300 mm) in the same direction, at the same speed. This symmetrical welding prevented uneven cooling states in either face, and hence deformation of the plate. From the welded plate, specimens were cut using electro-discharge machining (EDM) that were 7 mm in thickness, see Fig. 8.1a. Initial trial cuts with slimmer specimens caused distortion through stress relaxation around the heat affected zone. The welded specimen was prepared for TSA testing by applying a thin layer of matt black paint (<25 μm thick Electrolube Matt Black) to the surface of the specimen [6]. This increases the emissivity of the specimen surface improving the thermal emission and hence reducing noise in the thermoelastic response. An IR detector (Cedip Silver 480 M) was used to collect the data for TSA. The specimen was installed into a servo-hydraulic test machine (Instron 8800 Servo-Hydraulic) and the detector set up in front as shown in Fig. 8.1b. The specimen was cyclically loaded with a sine wave load of 7.5 ± 6.5 kN at a frequency of 15 Hz.

To produce a resultant bitmap of the change in thermoelastic response due to loading a reference dataset was required. To generate the reference dataset, a high resolution macro image was taken of the weld geometry, see Fig. 8.2a, and a model accurately portraying the weld geometry was created in Solidworks 2013. This geometry was converted to a 3D elastic FE model in ABAQUS 6.13 for pre-processing, simulation, and post-processing. In the pre-processing, experimentally derived material properties were applied to the FE model (E = 197 GPa, ν = 0.3) [3] and the mesh, loads, and boundary conditions

a)

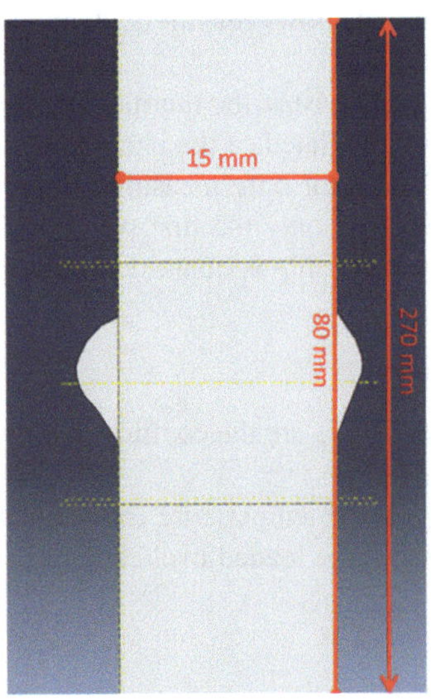

b)

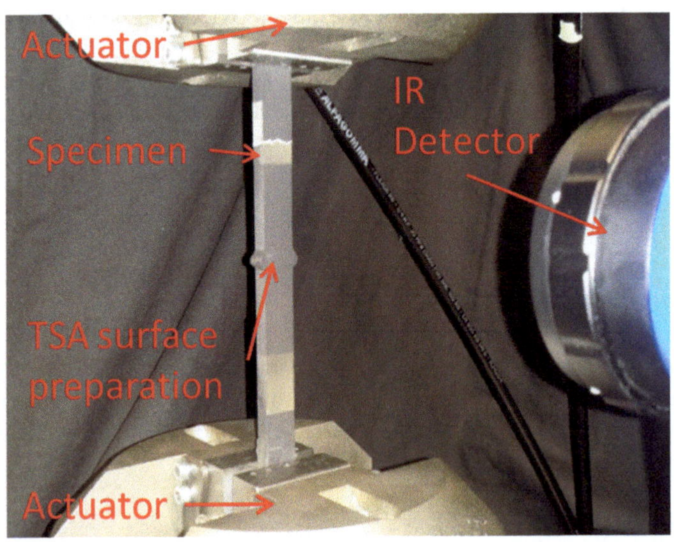

Fig. 8.1 (**a**) Specimen design. Eighty millimeter is the region prepared for TSA, 270 mm is the total length of specimen; (**b**) Experimental set up with specimen prepared for TSA using black paint and Cedip Silver 480 M IR detector in position

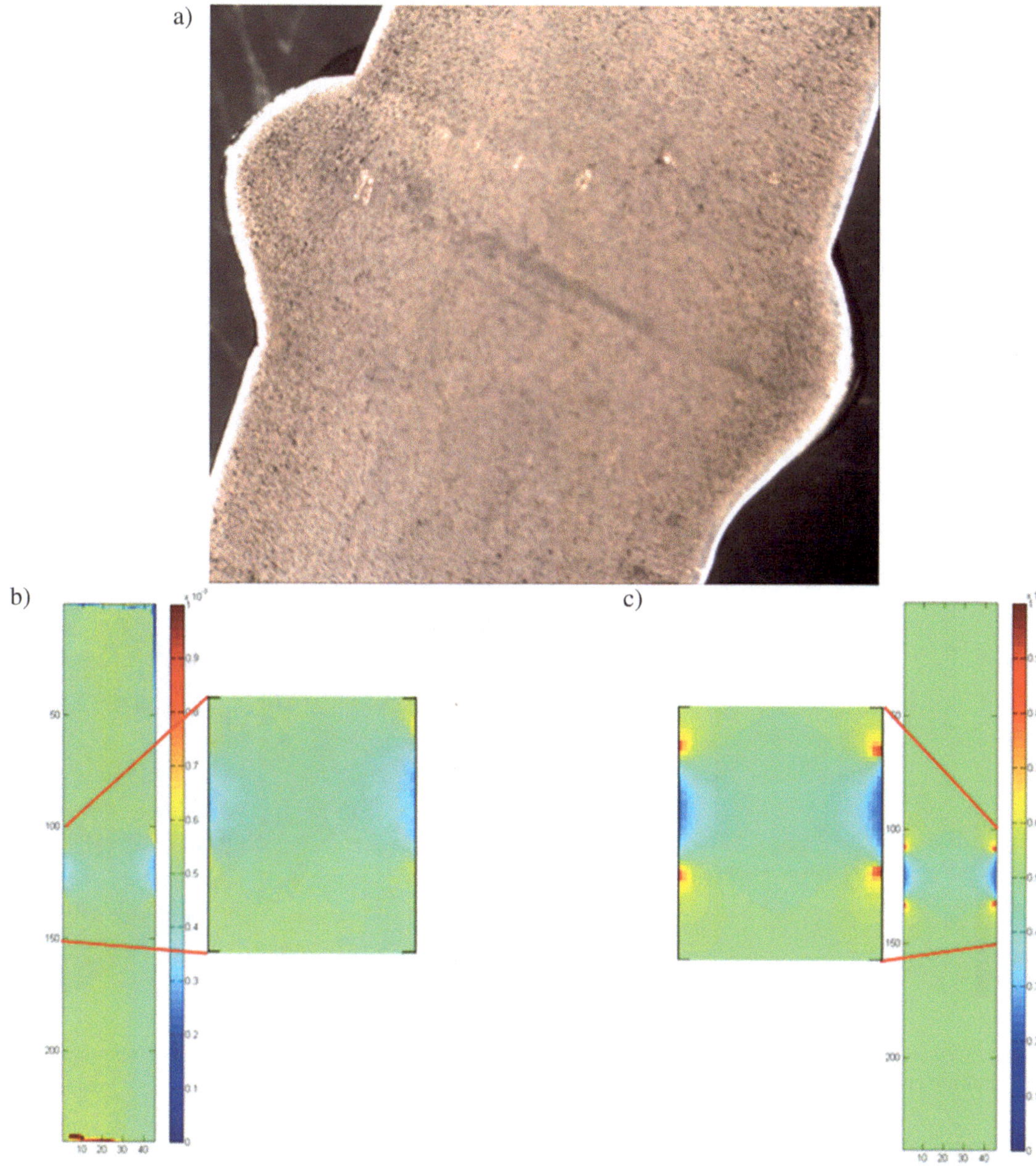

Fig. 8.2 (**a**) Image of weld geometry that was used to create the FE model for the synthetic bitmap; (**b**) ΔT/T experimental bitmap from TSA testing with expanded view of weld toe regions; (**c**) ΔT/T synthetic bitmap from FEA with expanded view of weld toe regions

were applied. A uniform mesh of quadratic solid elements, size 0.5 mm, was input to the model, and a single step load equivalent to the loading range of the experimental cyclic loading was applied to the lower end of the specimen giving a tensile stress of 123 MPa in the vertical direction. The simulation was completed and the stress results σ_1 and σ_2 were output from each node with the node coordinates. These values were combined to give a $\Delta(\sigma_1 + \sigma_2)$ at each node coordinate, and an interpolation routine was run to get the output stress map in a dataset matrix that was the same size and shape as that provided by the TSA, see Fig. 8.2b, c. The thermoelastic constant was used to convert the stress values into values of $\frac{\Delta T}{T}$ and these can then be compared to the experimental datasets.

8.3 Results and Discussion

When the reference dataset from the FE model, known as the synthetic bitmap, is taken from the experimental TSA results, known as the experimental bitmap, the values left over reveal the regions that have undergone plastic strain. This resultant dataset, Fig. 8.3a shows regions underneath the toe of each weld where there is significant departure from the elastic, reference model. The regions of high plastic strain predicted when a model is used to simulate the welding process [4] and can be seen in Fig. 8.3b. The regions appear negative because the results from the synthetic bitmap were taken from the results of the experimental bitmap. An improvement to this procedure would be to normalise the results through division rather than subtraction, and avoid the potential confusion of positive and negative strain results.

In Fig. 8.4, the thermoelastic response from beneath the weld through the centre of the specimen is compared from the synthetic bitmap, the experimental bitmap, and the resultant bitmap. The difference in thermoelastic response due to the

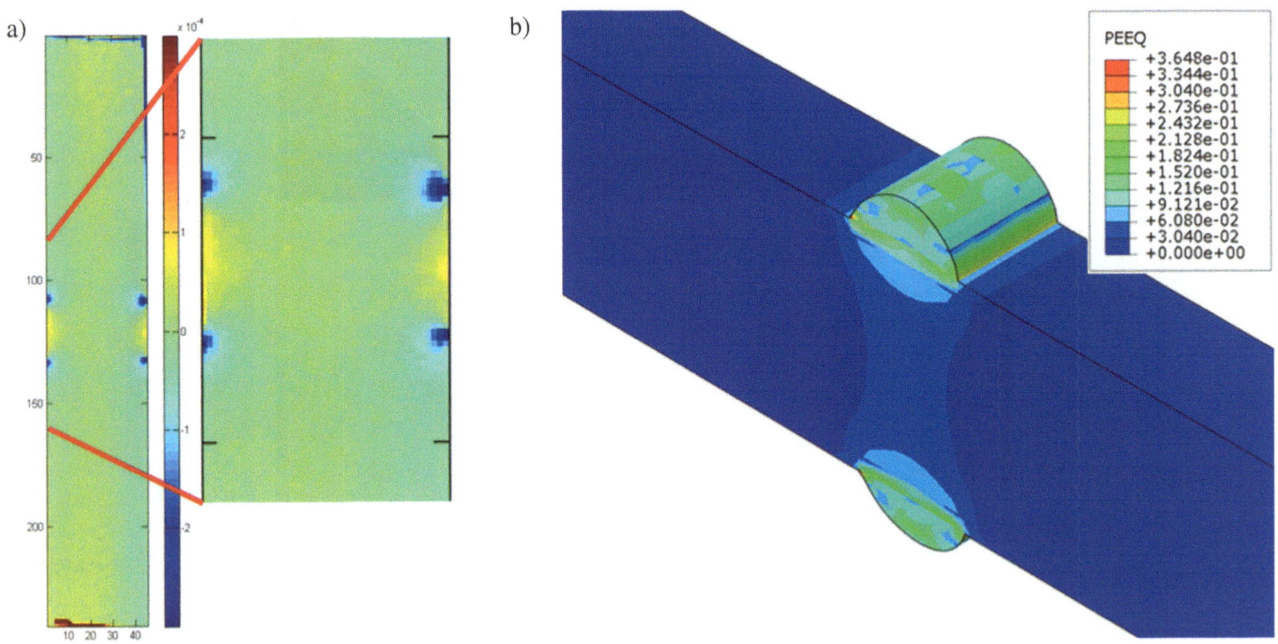

Fig. 8.3 (**a**) ΔT/T resultant bitmap with expanded view of weld toe regions; (**b**) Equivalent plastic strain results from thermal model of the welding process showing peak plastic strain directly under weld toe [4]

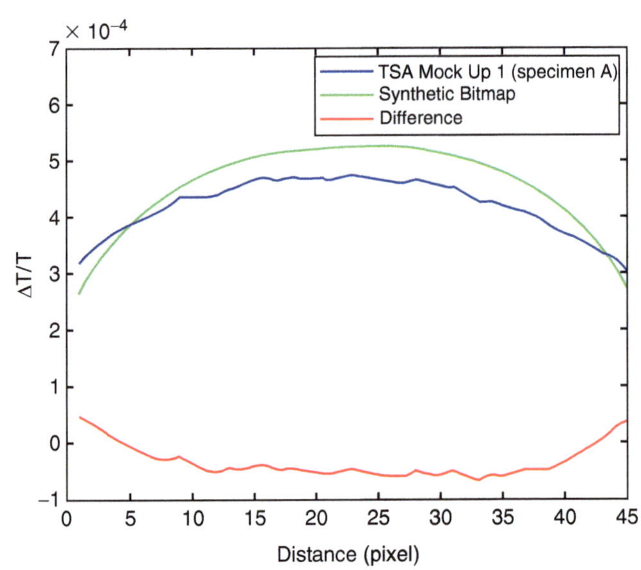

Fig. 8.4 Results of from experimental bitmap, synthetic bitmap, and resultant bitmap across the centre of the specimen beneath the welded region

plastic strain in the heat affected zone is clear to see, and the reduced response in the centre of the specimen where the heating affects are minimal are significantly smaller and are below the noise threshold of TSA testing (10^{-5} magnitude response) and so dominated by noise effects.

8.4 Conclusions and Future Work

The present paper has shown that through the use of a synthetic reference bitmap it is possible to identify regions that have undergone plastic strain without the need for a physical reference specimen. The results of the experiments have been compared to results from thermal models simulating the welding process and these have shown a good level of agreement between the expected results and the actual results. Already discussed have been the errors between the two datasets, with this attributed to the 8 % thermal reflectivity of the specimen surface during the experimental testing.

To further corroborate these results a further experiment is being undertaken, a specimen has been water-jet cut from 316 L stainless steel to identical dimensions to provide an experimental reference. This will enable a further validation of the synthetic bitmap approach. There is also a need to investigate localised plasticity due to specimen geometry; a second synthetic bitmap is being created that contains elastic and plastic material properties so any stress concentrations cause large stresses in the FE model can be accounted for in the thermoelastic response.

References

1. Withers, P.J., Bhadeshia, H.K.D.H.: Overview: residual stress part 1—measurement techniques. Mater. Sci. Technol. **17**, 355–365 (2001)
2. Dulieu-Barton, J.M., Stanley, P.: Development and applications of thermoelastic stress analysis. J. Strain Anal. **33**, 93–104 (1998)
3. Quinn, S., Dulieu-Barton, J.M., Landlands, J.M.: Progress in thermoelastic residual stress measurement. Strain **40**, 127–133 (2004)
4. Chevallier, E.: Private communication (2015)
5. Robinson, A.F.: Assessment of Residual Stress using Thermoelastic Stress Analysis. University of Southampton, Southampton (2011)
6. Robinson, A.F., Dulieu-Barton, J.M., Quinn, S., Burgeute, R.L.: Paint coating characterization for thermoelastic stress analysis of metallic materials. Meas. Sci. Technol. **21**(8), 1–11 (2010)

Chapter 9
Finite Element Modelling of a Series of Austenitic Stainless Steel 316 L Weldments to Inform Thermoelastic Stress Analysis Residual Stress Assessment

E.C. Chevallier, S. Blackwell, and J.M. Dulieu-Barton

Abstract The RESIST (Residual Stress and Structural Integrity Studies using Thermography) project aims to develop a non-contact, non-destructive, full-field measurement and portable residual stress assessment technique based on thermoelastic stress analysis (TSA). The TSA residual stress assessment (RSA) technique relies upon a change in the thermal expansion coefficient when a material is subjected to plastic straining. TSA RSA has been successfully applied to non-welded materials, so the next stage is to assess the feasibility of application on weldments. The study focuses on establishing the validation of the technique on austenitic stainless steel AISI 316 L (EN 1.4404), which is commonly used in power generation industry. A series of increasingly complex 'weld mock-ups' made out of 316 L are designed to test the TSA RSA approach. Finite element (FE) simulations of the welded mock-ups were produced. The design, FE modelling and manufacture of two of the mock-ups are presented. The expected amount of plastic strain is compared with the TSA assessment in another part of the RESIST project, and thus, informs the validation of the technique on weldments.

Keywords Welded structures • Finite element modelling • Plastic strain • Residual stress • Thermoelastic stress analysis

9.1 Introduction

The RESIST project aims to develop a non-contact, non-destructive, full-field measurement and portable residual stress assessment technique based on TSA [1]. TSA is an experimentally based technique in which the change in temperature of a material or component subject to a cyclic load, ΔT, is related to the change in the sum of the principal stresses $\Delta(\sigma_1 + \sigma_2)$, as follows:

$$\Delta T = -KT\Delta(\sigma_1 + \sigma_2) \tag{9.1}$$

where T is the temperature of the surface of the component and K is the thermoelastic constant, which is a material parameter defined as $K = \alpha/\rho C_p$; α is the coefficient of thermal expansion, ρ is the material density and C_p is the specific heat at constant pressure.

The TSA RSA technique relies upon the change of the thermoelastic constant K of the material, which occurs after a material has been subjected to plastic straining [2]. An infra-red detector measures T and ΔT of a component under a linear elastic cyclic loading. If K changes then ΔT will change, making it possible to identify areas that have been subject to plastic straining. If the change in K can be quantified then the amount plastic straining experienced can also be obtained. Since weld residual stresses are initiated by plastic deformation, the residual stresses could be derived from the amount of plastic strain experienced. The regions that have experienced plastic straining have been identified in non-welded materials [3], however, the next stage is to assess the feasibility of application on weldments.

E.C. Chevallier (✉)
TWI Ltd, Granta Park, Great Abington, Cambridge CB21 6AL, UK

Faculty of Engineering and the Environment, University of Southampton, Southampton SO17 1BJ, UK
e-mail: ecc1e13@soton.ac.uk

S. Blackwell
TWI Ltd, Granta Park, Great Abington, Cambridge CB21 6AL, UK

J.M. Dulieu-Barton
Faculty of Engineering and the Environment, University of Southampton, Southampton SO17 1BJ, UK

S. Quinn, X. Balandraud (eds.), *Residual Stress, Thermomechanics & Infrared Imaging, Hybrid Techniques and Inverse Problems, Volume 9*, Conference Proceedings of the Society for Experimental Mechanics Series, DOI 10.1007/978-3-319-42255-8_9

9.2 Methodology

To validate the TSA RSA technique, a series of weld mock-ups are designed and assessed using FE analysis. The geometry and welding method are investigated so that distortion is avoided, which may affect the thermoelastic response. Preliminary FE model of the mock-ups are produced to determine if enough plastic strain is generated during the welding process to provide a measurable change in the thermoelastic response. To calibrate the models thermocouples are used during the welding process to measure the heat generation and centre hole drilling (CHD) and X-ray diffraction residual stress measurements are made after the welding and manufacturing processes. The FE models are created and calibrated using both fusion line and temperature matching. Validation of the models against the experimental residual stress measurements is then performed. The residual stress and plastic strain are extracted for comparison with TSA RSA results.

The first step in the process is to design a welded component that contains a sufficient amount of plastic strain to produce a detectable change in the thermoelastic response. It was therefore decided to adopt an approach used in industry and create a 'mock-up'. Mock-ups are used in industry for the validation of FE models, when it is not possible to obtain measurements from the actual structure (e.g. in the nuclear industry). The structure or a part of the structure is manufactured, at the same dimension or at a different scale and measurements are performed on the mock-up, e.g. to give information to calibrate and validate the FE model. Once the FE model is validated, it can be used to predict distortion, stresses, fatigue resistance etc.

It is envisaged that a number of mock-ups that contain increasing complexity in terms of specimen and weld geometry will be made. The overall design criteria for the mock-ups are as follows:

1. to contain a weld in which all welding parameters are known so that the plastic straining and the residual stresses can be modelled accurately,
2. the mock-up must be designed to be cyclically loaded in a servo-hydraulic test machine with the weld positioned so it has a clear line of sight to the infra-red camera to enable images to be captured for the purpose of TSA.

The aim is to define a procedure to assess the potential of using TSA for residual stress assessment using mock-ups. The proposed procedure is shown schematically in Fig. 9.1. A mock-up is defined. Any welding and machining procedures are modelled using FE modelling and the plastic strain and residual stress fields are predicted. The thermoelastic response $\frac{\Delta T}{T}_{TSA}$ of the mock up is obtained. A linear elastic finite element (LEFE) model of the loaded mock-up (i.e. the effects of welding

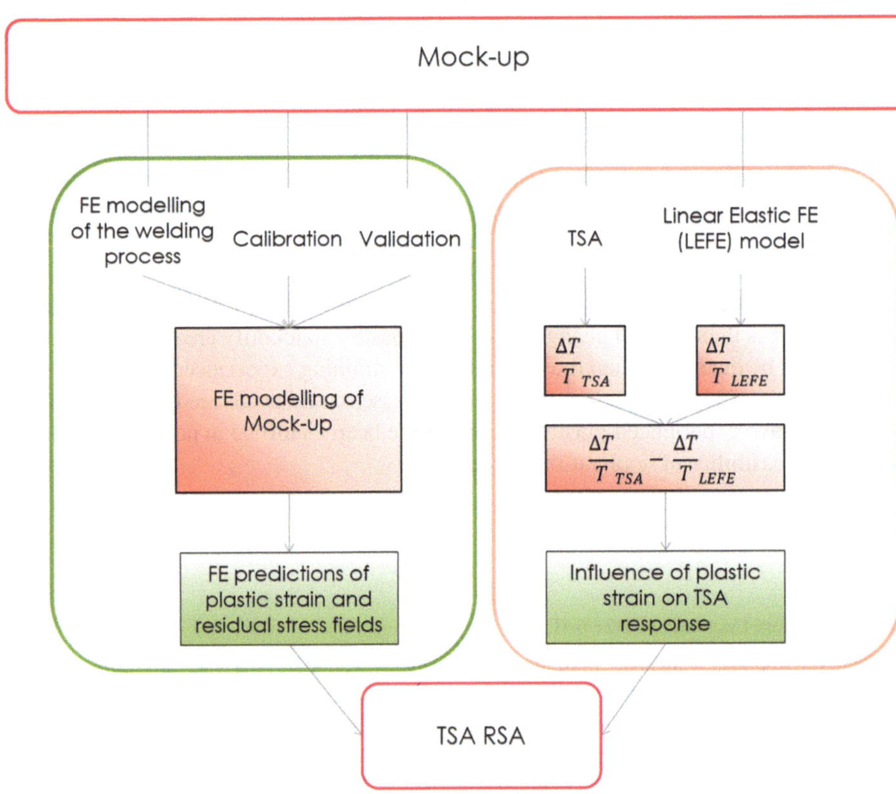

Fig. 9.1 Procedure to compare the plastic strain prediction from FE model of the mock-up with TSA results

are not modelled) is created and the calculated thermoelastic response $\frac{\Delta T}{T}_{LEFE}$ subtracted from the experimental thermoelastic response to obtain the influence of weld induced plastic strain on the thermoelastic response. The influence of the plastic strain on the thermoelastic response is then compared with the predicted plastic strain and residual stress field prediction to define the TSA response of a welded component.

9.3 Illustration Case: Mock-Up A

Mock-up A is a strip of material with a welded section at the centre as presented in Fig. 9.2. It is essentially a slice through a weld, where the weld penetration in the base material can be imaged. The proposed imaged region is delimited in green on Fig. 9.2 which is a zoom-in on the weld.

To create Mock-up A, a 316 L stainless steel 321 × 270 × 15 mm plate had a weldment deposited on both side simultaneously to create a 'bead-on-plate' specimen. The welding process was manual metal arc (MMA) welding with a heat input of 2310 W at a welding speed of 3.7 mm/s and the deposited weldment material is 308 stainless steel. To monitor the temperatures developed in the welding process eight thermocouples were attached on both sides of the plate. After complete cooling of the plate, residual stress measurements were calculated from centre hole drilling (CHD) measurements. The positions of the thermocouples and CHD are presented in Fig. 9.3 and were taken on both sides of the plate, away from the middle of the plate, not to impede with cutting of the mock-up. The double bead-on-plate specimen was subsequently cut using electrical discharge machining (EDM) to create three 7 mm thick slices, i.e. the mock-up in the direction transverse to the weld so the weld could be imaged with TSA.

The FE analysis software Abaqus was used to create the model of Mock-up A. The dimensions of the plate depicted in Fig. 9.3 were used to create the geometry of the bead-on-plate model. Material properties from [4] were used to model the

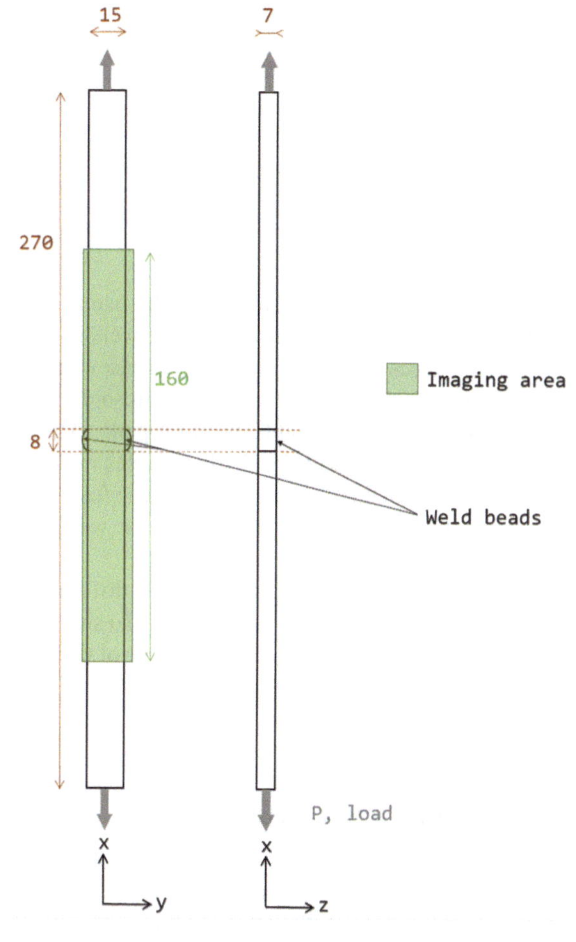

Fig. 9.2 Mock-up A dimensions (in mm), loading and imaging area

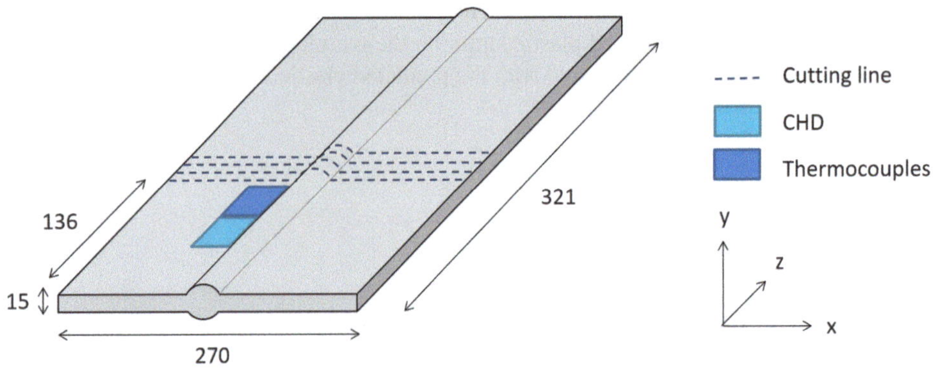

Fig. 9.3 Double bead on plate dimensions with CHD measurements, thermocouples and cutting lines positions (*right*)

Fig. 9.4 Mesh and boundary conditions during the welding process, the displacements of the *red lines* are set to zero in x, y and z directions

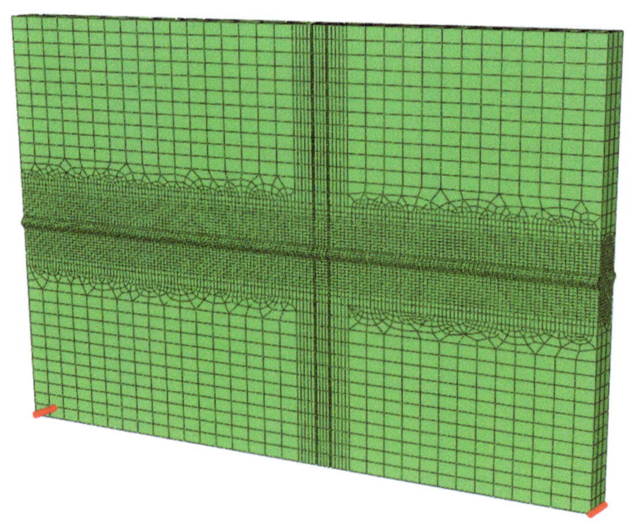

thermal and mechanical behaviour of the parent material. The mesh of the FE model is presented in Fig. 9.4. Tetrahedral elements of 2 mm^3 were used to map the weld and heat affected zone (HAZ). The mesh is fine in the region of the weld and around the slices to remove, and coarser elsewhere.

To model the heat source, a double ellipsoidal moving heat source model, developed by Goldak [5], was used to simulate the deposition of the double bead-on-plate. The efficiency and geometrical parameters of the double ellipsoid were calibrated against two of the thermocouples measurements, at 10 mm and 15 mm away from the weld bead. Satisfactory comparison between thermocouples and the fusion lines were obtained as it can be seen respectively in Figs. 9.5 and 9.6. The mechanical boundary conditions during the welding process were modelled by setting displacement to zero on the two red lines in Fig. 9.4.

A total of eight locations of centre hole drilling measurements were compared with the predicted residual stresses and the results were satisfactory for half of them. To simulate the cutting of the plate, small elements (1 mm^3) were removed around the slice. The resulting slice represents one of the Mock-up A specimens, and the plastic strain magnitude was predicted as it can be seen in Fig. 9.7.

9.4 Results and Discussion

Mock-up A, the first weldment designed for TSA was presented. It was manufactured from a 316 L bead-on-plate welded on both side and subsequently cut in the direction transverse to the weld, to produce a 7 mm slice, Mock-up A. The FE model of Mock-up A was created and good calibration of the heat source was obtained. The validation of the stress prediction against

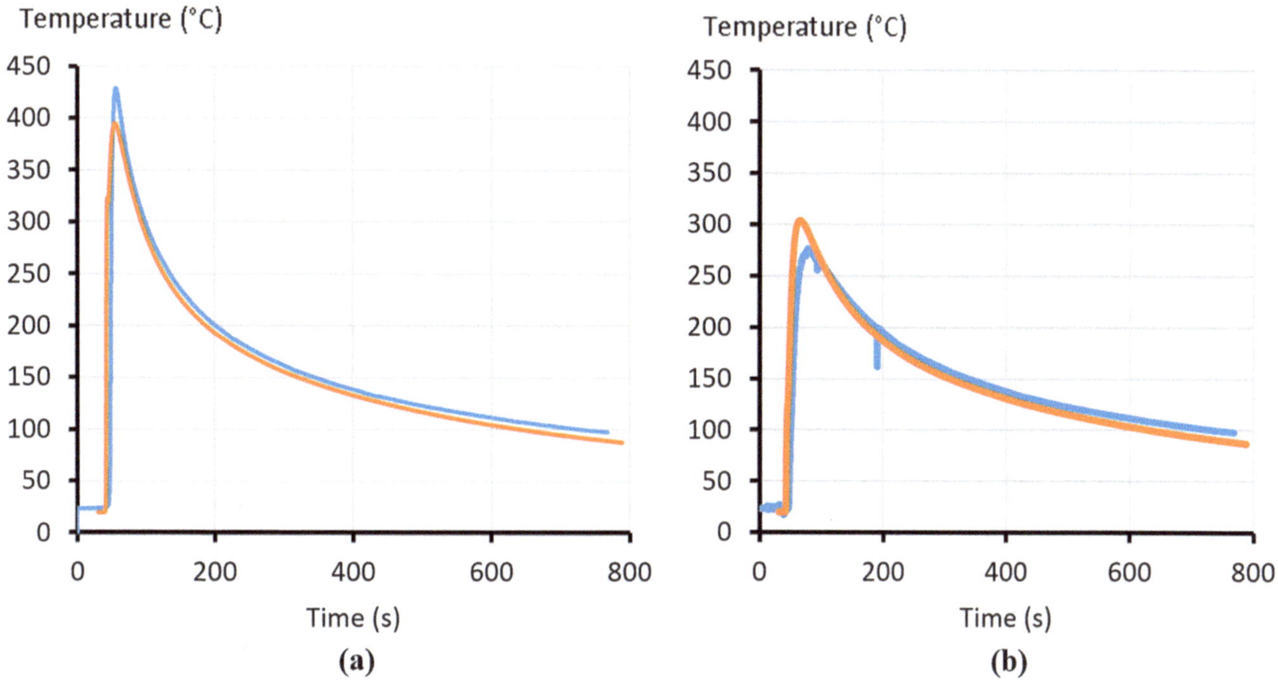

Fig. 9.5 Comparison between temperatures experienced at 10 mm (**a**) and 15 mm (**b**) away from weld bead from thermocouples (*blue*) and from FE prediction (*orange*)

Fig. 9.6 Comparison of the fusion line in the FE model (*left*) and macrograph (*right*)

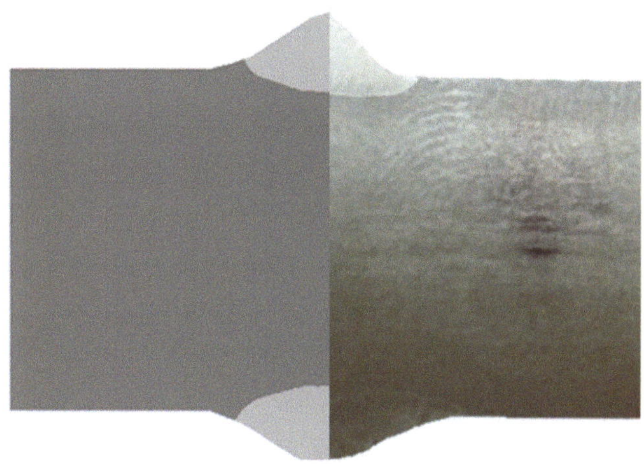

Fig. 9.7 Plastic strain magnitude prediction in Mock-up A

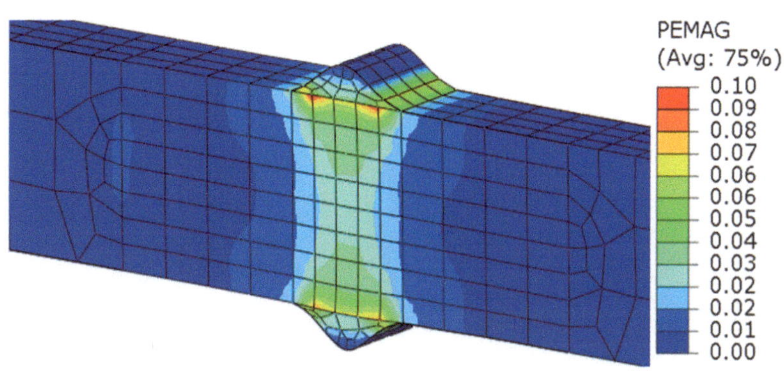

eight CHD measurements was satisfactory for 50 % of them. The difficulty in validating the residual stress prediction can be attributed to the manual aspect of the welding process, which involves many unknowns that cannot be taken into account in the modelling. Some of the CHD measurements values can be impacted by the large stress gradient in the HAZ as it is recommended CHD should indeed be located in regions of small stress gradients [6].

When the mock-up was created, EDM cutting process was selected to cut the slices as this technique is commonly used in the contour technique, where no residual stress or plastic strain should be induced [7]. Therefore the first approach to model this cutting was to remove a thin slice of elements, to release residual stresses normal to the cutting direction. Nonetheless EDM cutting modelling was investigated in further details and it was found that EDM introduces residual stress and plastic strain in the very near surface (typically a few tens of microns) called the "recast layer" [8]. TSA was applied on the cutting surface, so, if polishing does not remove the whole thickness of recast layer, results will be influenced by plastic strain induced by EDM cutting. The FE modelling of EDM is computationally difficult so it was decided that the next mock-ups will not be imaged on a surface resulting from an EDM cut.

9.5 Conclusions and Future Work

In Mock-up A one of the major difficulties was the uncertainties produced by the EDM cutting of the mock-ups from the large bead-on-plate specimen, particularly in predicting the residual stress relaxation during the cutting. It was therefore decided that Mock-up B should be such that the cutting stage is eliminated. A further consideration was controlling the shape and size of the weld bead in Mock-up A so Mock-up B contains a double autogenous weld, meaning that no bead is created, so it is not necessary to include the filler metal or the bead geometry in the model. It has been decided to use the TIG welding process for the autogenous weld using and automated process with the electrodes fixed to a support at a set distance from the plate, which will move across the welding torches. In this way the weld trajectory and the welding speed are controlled, unlike in Mock-up A which relied on the two welders.

References

1. Dulieu-Barton, J.M.: Thermoelastic stress analysis. In: Rastogi, P., Hack, E. (eds.) Optical Methods for Solid Mechanics. Wiley, Weinheim (2012)
2. Quinn, S., Dulieu-Barton, J.M., Langlands, J.M.: Progress in thermoelastic residual stress measurement. Strain **40**, 127–133 (2004)
3. Robinson, A.F., Dulieu-Barton, J.M., Quinn, S., Burguete, R.L.: The potential for assessing residual stress using thermoelastic stress analysis: a study of cold expanded holes. Exp. Mech. **53**, 299–317 (2013)
4. Bate, S.K., Charles, R., Warren, A.: Finite element analysis of a single bead-on-plate specimen using SYSWELD. Int. J. Press. Vessel. Pip. **86**, 73–78 (2009)
5. Goldak, J., Chakravarti, A., Bibby, M.: A new finite element model for welding heat sources. Metall. Mater. Trans. B **15**(2), 299–305 (1984)
6. Scaramangas, A.A., Porter Goff, R.F.D., Legatt, R.H.: On the correction of residual stress measurements obtained using the centre-hole-drilling method. Strain **18**(3), 88–97 (1982)
7. Hosseinzadeh, F., Kowal, J., Bouchard, P.J.: Towards good practice guidelines for the contour method of residual stress measurements. J. Eng (2014)
8. Cheng, W., Finnie, I., Gernaud, M., Prime, M.B.: Measurement of near surface residual stress using electric discharge wire machining. J. Eng. Mater. Technol. **116**(1), 1–7 (1994)

Chapter 10
Residual Stress Measurement of Full-Scale Jet-Engine Bearing Elements Using the Contour Method

Daulton D. Isaac, Michael B. Prime, and Nagaraj Arakere

Abstract Compressive residual stresses provide a well-known advantage to the fatigue life of bearing materials under rolling contact fatigue (RCF), but the stresses change under fatigue loading and may later contribute to failures. Previous measurements of the depth-wise distribution of residual stresses in post-fatigue bearings with X-rays involved the time consuming process of etching to determine subsurface stresses and only in limited locations. By contrast, the contour method determines the 2D residual stress map over a full cross section. The method involves the sectioning of the part using Electrical Discharge Machining, measuring the out of plane displacements of the exposed cross section, and using the afforded field as boundary conditions on a finite element model of the component to back calculate the causative residual stress. For this investigation, the residual hoop stresses in the split inner rings of the main shaft bearing assembly of an aircraft jet engine was mapped using the contour method. Prior to measurement, the full-scale bearing made of hardened AISI M50 was subjected to RCF during engine operation. In this talk, the unique challenges of the particular measurements are discussed. The tested bearings showed effectively no residual stresses induced by the RCF, probably because they were conservatively removed from service prior to sufficient cyclic loading. A more highly loaded bearing will be measured in future work.

Keywords Rolling contact fatigue • Bearings • Contour method • Residual stress • Finite element model

10.1 Introduction

Residual stresses play a significant role in many material failure processes like fatigue, fracture, fretting fatigue, and stress corrosion cracking [1–3]. Residual stresses are the stresses present in a part free from external load, and they are generated by virtually any manufacturing process. Because of their important contribution to failure and their almost universal presence, knowledge of residual stress is crucial for prediction of the life of any engineering structure. However, the prediction of residual stresses is a very complex problem. In fact, the development of residual stress generally involves nonlinear material behavior, phase transformation, coupled mechanical and thermal problems and also heterogeneous mechanical properties [4–20].

10.1.1 Residual Stresses and Roller Bearings

In this subsection, we make the points that residual stresses in general, and the radial component of residual stress in particular, (1) are important in the fatigue life of rolling bearings, (2) are usually ignored, and (3) are difficult to measure.

D.D. Isaac
Department of Mechanical and Aerospace Engineering, University of Florida, Gainesville, FL, USA

Los Alamos National Laboratory, Los Alamos, NM, USA

M.B. Prime (✉)
Los Alamos National Laboratory, Los Alamos, NM, USA
e-mail: prime@lanl.gov

N. Arakere
Department of Mechanical and Aerospace Engineering, University of Florida, Gainesville, FL, USA

S. Quinn, X. Balandraud (eds.), *Residual Stress, Thermomechanics & Infrared Imaging, Hybrid Techniques and Inverse Problems, Volume 9*, Conference Proceedings of the Society for Experimental Mechanics Series, DOI 10.1007/978-3-319-42255-8_10

Bearing steels are a family of ultra-high-strength structural materials subjected to the highest loading conditions in engineering systems throughout their life cycle. Some grades have tensile strengths exceeding 2.7 GPa and hardness values beyond 900 HV (~9 GPa). Industries have been aggressively pursuing development of new bearing steels to meet the increasingly challenging requirements for high power density, operation under transient adverse conditions and reliability of bearings for applications accumulating over 100 billion (10^{11}) cycles in aircraft, wind turbine, high-speed railway and other systems. The requirement for the L10 life, defined as the number of cycles at which 90 % of bearings survive under a rated load and speed, can be in excess of 30,000 h, about 100 billion (10^{11}) cycles of contacts. Properly installed and maintained bearings eventually fail under the influence of Rolling Contact Fatigue (RCF). RCF manifests as highly localized damage, accumulated under the contact surface, typically at a depth of 200–400 μm, eventually leading to the nucleation of subsurface cracks. Carbide precipitates ranging from tens of nm to 1–5 μm are largely responsible for material hardening.

Bearing material is subjected to a radically different fatigue loading under RCF compared to conventional high cycle fatigue (HCF). The salient features of RCF are:

1. The volume of cyclically stressed subsurface material in the RCF-affected zone is only a few mm^3 and highly localized [21].
2. Nature of cyclic fatigue loading is triaxial and nonproportional, over a very large number of cycles ($\sim 10^{11}$). The ability of material to work harden is much greater under these conditions.
3. Although loading is nominally elastic, heterogeneous cyclic plasticity at the scale of microstructure (i.e., microplasticity) is induced [22–24].
4. Material microstructure, residual stress and micro-hardness in the RCF-affected zone evolve continuously with cycles, with progressive fatigue damage accumulation [21, 22]. Tensile residual stresses in the radial direction are known to develop that can drive Mode I subsurface crack growth parallel to the running surface.

Bearing fatigue life estimation is still largely based on the seminal probabilistic life model by Lundberg and Palmgren (LP) [25–27], proposed in 1945. Despite many improvements to the LP model [21] current bearing life prediction methods are based on the ISO standard [28] and continue to rely on extensions to the LP model, are empirical in nature, and include variables that are obtained from extensive experimental testing. They do not directly consider the subsurface residual stress evolution, details of the constitutive behavior of materials under contact loading or localized microplasticity. Generally, LP models significantly underpredict bearing life.

Bearing life, based on LP models, can expressed as [25]

$$L \propto \frac{1}{[F_e]^p} \propto \frac{1}{S^n} \tag{10.1}$$

Where, L, F_e, and S represent bearing fatigue life, equivalent radial load, and maximum Hertzian contact pressure, respectively. In Eq. (10.1), p and n are the load-life and stress-life exponents respectively. Based on LP models, the stress-life exponent is 9 for ball bearings. The reevaluation study by Parker and Zaretsky [29] indicates that the load-life exponent and stress-life exponent for vacuum-processed steels are 4 and 12 respectively. However, a recent study [30] has shown that the load life exponent for ball bearings should be 4.1 (instead of 3 or 4), and the corresponding stress-life exponent should be 12.3. This shows that bearing life is highly sensitive to the small changes in the subsurface Hertzian stress due to the large exponent value of ~12.

Peak Hertzian contact stresses in aerospace bearings are typically 2.0–2.5 GPa (290–370 ksi) [21]. The peak orthogonal shear stress for this condition is about 600 MPa. The residual stress development is primarily the result of the accommodation of shape change of the small volume of RCF-induced plastic zone upon loading. SAE 52100 steel ball bearings (10–15 % retained austenite) tested at peak Hertzian stress of 3.3 GPa (485 ksi) developed a maximum compressive circumferential residual stress of 900 MPa after 2×10^9 inner ring revolutions, ~250 μm below the surface [31]. Axial compressive residual stresses were of similar magnitude while radial *tensile* residual stress of about 300 MPa persisted to greater depths of ~1.5 mm in the radial normal direction [31]. This indicates that the residual stresses developed in long-life bearings can be substantial, resulting in a perturbation of the Hertzian contact stress field, leading to significant variability in life prediction via Eq. (10.1).

The presence of subsurface residual stresses in bearings will certainly influence life calculations from Eq. (10.1). However, residual stress measurement [32] in bearing raceways is not easy and they evolve with RCF cycles. The important stresses are generally localized within a few 100 μm of the surface, which makes bulk measurement methods like neutron diffraction [33] impractical. Conventional hole drilling [34] is not possible because the steels are too hard to drill. X-ray diffraction has been used to measure residual stresses in bearings [31], but requires material removal to get subsurface stresses and typically only measures the circumferential residual stress. Hence, residual stresses are not adequately accounted for in life calculations. Compressive residual stresses that exist in case hardened bearings from manufacturing processing are known to greatly enhance life.

The influence and role of residual stress in the operation and life of RCF components being thus highlighted, it is understandable that such a parameter needs to be more fully understood. This paper presents a comprehensive method for residual stress measurement in rolling element bearings.

10.1.2 Bearing Residual Stress Measurements with the Contour Method

In this subsection, we describe how the residual radial stresses could be measured using a combination of the contour method with X-ray diffraction using superposition [35]. In this paper, only the results from the initial contour method measurements are reported.

The Contour Method [36] is a relatively new technique for measuring residual stresses [37, 38] but well proven and validated [39–48]. Figure 10.1 illustrates the ideas behind the technique. Step A in the figure represents a part containing residual normal stresses σ_x and σ_z that vary over the height of the beam. If the part is cut, the surface exposed will deform

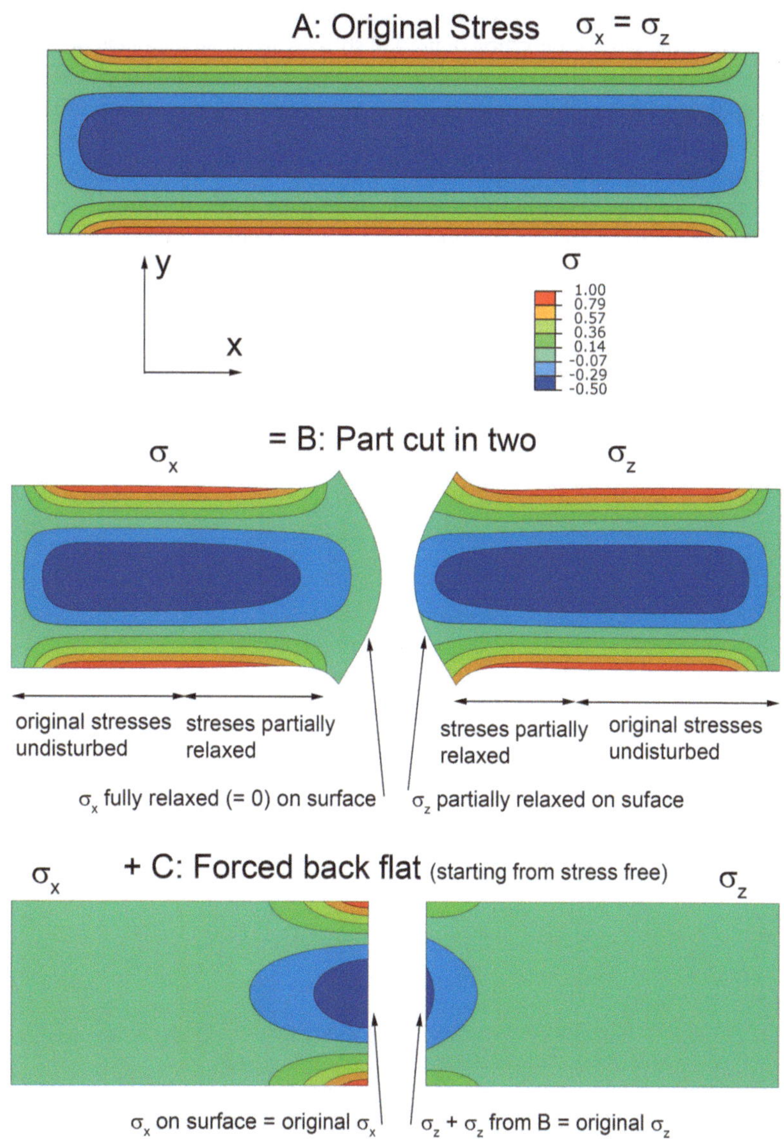

Fig. 10.1 Superposition principle behind the contour method (from [35])

elastically owing to the relaxation of the residual stresses as shown in Step B. Notice that σ_x must fully relax because it is normal to the newly created free surface. Because the deformation was elastic, forcing the deformed face back flat would reintroduce the stresses back into the part. Thus, if the deformations are measured and their reverse are applied as displacements on the surface of an unstressed part of similar geometry (as the deformed part), the residual stresses that relaxed during the initial deformation will be induced in this part as shown in Step C of the figure. In the contour method, a part containing residual stress is sectioned allowing the stresses to deform the exposed face. The out of plane displacement of this surface is then measured, for example with a coordinate measuring machine. In a linear elastic finite element analysis on the sectioned part, the opposite of the obtained contour is applied to an unstressed model of the part, and the resulting stresses are the stresses that changed as a result of the relaxation caused by the sectioning, i.e. $C = A - B$. Since the normal stress σ_x is zero in the free surface in B, this calculation gives the original (A) 2-D residual σ_x distribution on the full cross-section of the cut plane.

In order to get the radial stress, a hybrid variation of the contour method [35] is of particular interest for the case of RCF in bearings. The most logical cut to make when applying the method to a bearing race is one that exposes a radial-axial cross section (see Fig. 10.2) because in an axisymmetric part most of the variations are expected in that plane. The residual hoop stresses act normal to that section and thus is the only component that can be fully obtained through the standard contour method. Referring back to Fig. 10.1, we consider in particular the right side of the figure which depicts the behavior of the in-plane normal stress σ_z, which is analogous to the radial stress for our bearing measurements. Upon sectioning, the distribution of σ_z indeed changes as shown in Step B, but it is not fully relaxed. The amount that σ_z relaxed is actually quantifiable through the same operation of pushing back flat the deformed face as can be seen in Step C. This shows that although the contour method can only fully determine the stress component acting normal to the exposed surface, it can also determine the partial relaxation of normal stresses acting in the plane of the surface. In the case of a bearing, exposing the radial axial cross section allows for full determination of the residual hoop stresses and for quantifying how much the radial and axial components relaxed due to sectioning. Using a residual stress measurement technique that can measure in-plane stresses, such as X-ray diffraction, the unrelaxed portion of stress can be measured and combined with the relaxed portion to fully determine the in-plane normal stresses present in the bearing ring before sectioning. This is of particular interest seeing the residual radial stress in bearings, thought to be tensile, may play a role in Mode I crack growth. This new multiple-method superposition technique with the contour method [35], has been demonstrated a few times in the literature [40, 49, 50].

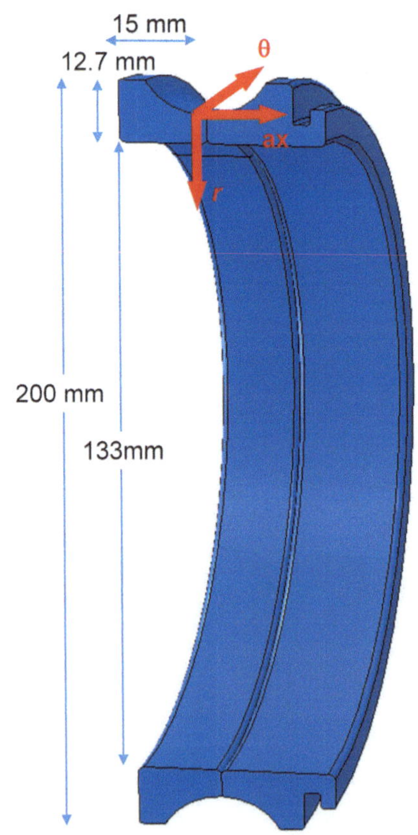

Fig. 10.2 Showing basic dimensions of the bearing and the radial-axial cross section with the accompanying coordinate system

Special attention has to be paid when utilizing the Contour Method to measure hoop stresses in cylindrical bodies such as pipes or as in this case a bearing ring [49, 51–54]. Assuming axisymmetry, the residual stresses on each cross section along the circumference of such geometries would be in force equilibrium, but there can be a net bending moment acting to either open up or close the part. Cutting through a section with a bending moment can result in a large load on the cut tip and local yielding [51, 53, 55], which clearly violates the elastic deformation assumption of the method. To reduce or eliminate plasticity, this bending residual stress is first relieved by cutting open the cylindrical part, and the contour method measurement is then made at another section on the circumference. However, the bending stresses relieved are of interest since they act in concert with the force equilibrium stresses to produce the net residual stress field, so measures are taken to calculate these stresses [51, 54] as discussed later in the paper.

10.2 Purpose

The work reported seeks to make a novel application of the contour method to determine the distribution of the residual stress present in the rings of a through hardened inner bearing race. Ideal cutting parameters, profiling techniques, data processing steps, and finite element modeling methods are sought after in an attempt to uncover the contour method particulars best adapted to measurement in bearings. Thus insight is offered on the state of residual stress in a bearing taken out of operation and on best practices to enable and continue a broad study of the phenomenon in high performance bearings using the contour method.

10.3 Methods

10.3.1 Experimental

The inner bearing rings under investigation are from the main shaft bearing assembly taken out of operation in an airplane jet engine. The bearing as shown in Fig. 10.2 is of the split inner ring type, and has a bore diameter of about 133 mm and an outer diameter of about 200 mm. The through hardened M-50 bearing steel typically has a uniform carbon content of 0.85 wt.% throughout its depth with a carbide volume fraction of 20 %. The primary carbides range in size from 1 to 5 μm. This high strength bearing steel has a hardness of about 8 GPa (63–65 HRC) with a yield strength of 2.8 GPa.

No spalls or other defects were observed on either of the rings taken from service. The stress level and cycles experienced by the bearing is unknown. Since this is a thrust bearing, one inner ring bears significantly more of the load. As such, the races are referred to as the thrust-loaded and unloaded races and their cross-sectional geometries are differentiable. The work described below is applicable to both rings. However, the contour data was processed into a residual map only for the unloaded race for reasons addressed in the Results and Discussion section.

10.3.2 Particulars of the Cut Plan and Reasoning

In order to determine the residual hoop stresses present, a cut in the radial-axial plane, shown in Fig. 10.2, into the bearing is required. Figure 10.3 illustrates the location of the various cuts made into the rings of the inner race. In general, the operations carried out on the two rings were similar. First the stress release cut is made, which results in the relieving of the bending moment stresses present in the ring. A strain gage is used at the location shown to capture the strain from the ring springing open when sectioned. The strain value is used in a finite element model of each ring to determine the bending residual stresses that were relieved. 180° from the location of the stress release cut, the contour cut is performed after the strain gage is removed. The out of plane displacement of the newly exposed surfaces of this cut is later used for determining the force equilibrium residual stresses. The measured displacement data from both sides of the cut is aligned on a common grid before being averaged. The averaging of the data from both sides removes the effects of cutting artifacts and shear stresses.

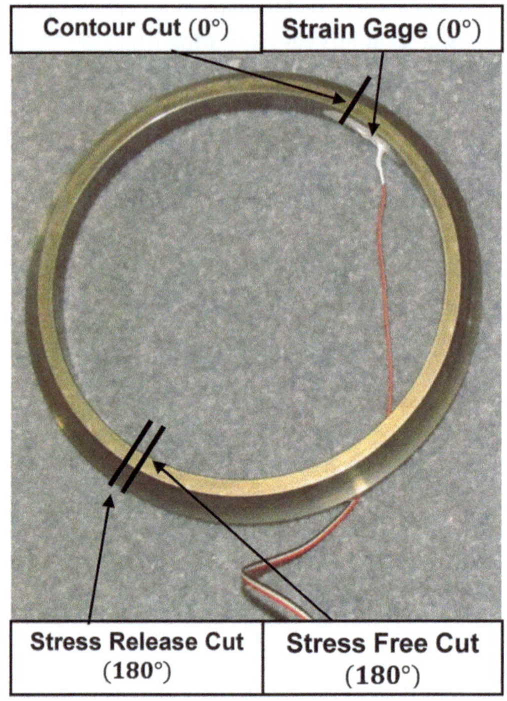

Fig. 10.3 Various cuts made into the bearing. *Cut 1* releases the geometric stresses, *Cut 2* reveals the surface whose contour will be measured, and *Cut 3* serves to check the validity of the surface obtained in the contour cut (*Cut 2*)

To remove the effects of noisy data, the averaged data is fitted to a smooth 3D surface using cubic splines, and the best fit is evaluated on the basis of a least squares analysis. The fitted surface is then evaluated at coordinates corresponding to the nodes of the finite element mesh of the ring model, and the analysis continues as discussed in the Finite Element Modeling section below.

The third and final cut is the stress-free test cut [36, 52, 56, 57] and is made by slicing a thin wafer off of one of the faces exposed by the stress release cut. The stress free cut should result in a flat surface seeing that the normal residual stresses were already released by the stress release cut. Any displacement measured from this last cut can be regarded as connected to the cutting of the part, and they can be used to make corrections to the displacement from the contour cut.

A 100 μm wire was used to perform the contour and stress free cuts using the skim cutting setting. The normal displacements of the contour faces were measured using a Taylor Hobson Talyscan 250 with a laser triangulation probe, and a grid of 50 × 50 μm was used to map the contour of the surface.

10.3.3 Finite Element Modeling

The value for the average strain obtained from the bending-moment stress release cut can be used to scale a model of each ring to determine the bending stresses that caused the ring to deform at this cut. The average strain at the location corresponding to that of the strain gage in the actual ring could then be calculated. The deformation being linear, the forces in the model can be scaled until the strain in the model matches that obtained during the stress release cut. The bending stresses thus determined can later be superimposed with those determined through the surface profiling of the contour cut.

A 180° portion model with the deformations from the contour cut applied as boundary conditions is used to obtain the remaining residual stresses. The displacement boundary conditions are generated for the normal direction only. All other degree of freedoms are left unconstrained on the model, except for those necessary to prevent rigid body translation or rotation of the model.

10.4 Results and Discussion

10.4.1 Strains from the Bending Moment Release Cut

During the stress release cut, a value of -42 με was recorded by the strain gage for the thrust-loaded race, and -15 με for the unloaded race. The magnitude of the bending moment stresses is estimated at under 7 MPa, small enough to be ignored in the remainder of the calculations.

10.4.2 Results for the Measured Contours

The raw data indicates that the residual stresses in the bearing races are quite small and that the in-service loading did not induce any significant stresses. Figures 10.4 and 10.5 show the measured contours for the thrust-loaded and unloaded bearing races, respectively. As is standard [36], the data is plotted relative to the best fit plane. The data all falls within ± 10 μm, not much greater than the noise level of the data, indicating relatively small deformations caused by residual stress release. There are some high and low features in the contours that appear precisely horizontal. These are all oriented in the wire direction and are almost certainly EDM-cutting artifacts [57]. At first glance, there appears to be a high spot near the surface of the curved bearing surface which might indicate some local stresses. For two reasons, further consideration indicates that such stresses are not evident or likely. First, the high spot appears in both the thrust-loaded race and the unloaded race at similar magnitudes, which would not be the case if the loading caused residual stresses. Second, the stress-free test cuts, Fig. 10.6, show very similar high regions, further indicating that the feature is a minor EDM-cutting artifact. When an EDM wire cuts through a non-uniform cross-section, there are small deviations from a perfectly constant width cut, which show up as such artifacts [52]. The stress-free test cuts also show linear features, confirming that the similar features in Figs. 10.4 and 10.5 are cutting artifacts.

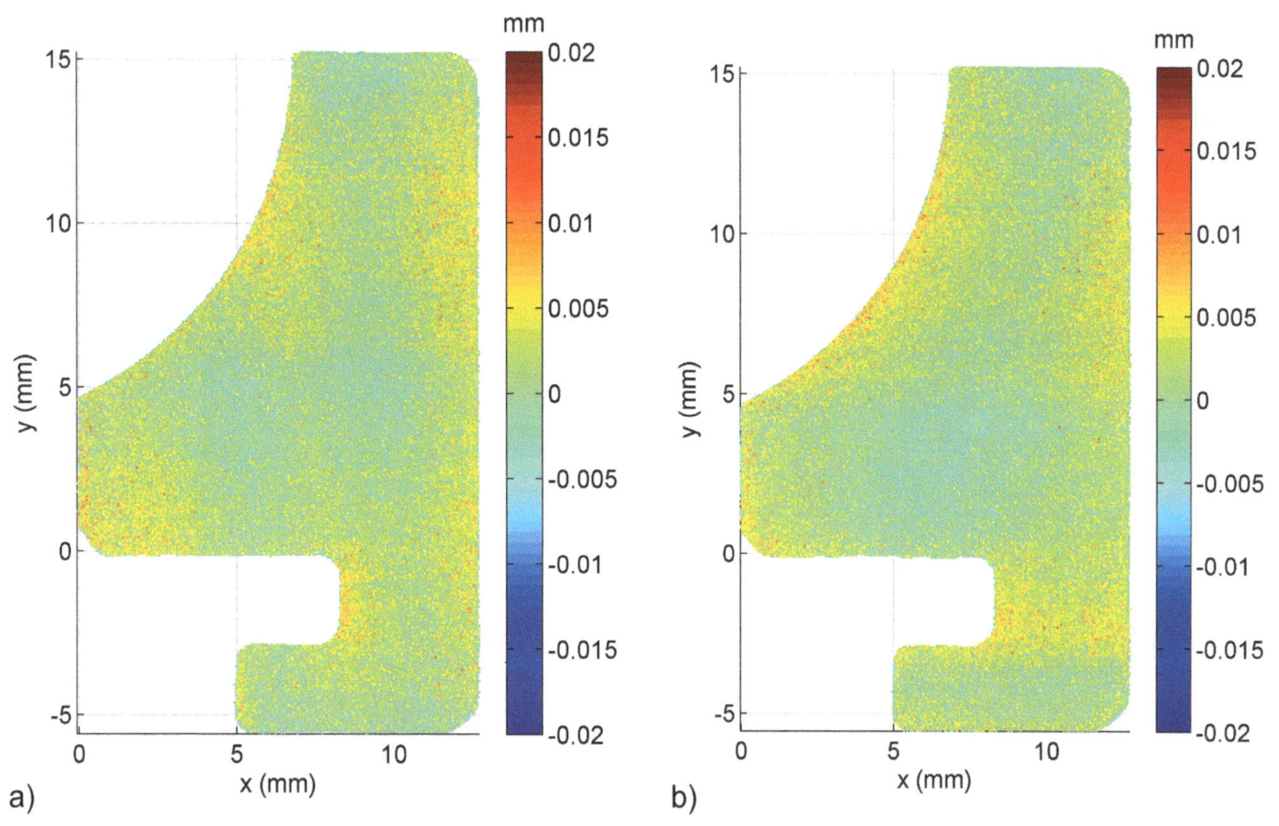

Fig. 10.4 The surface height maps for the two halves of the thrust-loaded race show relatively small deformations

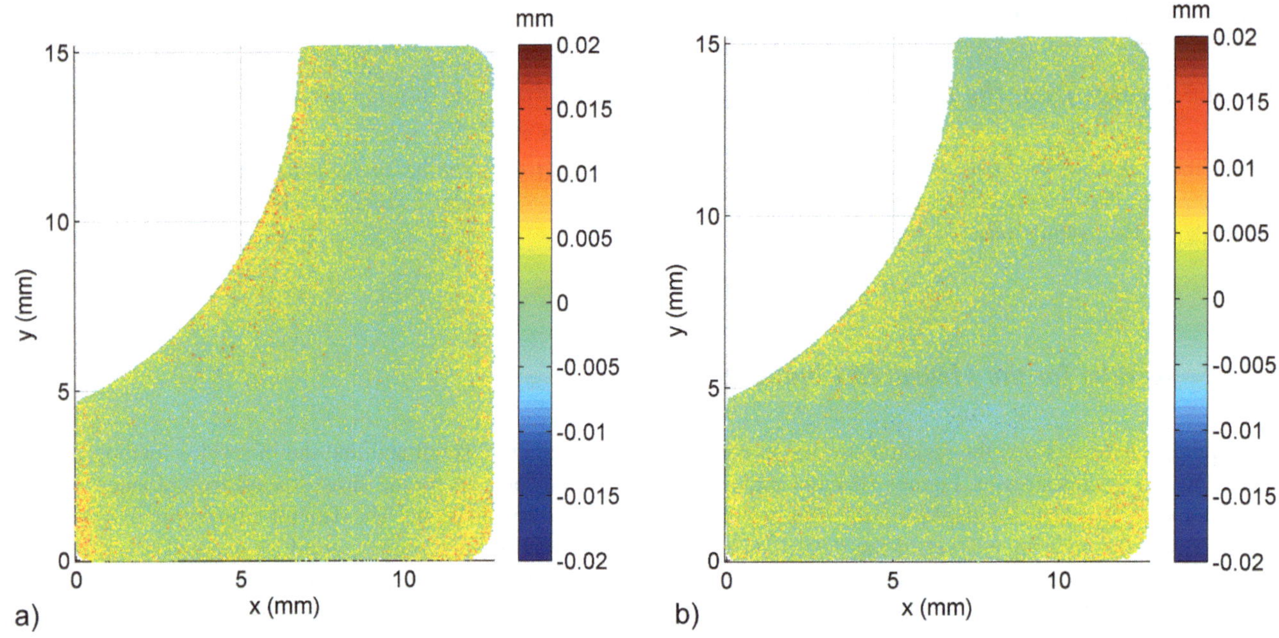

Fig. 10.5 Surface height maps for the two halves of the unloaded race show similar features to the loaded race

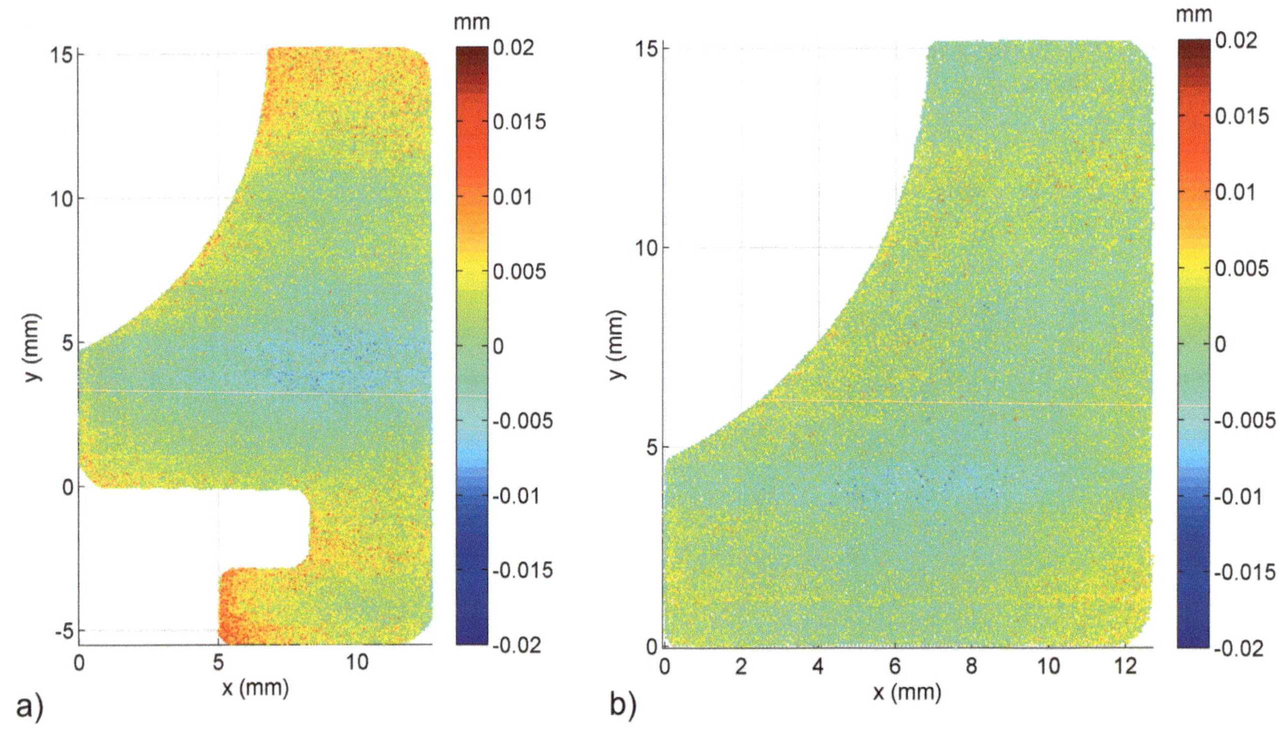

Fig. 10.6 Surface height maps for the stress-free test cut for the (**a**) loaded and (**b**) unloaded races show similar deformation magnitudes to the measurements in Figs. 10.5 and 10.6

In fact, the test cut for the thrust-loaded race, Fig. 10.6a, shows larger features than the measurement data in Fig. 10.4, which only indicates that there might have been some problem with the test cut. In all, the lack of significant differences between the measurement cuts and the stress-free test cuts indicate a lack of significant residual stresses.

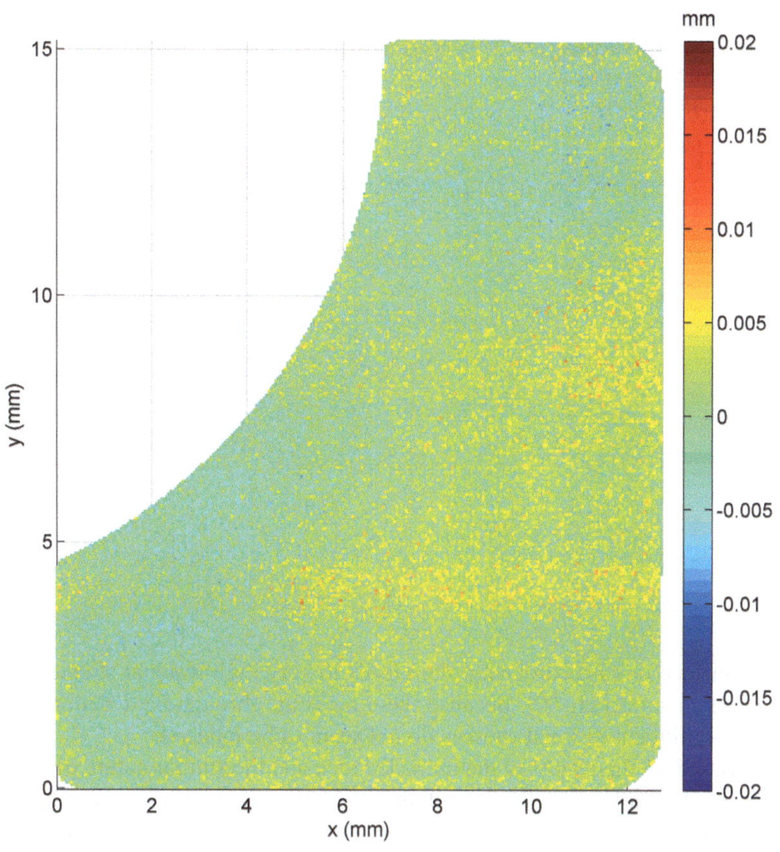

Fig. 10.7 After being averaged between the two sides then corrected for the stress-free test cut, the measured contours from the unloaded race show little evidence of residual stress

The raw data should not be taken to indicate a significant deficiency in the contour method itself. Had significant stresses been induced by the in-service loading of the bearing, the measurement cuts would have given much larger contours than those in Figs. 10.4 and 10.5, and would have been large compared to the test cuts.

Because the unloaded race seemed to have the most representative stress-free test cut, that data was analyzed first. Following standard procedure [36, 58], the measured contours, Fig. 10.5, were interpolated onto a regular grid and then averaged point-by-point. The test cut, Fig. 10.6b, was then interpolated onto the same grid. The test cut data was then subtracted from the averaged contour to reveal the deformations caused by residual stress relaxation [36, 38, 56, 57]. The resulting contour shown in Fig. 10.7 shows very little evidence of residual stress. There is no significant shape near the curved bearing surface. Linear horizontal features, indicative of the small EDM cutting artifacts, remain because they are not exactly duplicated in the test cut so are not corrected out. Some slightly low regions in the lower left and upper right regions of the figure might indicated tensile residual stress regions that are balanced by compressive stresses in the lower right, but the features may not be experimentally significant.

10.4.3 Residual Stresses

As expected from the data, no significant stresses are observed in the unloaded bearing race. When the deformations from Fig. 10.7 are processed into boundary conditions and applied to an unstressed model, the results shown in Fig. 10.8 are obtained. The stress uncertainty is estimated [58, 59], to be at least ±50 MPa. The stresses are not plotted within 300 μm of the edges because of inherent contour-method limitations getting stresses close to edges [36]. The results show some indication of subsurface tensile stress regions that might arise from the hardening and other thermal processing of the parts. However, the stresses lack significance in two ways. First, the peak stresses are barely larger than the predicted uncertainties, and it is difficult to account for all uncertainties so they may be lower than the actual uncertainties. Second, and more important, such stress magnitudes are not especially significant for the fatigue behavior of the races.

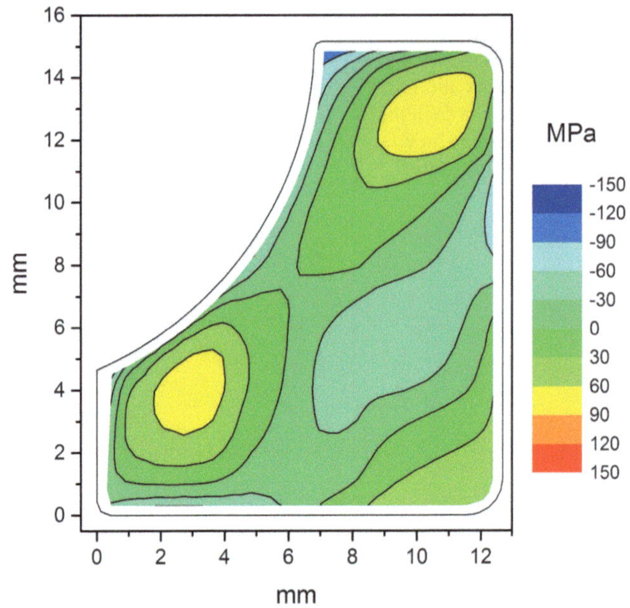

Fig. 10.8 The measure residual stresses in the unloaded bearing race are mostly insignificant

Surface or case hardening often leads to very high residual stresses because of non-uniform plastic deformation and/or phase changes [11, 60]. Because these races are through-hardened, the thermo-mechanical history is sufficiently uniform that significant residual stresses do not seem to be induced.

Based on the results of the unloaded race and the similar nature of the data on the thrust-loaded race, combined with the poorer data quality for the test cut, the effort was not made to process the data on the loaded race. Based on the data, it is clear that the same conclusion can be made: there are no significant residual stresses in the thrust-loaded race.

Upon further consideration, the lack of service-induced residual stress is less of a surprise. Subsurface residual stresses in bearings typically develop after a significant number of RCF cycles, from localized microplasticity, at elevated Hertzian stress. The small surface deformation and stress uncovered by the contour method thus implies that the hardened steel raceways of this bearing were not subjected to high Hertzian stresses or a substantially great amount of cycles. Because the bearing was conservatively removed from service prior to sufficient cyclic loading, such stresses were not induced.

10.5 Conclusions and Future Plans

Contour method measurements showed no significant residual stresses in two races taken from a main-shaft ball bearing removed from service from an aircraft jet engine. The lack of stress in the race that did not see a thrust load indicates no residual stresses from the bearing manufacture, which is attributed primarily to the races being through-hardened as compared to case hardened. The lack of stress in the thrust-loaded counterpart race indicates that the in-service loading did not induce any new residual stresses. Since the precise in-service loading history of this race in not currently known, we cannot compare this result to expectations based on the details of the loading. However, the bearing was probably conservatively removed from service prior to sufficient cyclic loading to induce stresses.

The data is encouraging for future measurements on races and other test parts where higher residual stresses are expected. The stress-free test cuts showed contours generally under $\pm$ 10 μm in amplitude, which should allow for reasonable resolution of stress-induced contours. Even more encouraging is that once the data was averaged between the two halves of a measurement cut and then corrected for the stress-free cut, see Fig. 10.7, it shows even flatter data in the region near the curved bearing contact surface, which is the region of interest where we hope to resolve loading induced stresses in bearings that have seen more significant loads. We should be able to resolve localized stresses in such a region if they exist.

This novel application of the contour method has paved the way and generated insights for future applications of the method on RCF components. Further investigations are anticipated using laboratory RCF-tested ball and rod specimens of both through and case-hardened bearing steels that have experienced a higher range of Hertzian stress and cycles.

Acknowledgements The authors would like to thank Adrian DeWald and Hill Engineering, LLC for the expert experimental work, both EDM cutting and surface scanning, for the measurements reported in this paper. Gratitude is also expressed to Lewis Rosado, Kevin Thompson, Hitesh Trivedi and their colleagues in the Mechanical Systems Branch of the Air Force Research Laboratory for providing the bearing and their valuable support of the project thus far.

Los Alamos National Laboratory, an affirmative action/equal opportunity employer, is operated by the Los Alamos National Security, LLC for the National Nuclear Security Administration of the U.S. Department of Energy under contract DE-AC52-06NA25396. By approving this article, the publisher recognizes that the U.S. Government retains nonexclusive, royalty-free license to publish or reproduce the published form of this contribution, or to allow others to do so, for U.S. Government purposes. Los Alamos National Laboratory requests that the publisher identify this article as work performed under the auspices of the U.S. Department of Energy. Los Alamos National Laboratory strongly supports academic freedom and a researcher's right to publish; as an institution, however, the Laboratory does not endorse the viewpoint of a publication or guarantee its technical correctness.

References

1. Withers, P.J.: Residual stress and its role in failure. Rep. Prog. Phys. **70**(12), 2211–2264 (2007)
2. James, M.N.: Residual stress influences on structural reliability. Eng. Fail. Anal. **18**(8), 1909–1920 (2011)
3. Liu, K.K., Hill, M.R.: The effects of laser peening and shot peening on fretting fatigue in Ti-6Al-4V coupons. Tribol. Int. **42**(9), 1250–1262 (2009)
4. Edwards, L., Smith, M.C., Turski, M., Fitzpatrick, M.E., Bouchard, P.J.: Advances in residual stress modeling and measurement for the structural integrity assessment of welded thermal power plant. Adv. Mater. Res. **41–42**, 391–400 (2008)
5. Aydiner, C.C., Ustundag, E., Prime, M.B., Peker, A.: Modeling and measurement of residual stresses in a bulk metallic glass plate. J. Non Cryst. Solids **316**(1), 82–95 (2003)
6. Tanner, D.A., Robinson, J.S.: Modelling stress reduction techniques of cold compression and stretching in wrought aluminium alloy products. Finite Elem. Anal. Des. **39**(5/6), 369–386 (2003)
7. Yaghi, A.H., Hilson, G., Simandjuntak, S., Flewitt, P.E.J., Pavier, M.J., Smith, D.J., Hyde, T.H., Becker, A.A., Sun, W.: A comparison between measured and modeled residual stresses in a circumferentially butt-welded P91 steel pipe. J. Press. Vessel. Technol. **132**(1), 011206 (2010). doi:10.1115/1.4000347
8. Rolph, J., Preuss, M., Iqbal, N., Hofmann, M., Nikov, S., Hardy, M.C., Glavicic, M.G., Ramanathan, R., Evans, A.: Residual stress evolution during manufacture of aerospace forgings. In: Huron, E.S., Reed, R.C., Hardy, M.C., et al. (eds.) Superalloys 2012, pp. 881–891. Wiley, Hoboken (2012). doi:10.1002/9781118516430.ch97
9. Dai, H., Francis, J.A., Withers, P.J.: Prediction of residual stress distributions for single weld beads deposited on to SA508 steel including phase transformation effects. Mater. Sci. Technol. **26**, 940–949 (2010). doi:10.1179/026708309x12459430509454
10. DeWald, A.T., Hill, M.R.: Eigenstrain based model for prediction of laser peening residual stresses in arbitrary 3D bodies. Part 1: model description. J. Strain Anal. Eng. Des. **44**(1), 1–11 (2009)
11. Prime, M.B., Prantil, V.C., Rangaswamy, P., Garcia, F.P.: Residual stress measurement and prediction in a hardened steel ring. Mater. Sci. Forum **347–349**, 223–228 (2000)
12. Muránsky, O., Hamelin, C.J., Smith, M.C., Bendeich, P.J., Edwards, L.: The role of plasticity theory on the predicted residual stress field of weld structures. Mater. Sci. Forum **772**, 65–71 (2014)
13. Carlone, P., Palazzo, G.S., Pasquino, R.: Finite element analysis of the steel quenching process: temperature field and solid–solid phase change. Comput. Math. Appl. **59**(1), 585–594 (2010). doi:10.1016/j.camwa.2009.06.006
14. Tanner, D.A., Robinson, J.S.: Residual stress prediction and determination in 7010 aluminum alloy forgings. Exp. Mech. **40**(1), 75–82 (2000). doi:10.1007/bf02327551
15. Ismonov, S., Daniewicz, S.R., Newman, J.J.C., Hill, M.R., Urban, M.R.: Three dimensional finite element analysis of a split-sleeve cold expansion process. J. Eng. Mater. Technol. **131**(3), 031007 (2009). doi:10.1115/1.3120392
16. Carlone, P., Palazzo, G.S.: Experimental analysis of the influence of process parameters on residual stress in AA2024-T3 friction stir welds. Key Eng. Mater. **504–506**, 753–758 (2012). doi:10.4028/www.scientific.net/KEM.504-506.753
17. Kaiser, R., Stefenelli, M., Hatzenbichler, T., Antretter, T., Hofmann, M., Keckes, J., Buchmayr, B.: Experimental characterization and modelling of triaxial residual stresses in straightened railway rails. J. Strain Anal. Eng. Des. **50**(3), 190–198 (2015). doi:10.1177/0309324714560040
18. Xie, P., Zhao, H., Wu, B., Gong, S.: Evaluation of residual stresses relaxation by post weld heat treatment using contour method and X-ray diffraction method. Exp. Mech. **55**(7), 1329–1337 (2015). doi:10.1007/s11340-015-0040-2
19. Zhang, Z., Yang, Y., Li, L., Chen, B., Tian, H.: Assessment of residual stress of 7050-T7452 aluminum alloy forging using the contour method. Mater. Sci. Eng. A **644**, 61–68 (2015). doi:10.1016/j.msea.2015.07.018
20. Vrancken, B., Cain, V., Knutsen, R., Van Humbeeck, J.: Residual stress via the contour method in compact tension specimens produced via selective laser melting. Scr. Mater. **87**, 29–32 (2014). doi:10.1016/j.scriptamat.2014.05.016
21. Sadeghi, F., Jalalahmadi, B., Slack, T.S., Raje, N., Arakere, N.K.: A review of rolling contact fatigue. J. Tribol. **131**(4), 041403 (2009). doi:10.1115/1.3209132
22. Voskamp, A.P., Österlund, R., Becker, P.C., Vingsbo, O.: Gradual changes in residual stress and microstructure during contact fatigue in ball bearings. Met. Technol. **7**(1), 14–21 (1980). doi:10.1179/030716980803286676
23. Österlund, R., Vingsbo, O.: Phase changes in fatigued ball bearings. Metall. Trans. A **11**(5), 701–707 (1980). doi:10.1007/bf02661199
24. Voskamp, A.P.: Material response to rolling contact loading. J. Tribol. **107**(3), 359–364 (1985). doi:10.1115/1.3261078

25. Palmgren, A.: Ball and roller bearing engineering. SKF Industries, Philadelphia (1959)
26. Lundberg, G., Palmgren, A.: Dynamic capacity of roller bearings, vol. 196. Generalstabens litografiska anstalts förlag, Stockholm (1947)
27. Lunberg, G., Palmgren, A.: Dynamic capacity of roller bearings. Acta Polytech. Mech. Eng. Ser. **2**, 96 (1952)
28. ISO. Rolling bearing–dynamic load ratings and rating life. ISO (2007)
29. Zaretsky, E.V.: STLE life factors for rolling bearings. STLE Special Publication SP (1992)
30. Londhe, N.D., Arakere, N.K., Haftka, R.T.: Reevaluation of rolling element bearing load-life equation based on fatigue endurance data. Tribol. Trans. **58**(5), 815–828 (2015). doi:10.1080/10402004.2015.1021943
31. Voskamp, A.P., Mittemeijer, E.J.: State of residual stress induced by cyclic rolling contact loading. Mater. Sci. Technol. **13**(5), 430–438 (1997). doi:10.1179/mst.1997.13.5.430
32. Schajer, G.S.: Practical residual stress measurement methods. Wiley, Chichester (2013)
33. Holden, T.M.: Neutron diffraction. In: Schajer, G.S. (ed.) Practical Residual Stress Measurement Methods, pp. 195–223. Wiley, Chichester (2013). doi:10.1002/9781118402832.ch8
34. Schajer, G.S., Whitehead, P.S.: Hole drilling and ring coring. In: Schajer, G.S. (ed.) Practical Residual Stress Measurement Methods, pp. 29–64. Wiley, Chichester (2013)
35. Pagliaro, P., Prime, M.B., Robinson, J.S., Clausen, B., Swenson, H., Steinzig, M., Zuccarello, B.: Measuring inaccessible residual stresses using multiple methods and superposition. Exp. Mech. **51**(7), 1123–1134 (2011). doi:10.1007/s11340-010-9424-5
36. Prime, M.B., DeWald, A.T.: The contour method. In: Schajer, G.S. (ed.) Practical Residual Stress Measurement Methods. Wiley, Chichester (2013). doi:10.1002/9781118402832.ch5
37. Prime. M.B., Gonzales, A.R.: The contour method: simple 2-D mapping of residual stresses. In: The 6th International Conference on Residual Stresses, Oxford, IOM Communications, London, pp. 617–624 (2000)
38. Prime, M.B.: Cross-sectional mapping of residual stresses by measuring the surface contour after a cut. J. Eng. Mater. Technol. **123**(2), 162–168 (2001)
39. Hill, M.R., Olson, M.D.: Repeatability of the contour method for residual stress measurement. Exp. Mech. **54**(7), 1269–1277 (2014). doi:10.1007/s11340-014-9867-1
40. Toparli, M.B., Fitzpatrick, M., Gungor, S.: Determination of multiple near-surface residual stress components in laser peened aluminum alloy via the contour method. Metall. Mater. Trans. A **46**(9), 4268–4275 (2015). doi:10.1007/s11661-015-3026-x
41. Woo, W., An, G.B., Em, V.T., DeWald, A.T., Hill, M.R.: Through-thickness distributions of residual stresses in an 80 mm thick weld using neutron diffraction and contour method. J. Mater. Sci. **50**(2), 784–793 (2014)
42. Braga, D.F.O., Coules, H.E., Pirling, T., Richter-Trummer, V., Colegrove, P., de Castro, P.M.S.T.: Assessment of residual stress of welded structural steel plates with or without post weld rolling using the contour method and neutron diffraction. J. Mater. Process. Technol. **213**(12), 2323–2328 (2013). doi:10.1016/j.jmatprotec.2013.07.004
43. Elmesalamy, A., Francis, J.A., Li, L.: A comparison of residual stresses in multi pass narrow gap laser welds and gas-tungsten arc welds in AISI 316L stainless steel. Int. J. Press. Vessel. Pip. **113**, 49–59 (2014). doi:10.1016/j.ijpvp.2013.11.002
44. Rolph, J., Iqbal, N., Hoffman, M., Evans, A., Hardy, M., Glavicic, M., Preuss, M.: The effect of d0 reference value on a neutron diffraction study of residual stress in a γ/γ' nickel-base superalloy. J. Strain Anal. Eng. Des. **48**(4), 219–228 (2013). doi:10.1177/0309324713486273
45. Traore, Y., Paddea, S., Bouchard, P., Gharghouri, M.: Measurement of the residual stress tensor in a compact tension weld specimen. Exp. Mech. **53**(4), 605–618 (2013). doi:10.1007/s11340-012-9672-7
46. Pagliaro, P., Prime, M.B., Clausen, B., Lovato, M.L., Zuccarello, B.: Known residual stress specimens using opposed indentation. J. Eng. Mater. Technol. **131**, 031002 (2009)
47. Prime, M.B., DeWald, A.T., Hill, M.R., Clausen, B., Tran, M.: Forensic determination of residual stresses and KI from fracture surface mismatch. Eng. Fract. Mech. **116**, 158–171 (2014). doi:10.1016/j.engfracmech.2013.12.008
48. Mahmoudi, A.H., Saei, A.: Influence of asymmetrical cuts in measuring residual stresses using contour method. Int. J. Press. Vessel. Pip. **134**, 1–10 (2015). doi:10.1016/j.ijpvp.2015.08.004
49. Hosseinzadeh, F., Bouchard, P.: Mapping multiple components of the residual stress tensor in a large P91 steel pipe girth weld using a single contour cut. Exp. Mech. **53**(2), 171–181 (2013). doi:10.1007/s11340-012-9627-z
50. Olson, M.D., Hill, M.R.: A new mechanical method for biaxial residual stress mapping. Exp. Mech. **55**(6), 1139–1150 (2015). doi:10.1007/s11340-015-0013-5
51. Prime, M.B.: Contour method advanced applications: hoop stresses in cylinders and discontinuities. In: Proulx, T. (ed.) Engineering Applications of Residual Stress, Conference Proceedings of the Society for Experimental Mechanics Series, vol. 8, pp. 13–28. Springer, New York (2011). doi:10.1007/978-1-4614-0225-1_2
52. Hosseinzadeh, F., Kowal, J., Bouchard, P.J.: Towards good practice guidelines for the contour method of residual stress measurement. J. Eng. (2014)
53. Johnson. G.: Residual stress measurements using the contour method. Ph.D. Dissertation, University of Manchester (2008). http://www.lanl.gov/contour
54. Brown, D.W., Holden, T.M., Clausen, B., Prime, M.B., Sisneros, T.A., Swenson, H., Vaja, J.: Critical comparison of Two independent measurements of residual stress in an electron-beam welded uranium cylinder: neutron diffraction and the contour method. Acta Mater. **59**(3), 864–873 (2011). doi:10.1016/j.actamat.2010.09.022
55. de Swardt, R.R.: Finite element simulation of crack compliance experiments to measure residual stresses in thick-walled cylinders. J. Press. Vessel. Technol. **125**(3), 305–308 (2003)
56. Ahmad, B., Fitzpatrick, M.E.: Minimization and mitigation of wire EDM cutting errors in the application of the contour method of residual stress measurement. Metall. Mater. Trans. A **47**(1), 301–313 (2016). doi:10.1007/s11661-015-3231-7
57. Prime, M.B., Kastengren, A.L.: The contour method cutting assumption: error minimization and correction. In: Proulx, T. (ed.) Experimental and Applied Mechanics, Volume 6, Conference Proceedings of the Society for Experimental Mechanics Series, vol. 17, pp. 233–250. Springer, New York (2011). http://www.lanl.gov/contour/. doi:10.1007/978-1-4419-9792-0_40

58. Prime, M.B., Sebring, R.J., Edwards, J.M., Hughes, D.J., Webster, P.J.: Laser surface-contouring and spline data-smoothing for residual stress measurement. Exp. Mech. **44**(2), 176–184 (2004)
59. Olson, M.D., DeWald, A.T., Prime, M.B., Hill, M.R.: Estimation of uncertainty for contour method residual stress measurements. Exp. Mech. **55**(3), 577–585 (2015). doi:10.1007/s11340-014-9971-2
60. Savaria, V., Hoseini, M., Bridier, F., Bocher, P., Arkinson, P.: On the measurement of residual stress in induction hardened parts. Mater. Sci. Forum **681**, 431–436 (2011)

Chapter 11
ESPI Hole-Drilling of Rings and Holes Using Cylindrical Hole Analysis

T.J. Rickert and Wade Gubbels

Abstract The simplest geometry for hole-drilling analysis is a perfectly cylindrical hole and that geometry is the first option when generating coefficients that correlate surface displacements with stress. The actual hole geometry usually isn't quite perfect, though. Plunge drilling with square-end end mills, for instance, creates an inverted cone on the bottom of the hole. Also, core-drills may be the best option for drilling some materials, such as ceramics. This study was made to evaluate the error due to such geometry mismatch when coefficients for a cylindrical hole are used for the analysis. A shot-peened aluminum sample that is assumed to have a very homogeneous stress distribution was chosen for the study. Plunge drilling, orbital drilling and ring drilling with different sizes and diameter ratios was performed. All measurements were repeated twice to test measurement statistics.

Keywords Stress measurement • ESPI • Incremental hole-drilling • Residual stress • Hole shape

11.1 Background

Square-end end mills are common machining tools and can be used for making holes for stress measurement. The cutting edges on the front end have a small angle against the surface to make plunge drilling possible. After plunge drilling, the hole will have an inverted cone on its bottom. The cone height may be about 2–3 % of the tool diameter, which may not significantly affect the overall results. However, when stresses very close to the surface are of interest this error should be considered. By adding orbital motion to the drill flat bottom holes that more closely match the theoretical model can be created. When using the same square-end end mills, a flat-bottom cylinder would require a hole diameter twice that of the tool. Larger eccentricity results in a ring, which is what core-drill drilling and ring-coring generate. The ring-core method typically measures changes on the surface of the columnar volume remaining in the middle [1]. Here, measurements were always made on the surface outside of the ring.

When evaluating a relatively small effect like that of the inverted cone on the hole bottom, one also needs to consider other, relatively small error sources. One of these is the accuracy of zero-depth determination. The user basically assigns this depth when "test drilling" before measurement start and assessing images after the first contact. The task becomes more difficult when the surface is rough—like after shot-peening. The material volume removed will be lower than in the analysis model until the depth exceeds the surface roughness. This only affects the lowest depths. But when zero-depth is not determined consistently, the whole depth profile shifts to lower or higher depths. The range of the shift can be limited to a few microns, though, and is thus often negligible.

11.2 Experimental

One shot-peened aluminum plate (alloy 7075) was used for all measurements (Fig. 11.1). The sample was lightly spray painted to improve its optical properties. Similar plates had previously been found to have very uniform stress depth profiles across their surfaces. Square-end end mills of four sizes were used: 0.015″ (0.38 mm), 1/32″ (0.79 mm), 1/16″ (1.59 mm) and 1/8″ (3.18 mm). These sizes will be referred to as 0.4 mm, 0.8 mm, 1.6 mm and 3.2 mm, respectively, in the following. The tools were 2-fluted and TiN-coated (except for the 3.2 mm end mill, which was not coated). They were used to make holes

T.J. Rickert (✉) • W. Gubbels
American Stress Technologies, Inc., 540 Alpha Drive, Pittsburgh, PA 15238, USA
e-mail: Theo.Rickert@astresstech.com; Wade.Gubbels@astresstech.com

S. Quinn, X. Balandraud (eds.), *Residual Stress, Thermomechanics & Infrared Imaging, Hybrid Techniques and Inverse Problems, Volume 9*, Conference Proceedings of the Society for Experimental Mechanics Series, DOI 10.1007/978-3-319-42255-8_11

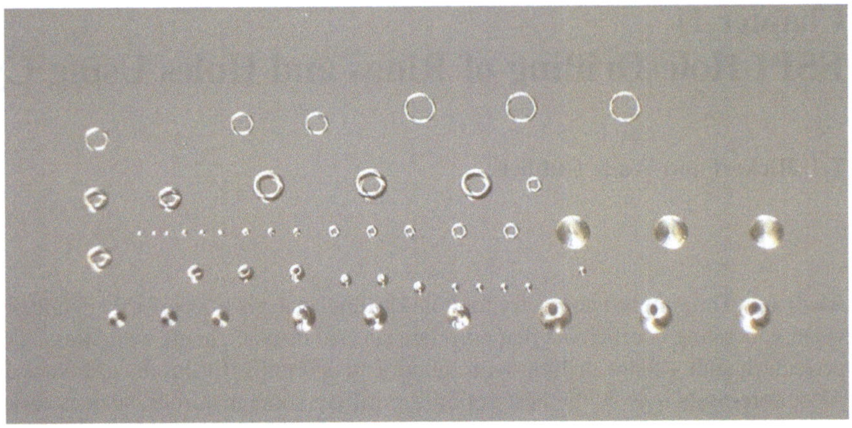

Fig. 11.1 Sample with the complete set of measurements

and rings of the following approximate sizes: 0.4, 0.6, 0.8, 1.2, 1.6, 2.4 and 3.2 mm. It is practical to also characterize the measurements by the ratio between hole or ring and tool diameter. In many cases, there are multiple tool sizes for the same ratio. The ratio ranges from 1 to 8.

All measurements were made with a Prism[1] residual stress measurement system, which combines incremental hole-drilling with ESPI (Electronic Speckle Pattern Interferometry) [2]. The stress calculation followed the Integral Method [3]. All depth profiles shown were calculated using modest regularization, a form of smoothing. Stress calculation and regularization are equivalent to those described in the ASTM standard for strain gage hole-drilling [4]. The analysis area included the ring-shaped area between 2 and 4 times the feature radius from its center point.

The orbital motion was generated by a separately controlled attachment. Unfortunately, the orbiting speed was much lower than the feed motion and the software doesn't permit adding a holding time at depth. The hole depth therefore wasn't identical everywhere even with orbital drilling for a hole twice the tool size. The feed was set to the minimum practical speed of 0.01 mm/s and the speed of the orbital motion maximized. The latter had to be reduced for the larger holes and ranged from about 95 to 57 rpm. The drill rotation speed was 18,000 rpm. The hole sizes were measured using images containing a ruler next to the hole. The hole diameters used in the calculations were on average 1.1 % off the expected values; the maximum ±2.4 %. The impact on the depth profiles is negligible.

11.3 Results

The repeatability is generally quite good. The highest variability is found for plunge drilling. For the smallest hole of 0.4 mm (Fig. 11.2a), the camera magnification was at its limits and the depth profiles deviate significantly from most others. The plunge drilling results for 3.2 mm holes and the results from rings with the largest diameter ratios are also clearly different. But the majority of measurements are similar to each other. Examples are shown in Fig. 11.2. The results in vertical direction are similar to those in horizontal direction and have only slightly lower repeatability.

Averages were calculated for each depth increment and condition. Some of these average depth profiles are shown in Fig. 11.3, organized by diameter ratio. Overall, these graphs show great similarity of the stress curves despite differences in the drilling. Yet the maximum compressive stress increases slightly with increasing hole or ring diameter. Averaging the curves for the hole sizes 0.8, 1.6 and 3.2 mm shows a characteristic difference between plunge and orbital drilling. The plunge drilling depth profiles are clearly shifted to greater depths (~20 μm) and less compressive stresses at the lowest depths (Fig. 11.4).

The measurement variability was also assessed by calculating standard deviations for each set of three measurements of the same condition. After averaging these values over all depths to obtain a single number for each drilling condition, one can make a direct comparison (Fig. 11.5). The average standard deviations are clearly largest for plunge drilling. Above a diameter ratio of 2, the values are essentially the same. Another description of the variabilities of plunge and hole drilling is given in Fig. 11.6. The averages of the standard deviations for the hole diameters of 0.8, 1.6 and 3.2 mm are consistently higher for plunge drilling up to depths where the stresses become small.

[1] ®Prism is a registered trade mark of American Stress Technology, Pittsburgh, PA.

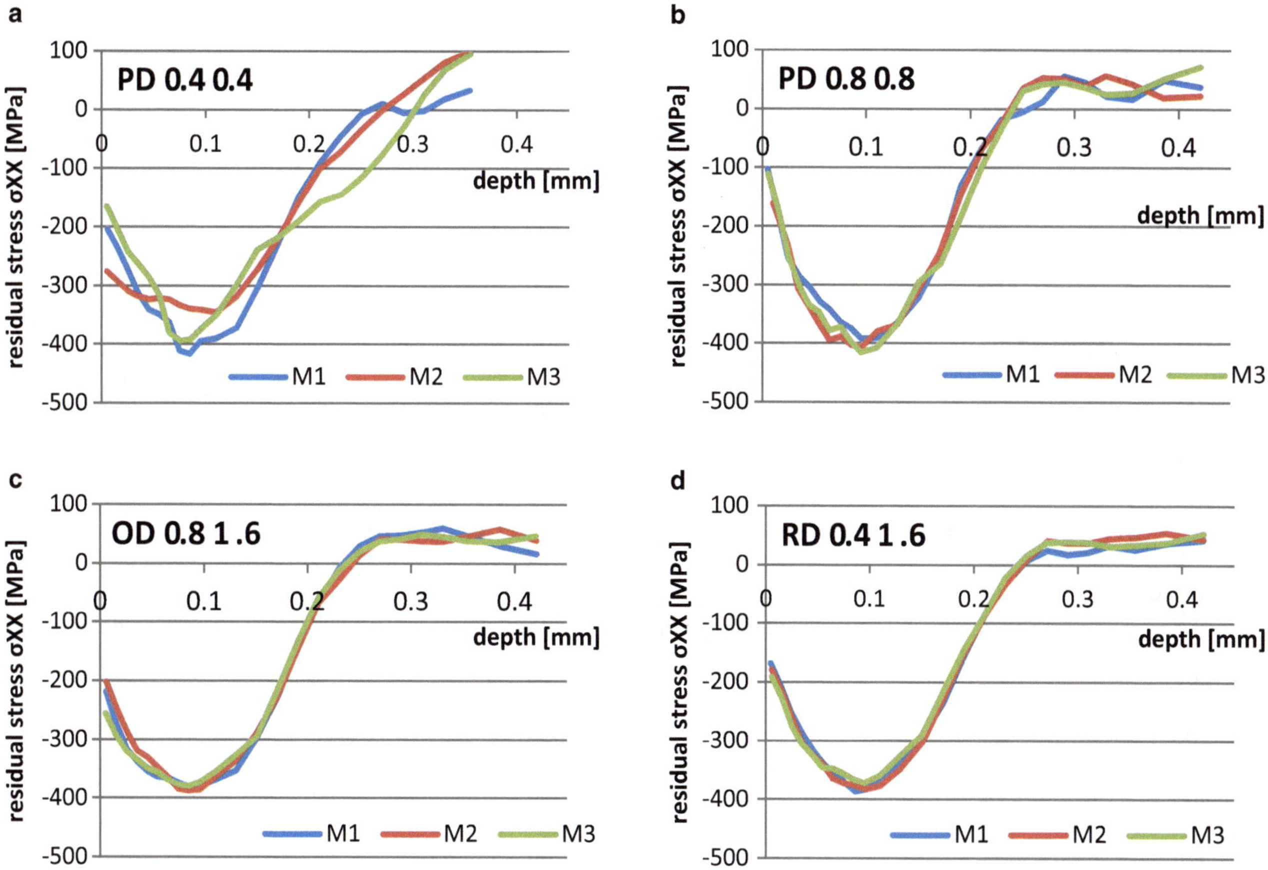

Fig. 11.2 Example results for plunge drilling: 0.4 mm hole (**a**) and 0.8 mm hole (**b**), orbital drilling: 0.8 mm tool, 1.6 mm hole (**c**); orbital drilling: 0.4 mm tool, 1.6 mm ring (**d**); three measurements each

The ring drilling depth profiles for the diameter ratios of 3 and 4 are very similar to those from orbital drilling for a ratio of 2. With increasing ratio, the stresses closest to the surface become less compressive, similar to the inverted cone effect in plunge drilling. At the highest ratios of 6 and 8, the whole curves become significantly different.

11.4 Discussion

A possible cause for the shift of the depth profiles between plunge and orbital drilling is simply a different assessment of when the zero-position is reached during measurement preparation. Figure 11.7 shows typical images from before measurement start. The zero-determination process is deemed precise enough that the estimated variability should be less than 10 μm despite the relatively high surface roughness. However, it should be of similar magnitude for both plunge and orbital drilling. Therefore, the shift would have to be due to a systematic error. It is possible that the start for orbital drilling is set consistently deeper in the material than for plunge drilling since it is a visual process. But the fact that the material removed is always a slice parallel to the inverted cone surface also causes a shift (Fig. 11.8a, bottom). The average depth to which one should attribute the stress for the interval shifts to lower depths. An estimation for a 3° angle of the cutting edges and a tool diameter of 1.6 mm suggests an 8–9 μm shift.

A second effect, described by the drawing in Fig. 11.8a (top), affects only the results near the surface. The actual volume of material removed is much smaller at the very beginning of drilling than the cylinder volume assumed for the coefficients. This reduces the magnitude of the stress calculated for the first depth increment substantially (depending on increment size).

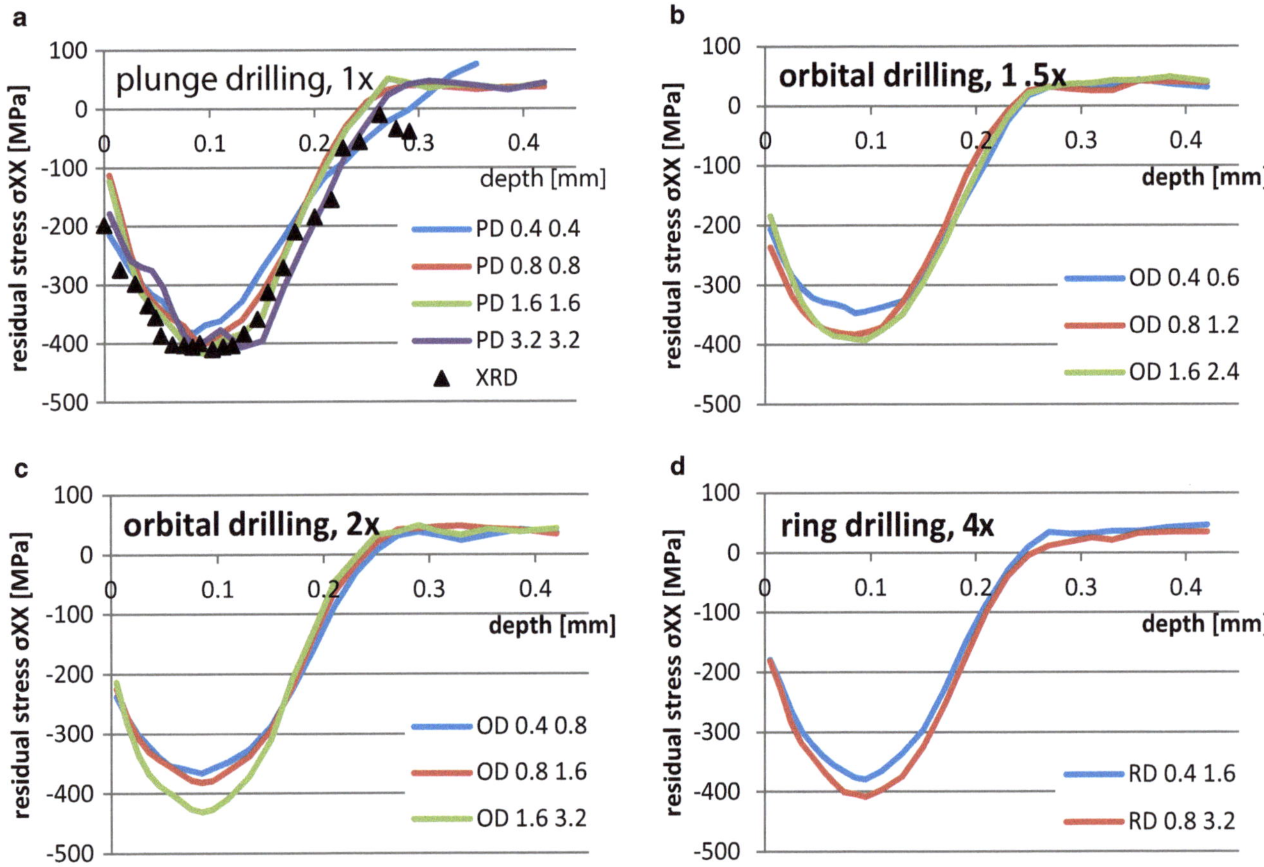

Fig. 11.3 Comparisons of averaged depth profiles for fixed diameter ratios

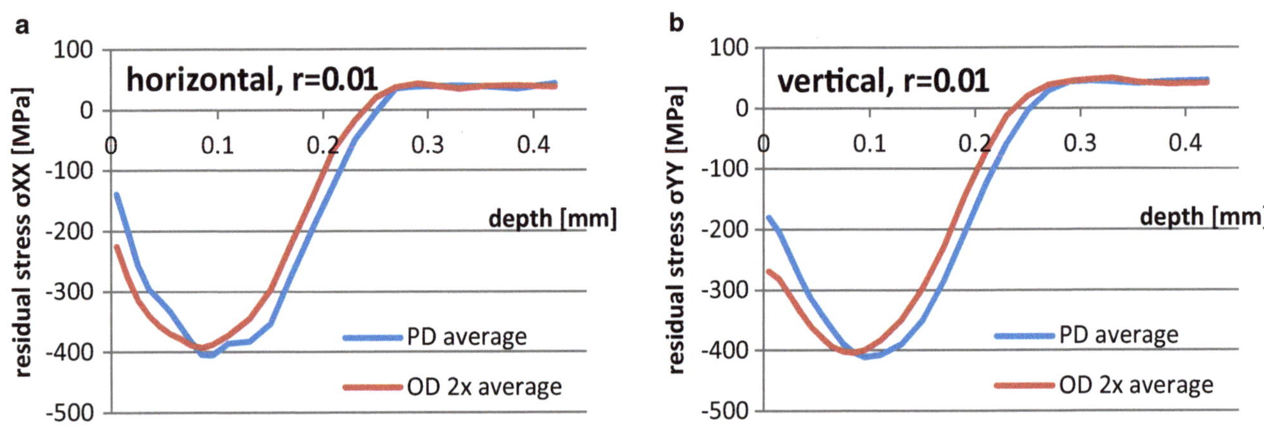

Fig. 11.4 Comparison of the averages over all plunge drilling and orbital drilling (2×) measurements for holes of 0.8 mm, 1.6 mm and 3.2 mm diameters; regularized: r = 0.01, *horizontal* (**a**) and *vertical* stresses (**b**)

To verify that, one needs to evaluate non-regularized data because regularization is a smoothing process. Figure 11.9 shows that the stresses calculated for plunge drilling are less compressive than those for orbital drilling at the lowest depths. But even the orbital drilling depth profiles have suspiciously low compressive stresses at the lowest depth. It is suspected that the actual stresses for the first increment are similar to those measured subsequently. Note that the orbital motion was too slow to reach the same depth in all locations so that there must be a volume error here, too. It should affect a much smaller depth range, though.

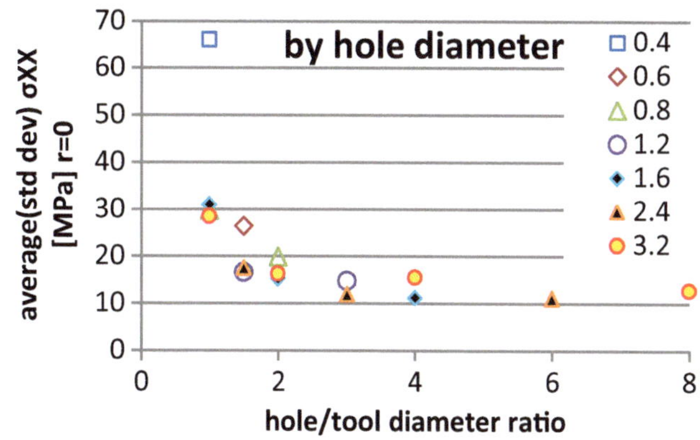

Fig. 11.5 Standard deviations averaged over all depths; all conditions, organized by hole diameter and plotted over the diameter ratio

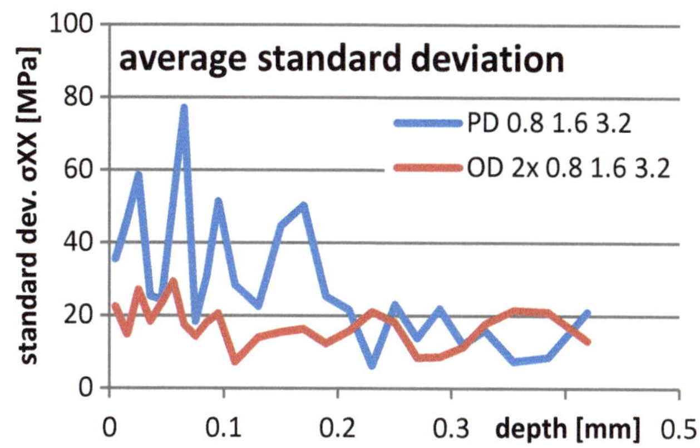

Fig. 11.6 Averages of the standard deviations for holes of 0.8 mm, 1.6 mm and 3.2 mm diameters; plunge drilling compared to orbital drilling

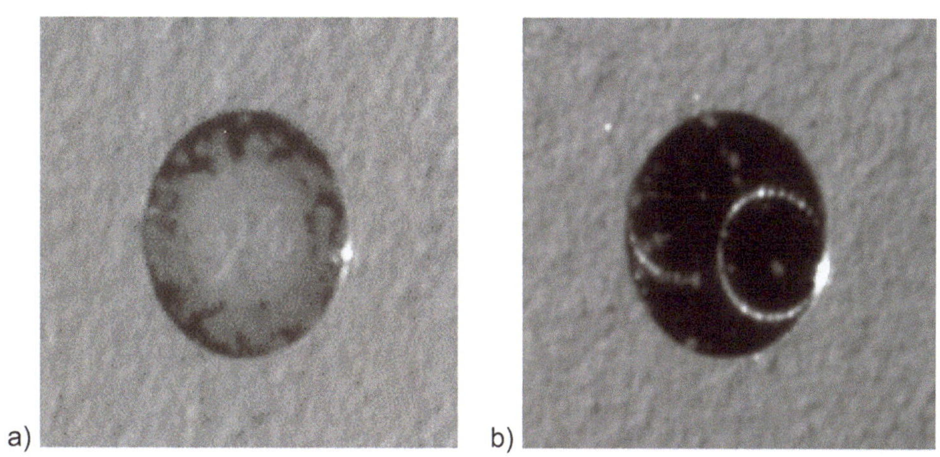

Fig. 11.7 Hole location after zero-depth definitions; spray painted surface; (**a**) 1.6 mm plunge drilling, (**b**) orbital drilling of 1.6 mm hole with 0.8 mm tool

The ring drilling process removes a much smaller volume than the full cylinder assumed for the coefficients (Fig. 11.8b). Yet for diameter ratios of 3 and 4, the stress profiles are very similar to orbital drilling with a ratio of 2. The reason is that any stresses in the columns left inside the ring affect the measurement site outside the ring only very little. Larger differences are found only for higher ratios—thinner rings. When looking a bit closer at the stresses at the lowest depths, a small effect similar to that of the inverted cone becomes visible, and it increases with the diameter ratio. This supports that the surface-near effects seen are related to reduced material removal.

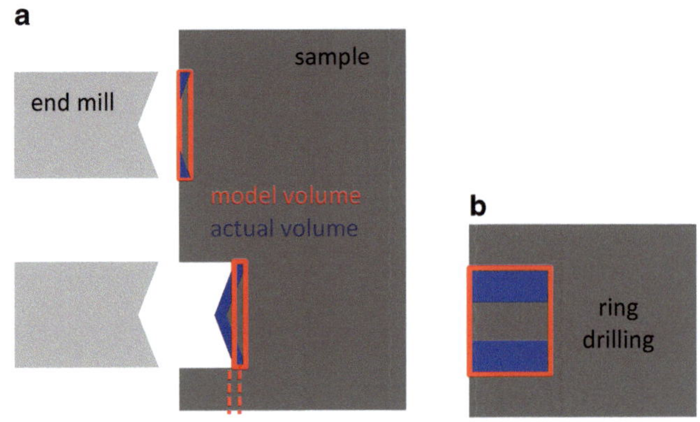

Fig. 11.8 Schematic material removal process for plunge drilling with a square-end end mill (**a**) and ring drilling (**b**), describing volume and depth errors

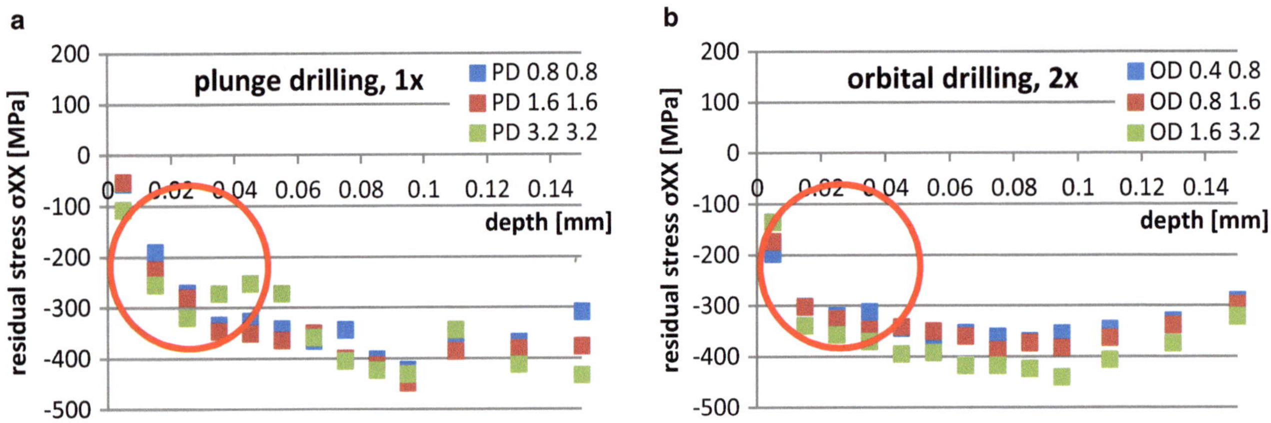

Fig. 11.9 Comparison of near-surface stress values determines for plunge drilling (**a**) and orbital drilling (**b**); no regularization

The plunge drilling results have clearly higher deviations than orbital drilling. This may be indicative of the advantage that orbital drilling is reported to provide [5]. It is possible though, that other drilling parameters, e.g. a different rotation speed would work better for plunge drilling. That was not investigated.

The maximum compressive stress of the depth profiles increases with hole and ring diameter. Increasing feature size means an increased magnitude of the displacements, which is what the instrument measures. Therefore, one might assume that data accuracy would increase and that the profiles determined from larger features are more meaningful. There is no confirmation for this, though. It is questionable whether comparison XRD measurements could be accurate enough to support or challenge this idea since the differences are so small.

11.5 Conclusions

Orbital drilling has been shown to decrease measurement variability compared to plunge drilling for an ESPI hole-drilling instrument. The high precision allowed analyzing relatively small effects that could be shown to be related to deviations from the perfectly cylindrical hole shape that was assumed for the generation of the coefficients. Analyzing rings as if they were full holes can produce very similar results as long as the ring is relatively thick.

References

1. Keil, S.: Experimental determination of residual stresses with the ring core method and an on-line measuring system. Exp. Tech. **16**(5), 17–24 (1992)
2. Steinzig, M., Ponslet, E.: Residual stress measurement using the hole drilling method and laser speckle interferometry, parts I–IV. Exp. Tech. **27** (3–6), 59–63 (2003)
3. Schajer, G.S., Steinzig, M.: Full-field calculation of hole drilling residual stresses from electronic speckle pattern interferometry data. Exp. Mech. **45**(6), 526–532 (2005)
4. American Society for Testing and Materials: Standard Test Method for Determining Residual Stresses by the Hole Drilling Strain-Gage Method, ASTM E837-13a. American Society for Testing and Materials, Conshohocken (2013)
5. Flaman, M.T., Herring, J.A.: Ultra-high-speed center-hole technique for difficult machining materials. Exp. Tech. **10**(1), 34–35 (1986)

Chapter 12
Preliminary Study on Residual Stress in FDM Parts

C. Casavola, A. Cazzato, V. Moramarco, and G. Pappalettera

Abstract The Fused Deposition Modelling (FDM) is nowadays one of the most widespread techniques for 3D object rapid prototyping. In recent years, the FDM evolved from rapid prototyping technique towards a rapid manufacturing method, changing the main purpose in producing finished components ready for use. However, as the parts are built as a layer-by-layer deposition of a feedstock wire, the FDM technique shows, during the building process, distortion and de-layering problems. This issue influences the shape and the final dimensions of the parts or it can prevent the finalization of the objects due to unsticking problems from the bed. Several techniques can be employed in order to obtain parts of correct shape and dimensions. Many of these, such as depositing glue on the bed, aim to constrain the object. As a consequence, the FDM parts could show residual strain and residual stress that could influence their mechanical behaviour. The aim of this work is to measure, by ESPI technique, the displacements around a hole drilled into the material. This can be considered as a preliminary indication of the level of residual stress inside the FDM parts.

Keywords 3D printing • Residual stress • Fused deposition modelling • Hole drilling • ESPI

12.1 Introduction

The Fused Deposition Modelling (FDM), invented in the early 1990s by Stratasys, is one of the most used techniques to 3D object rapid prototyping. The main advantage of this technology is the easy fabrication of complex 3D prototypes directly from a computerized solid model without any need for machining. Nowadays, the reliability of FDM parts has been widely tested and this technology is employed in many fields such as aerospace, medical, construction, cultural and other [1, 2]. Moreover, the low-cost desktop 3D printers such as RepRap, Maker-Bot, Cube, etc., have made this technology widely accessible for producing custom products at home and office. This wide diffusion and reliability have transformed the FDM from rapid prototyping technique in a rapid manufacturing method that produces finished components in single parts or in low volume [3]. In the FDM process, a filament feedstock is partially melted, extruded and deposited by a numerically guided heated nozzle. After the deposition, the material cools, solidifies and sticks to the surrounding material. The part is built bottom up, depositing layer by layer and once the building process is completed, the FDM part shows orthotropic material properties with a behaviour similar to a laminate composite structure [4]. Currently, most desktop 3D printers build parts using only thermoplastic materials such as polylactic acid (PLA) and acrylonitrile-butadiene-styrene (ABS) even if many others materials have been developed, such as bioresorbable polymer (PCL) [5], short fibre composites [6], ceramics [7], metal [8] and metal/polymers mixture materials [9]. The PLA offers better thermo-mechanical characteristics than ABS having a stronger mechanical resistance and a lower coefficient of thermal expansion that improves the printability of PLA reducing the effects of warping during the printing phase. Indeed, one of the most important problems in the FDM process is the distortion of the part during the printing phase. The heating and rapid cooling cycles of the feedstock material produces accumulation of residual stresses and strains during material building up [10]. This residual stress would lead to distortion and de-layering problems, which seriously affects the shape and the final dimensions of the parts or it can prevent the finalization of the objects due to unsticking problems from the bed. In order to reduce the distortion problems, the part is glued on a heated bed even if this does not resolve the residual stress problems.

Some studies have shown that the final quality of the object (i.e. strength, distortions, residual strains and stresses, etc.) is related to the process parameters such as layer thickness, deposition direction (raster angle), stacking sequence, bead width and overlap between beads. In particular, Mahesh et al. [11] evaluated the distortions in Rapid Prototyping Techniques showing that the deposition of the material leads to asymmetrical thermal gradients that caused residual stress and then

C. Casavola (✉) • A. Cazzato • V. Moramarco • G. Pappalettera
Dipartimento di Meccanica, Matematica e Management (DMMM)—Politecnico di Bari, Viale Japigia 182, Bari 70126, Italy
e-mail: katia.casavola@poliba.it

S. Quinn, X. Balandraud (eds.), *Residual Stress, Thermomechanics & Infrared Imaging, Hybrid Techniques and Inverse Problems, Volume 9*, Conference Proceedings of the Society for Experimental Mechanics Series, DOI 10.1007/978-3-319-42255-8_12

distortions. Wang et al. [12] introduced a mathematical model to analyse the effects of the chamber temperature and the number of deposited layer on the distortion of the parts, finding that the increase of the chamber temperature reduces the distortion of the part. Zhang and Chou [13], using a 3D finite element model of the FDM process, studied the effect of path patterns of the head on residual stresses and part distortions. Although many analytical models have been proposed to predict and reduce the distortion of FDM parts, these show some issues such as lack of accuracy in the prediction. Zhang and Chou, by a FE 3D numerical model, provide a better prediction of residual stress and strain changing the process parameters. However, their model has been validated only on the distortion of real specimens and not by comparing measured residual stress with the predicted one.

The aim of this work is to measure the displacements in FDM parts by ESPI-HDM (Electronic Speckle Pattern Interferometry—Hole Drilling Method) technique. Due to the orthotropic behaviour and the consequently difficulties in calculating the residual stress, this work is a preliminary study on the measurement of residual stress in FDM parts. Three different stacking sequences have been analysed.

12.2 Material and Methods

In this work ESPI technique has been used to measure displacement around a hole drilled inside the material. The HDM combined with the ESPI allows to avoid local reinforcement due to strain gauge rosette that is a serious issue in materials with low Young's module. A diode pumped solid state laser source ($\lambda = 532$ nm) was used in order to shine the sample and to generate the speckle pattern. The laser beam is divided in two parts by a beamsplitter and delivered by two optical fibres. The beam emerging out from the first fibre is collimated and then directed towards the sample at a given angle ($\alpha = 44°$). The beam emerging from the second fibre, instead, is directed towards the CCD matrix of the camera and acts as a reference beam. The CCD camera (640 × 480 pixel) itself is placed at a given angle with respect to the normal to the sample ($\beta = 24°$). Light diffused by the sample interferes on the CCD matrix with the reference beam. Four-step temporal phase shifting algorithm was adopted in order to obtain the phase [14, 15]. This means that four reference images are taken initially having a $\pi/2$ phase difference among each other. Another set of four images is, analogously taken for each drill increment. These intensity patterns were subtracted from the reference intensity pattern recorded on the sample before starting the drilling procedure. This operation allows obtaining fringe patterns encoding the information about the displacement experienced by the sample along the sensitivity vector . The hole was drilled by means of a high-speed turbine which is mounted on a precision travel stage. Turbine rotation speed was set to 5000 rpm after some preliminary tests that indicated that this speed allows obtaining good quality holes. The cutter is made by tungsten coated by TiN and it has a nominal diameter d = 1.59 mm. Hole was drilled to a depth of 0.6 mm through 30 drill increments. Compressed air was activated during the test to clean the surface of the sample by the formation of drilling chips.

A RepRap Prusa i3 equipped with a marlin firmware and a nozzle of 0.4 diameter has been employed for the production of the specimens. These have a rectangular shape and the dimensions of 80 × 40 × 7 mm. Three different stacking sequences have been analysed, i.e., the raster angles are ±30°, ± 45° and 0°/90°. A layer with a 0° raster angle have the deposited beads parallel to the major side of the specimen. The Fig. 12.1 shows two examples of specimens used in this work.

Some printing parameters, such as the layer thickness or the number of contour lines, have been kept constant for every specimens. These values have been reported in Table 12.1. Finally, the specimens have been fabricated with the minimum dimension of the part perpendicular to the build platform.

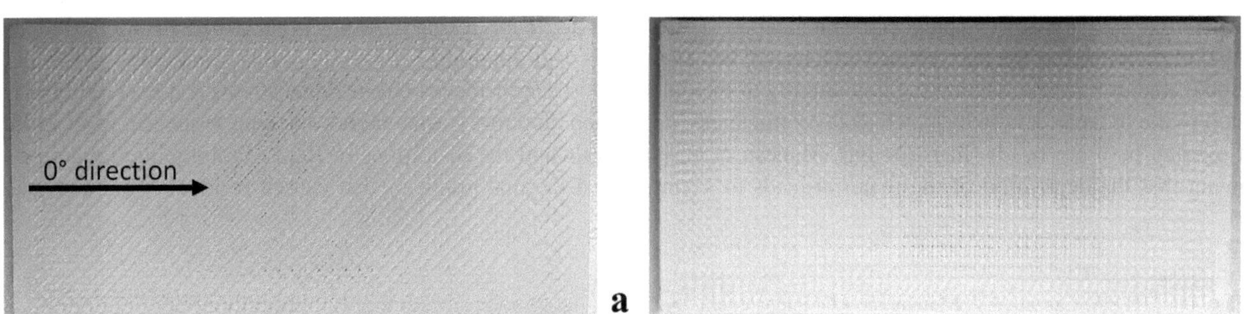

Fig. 12.1 Specimen examples with ±45° (**a**) and 0°/90° (**b**) stacking sequence

Table 12.1 Fixed printer parameters

Parameter	Value
Air gap [mm]	0
Layer thickness [mm]	0.2
Bead width [mm]	0.67
Number of contour lines	3
Bed temperature [°C]	90
Nozzle temperature [°C]	215

In Table 12.1, the air gap is the distance between two, adjacently deposited, beads of the same layer; the layer thickness and the bead width are respectively the height and the width of a deposited filament. The number of contours represents how many edges have been deposited before filling the inner part by inclined beads.

The solid model, created using a 3D CAD, has been sliced using the open source software Slic3r. On each specimen, five holes have been drilled on the top of the specimen, i.e. starting from the last layer deposited. In order to obtain a better knowledge of the residual strain in different zone of the sample also a hole has been drilled on the bottom of the specimens.

12.3 Results and Discussions

Preliminary tests have been carried out in order to determine the correct drilling parameters. In particular, the effects of the local increase of temperature due to drilling speed have been considered. In fact, because of the low melting point of the studied material, temperature can seriously influence the quality of the holes [16]. In Fig. 12.2, the hole macrography and thermograms comparison between 5000 rpm (Fig. 12.2a, b) and 50,000 rpm (Fig. 12.2c, d) are reported.

The temperatures comparison between Fig. 12.2b, d shows that increasing the drilling speed from 5000 to 50,000 rpm leads to a remarkable increase of temperature. Indeed, the maximum temperature for 5000 rpm is 31.9 °C whereas for 50,000 rpm is 77.9 °C. This could explain the difference in the hole quality between the two drilling speeds. In Fig. 12.2a the hole bottom is flat and regular whereas in Fig. 12.2c there is a central bulge that is probably due to a local boiling of the ABS material during the drilling process.

In Fig. 12.3 a scheme of the measurement area has been reported. In order to avoid local effect (e.g. plasticization or chips presence) the area considered for the measurements goes from two times to four times the hole radius. In this circular crown, a 7 × 7 pixels area has been selected to record the displacements after the drilling phase. In this paper, in order to simplify the results only the displacements every five step of drilling, i.e. 0.1 mm of depth, have been reported. Moreover, these values have been averaged in order to obtain one value for the whole area.

In Figs. 12.4 and 12.5, a comparison between the displacements recorded for the three specimens has been reported. The displacements for the ±45° specimen are higher than the ±30° and 0/90° specimens. This should be related to higher residual stress in the sample and this is confirmed by major difficulties encountered in printing this specimen type. Indeed, even if the sample is glued on the heated bed, this specimen tends to unstick from the bed and to warp more than the other specimen types. The sample 0/90° shows lower displacements that should be due to lower residual stress and also in this case this is confirmed by an higher printability.

12.4 Conclusion

In this paper, the measure of the displacements consequent to the hole-drilling process has been carried out by ESPI. This technique allows to avoid local reinforcement due to strain gauge rosette that is serious issue in materials, such as ABS, that have a low Young's module. In view of this aim, three different specimens with different stacking sequences, i.e., ±30°, ± 45° and 0°/90°, have been printed and studied. The specimens have been made of ABS and they have the dimensions of 80 × 40 × 7 mm.

Some preliminary tests have been carried out in order to determine the correct drilling parameters. These show that the local increase of temperature due to the drilling phase can seriously influence the quality of the hole. Indeed, whereas using 5000 rpm the hole bottom is flat and regular, with 50,000 rpm there is a central bulge that is probably due to a local boiling of

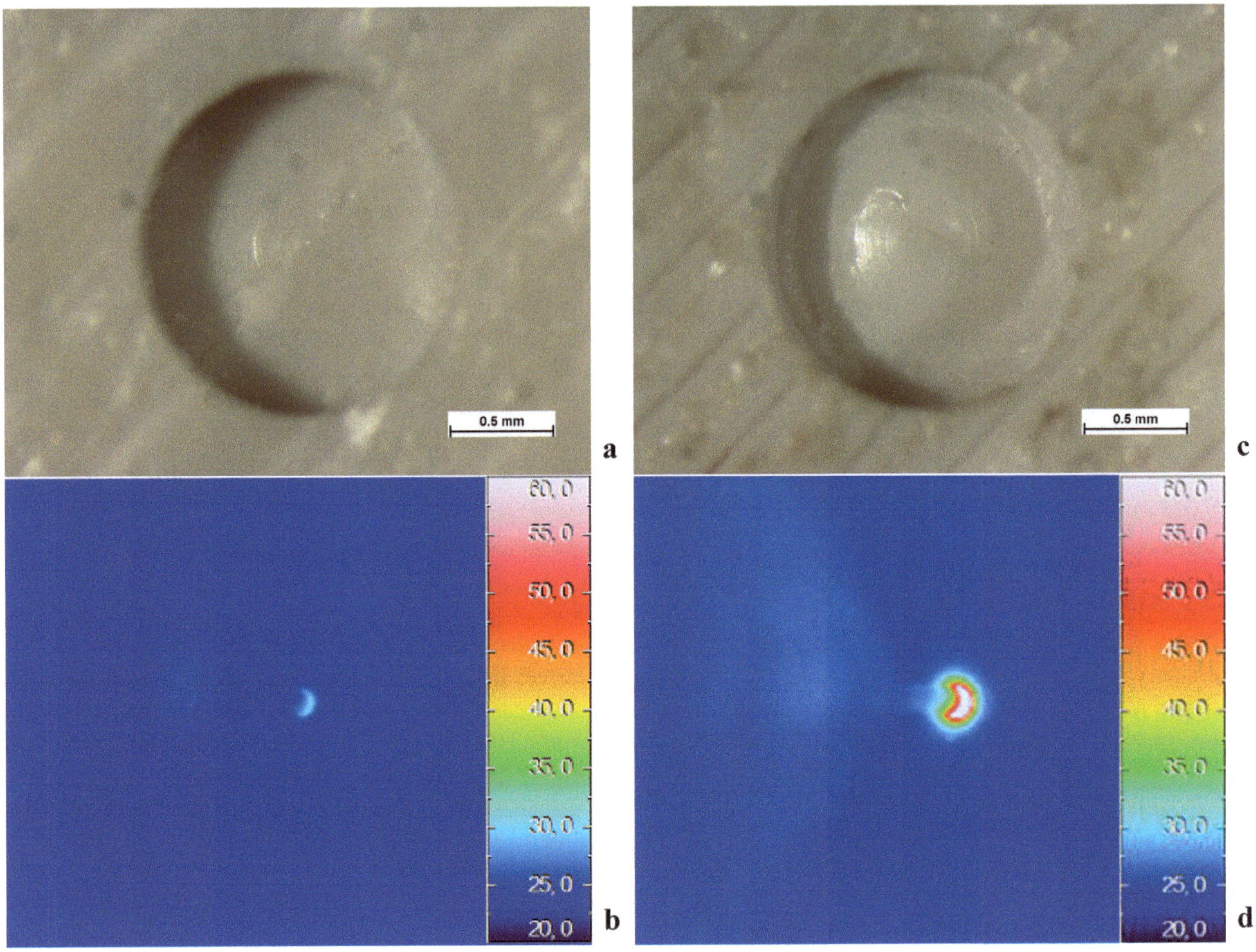

Fig. 12.2 Hole macrography and temperatures maps comparison between 5000 rpm (**a**, **b**) and 50,000 rpm (**c**, **d**)

Fig. 12.3 Scheme of the measurement area, the solid arrow indicates the direction of the component of displacement measured with the ESPI set-up

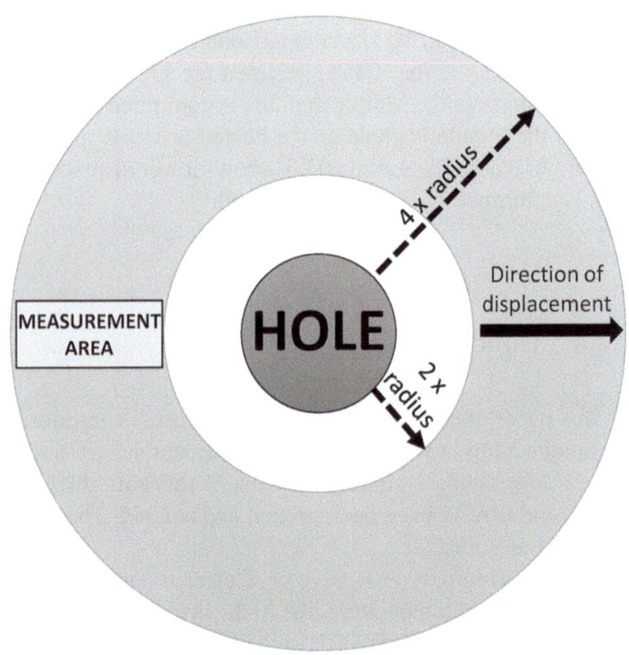

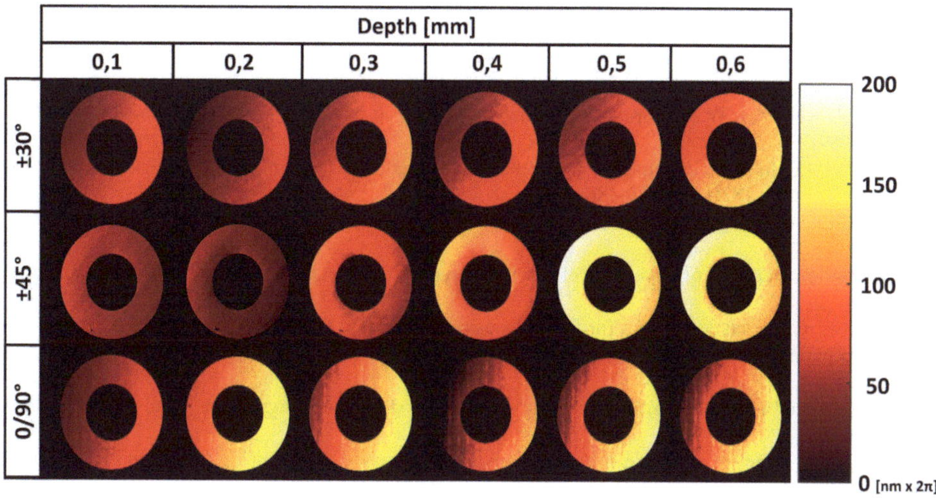

Fig. 12.4 Comparison between the displacements maps for the three specimens

Fig. 12.5 Comparison between the displacements of the three specimens

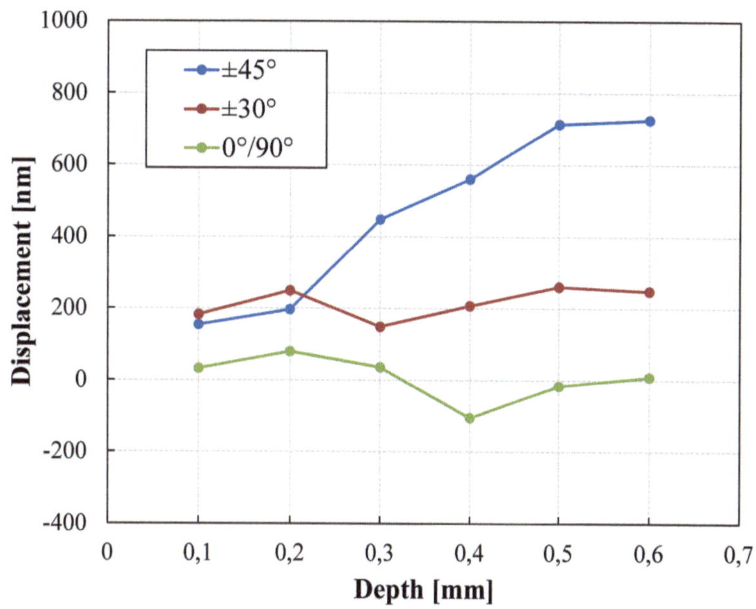

the material. This is confirmed also by thermographic analysis that show an increment of temperature from 31.9 °C for 5000 rpm to 77.9 °C for 50,000 rpm.

A 7 × 7 pixels area has been selected to record the displacements after the drilling phase. These values have been averaged in order to obtain one value of displacement for the whole area. The results show that the displacements for the ±45° specimen are higher than the ±30° and 0/90° specimens and this should be related to higher residual stress in the sample. This is confirmed by the higher printing difficulty of this particular specimen type. The sample 0/90° shows lower displacements that should be due to lower residual stress.

However, due to the orthotropic behaviour and the consequent difficulties to calculate the residual stress, this work is a preliminary study on the measurement of residual stress in FDM parts. A FE model will be the next step in order to determine the coefficients matrix that correlates the strain to the stress.

References

1. Yan, X., Gu, P.: A review of rapid prototyping technologies and systems. Comput. Aided Des. **28**(4), 307–318 (1996)
2. Petzold, R., Zeilhofer, H.F., Kalender, W.A.: Rapid prototyping technology in medicine—basics and applications. Comput. Med. Imaging Graph. **23**(5), 277–284 (1999)
3. Caulfield, B., McHugh, P.E., Lohfeld, S.: Dependence of mechanical properties of polyamide components on build parameters in the SLS process. J. Mater. Process. Technol. **182**(1–3), 477–488 (2007)
4. Casavola, C., et al.: Orthotropic mechanical properties of fused deposition modelling parts described by classical laminate theory. Mater. Des. **90**, 453–458 (2016)
5. Zein, I., et al.: Fused deposition modeling of novel scaffold architectures for tissue engineering applications. Biomaterials **23**(4), 1169–1185 (2002)
6. Zhong, W., et al.: Short fiber reinforced composites for fused deposition modeling. Mater. Sci. Eng. A **301**(2), 125–130 (2001)
7. Allahverdi, M., et al.: Processing of advanced electroceramic components by fused deposition technique. J. Eur. Ceram. Soc. **21**(10–11), 1485–1490 (2001)
8. Mireles, J., et al.: Development of a fused deposition modeling system for low melting temperature metal alloys. J. Electron. Packag. **135**(1), 011008 (2013)
9. Masood, S.H., Song, W.Q.: Development of new metal/polymer materials for rapid tooling using fused deposition modelling. Mater. Des. **25** (7), 587–594 (2004)
10. Kantaros, A., Karalekas, D.: Fiber Bragg grating based investigation of residual strains in ABS parts fabricated by fused deposition modeling process. Mater. Des. **50**, 44–50 (2013)
11. Mahesh, M., et al.: Benchmarking for comparative evaluation of RP systems and processesnull. Rapid Prototyp. J. **10**(2), 123–135 (2004)
12. Wang, T.M., Xi, J.T., Jin, Y.: A model research for prototype warp deformation in the FDM process. Int. J. Adv. Manuf. Technol. **33**(11–12), 1087–1096 (2007)
13. Zhang, Y., Chou, K.: A parametric study of part distortions in fused deposition modelling using three-dimensional finite element analysis. Proc. Inst. Mech. Eng. B J. Eng. Manuf. **222**(8), 959–967 (2008)
14. Kujawinska, M.: Use of phase-stepping automatic fringe analysis in moire interferometry. Appl. Optics **26**(22), 4712–4714 (1987)
15. Ghiglia, D.C., Pritt, M.D.: Two-Dimensional Phase Unwrapping: Theory, Algorithms, and Software, vol. 4. Wiley, New York (1998)
16. Barile, C., et al.: Analysis of the effects of process parameters in residual stress measurements on Titanium plates by HDM/ESPI. Measurement **48**, 220–227 (2014)

Chapter 13
Predicting Residual Stress on X-ray Tomographed Complex Bi-Layer Geometries using 3D Finite Element Analysis

Masoud Allahkarami, Leila Seyed Faraji, and Jay C. Hanan

Abstract Micro-Computed Tomography was performed on an artificial mandibular first molar crown. A complex bi-layer geometry finite element model representing a real dental crown shape was created by converting 3D reconstructed micro-tomographs into a meshed model. The distribution of thermal residual stresses, provided by finite elements, demonstrated how thickness of the veneer contributes to localized compressive residual stress at the external surface of the veneer layer. To inhibit fracture initiation, it is desirable to manufacture a crown with compressive residual stress on its external surface. In order to quantify the correlation between thicknesses of the veneer layer and compressive residual stress at the external surface of the veneer, the local thickness of each surface node is required. The method introduced here computes the minimum distance between two triangulated complex surfaces by finding the closest point on the facing surface for each surface vertex. Statistical correspondence between local thickness of the veneer layer and the calculated residual stress components from finite elements revealed a curvilinear relationship over the range of variables and suggests a preferred layer thickness for engineering future dental restorations.

Keywords Micro X-ray tomography • Residual stress • 3D Geometry • Zirconia • Finite Element Model • Thickness measurement

13.1 Introduction

Advances in computer technology and progress in numerical computation power has enabled 3D finite element analysis as a viable and important tool for many industries to evaluate and optimize performance [1, 2] and reliability of their designs and manufacturing processes [3]. It also helps to validate new methods developed for residual stress measurement [4, 5]. Along with advances in creating 3D model geometries using computer aided design (CAD), there are new methods to create finite element models resembling the real 3D geometry of an object [6]. Allahkarami et al. developed a method to include surface feature details into a finite element model of a single carbon fiber based on stochastic analysis of AFM Images [7, 8]. Prediction of residual stress due to thermal expansion mismatch between two complex shape ceramic layers in a bi-layer ceramic dental crown is an example of incorporating the actual geometry details in a 3D model. A real geometry model, with internal details could be obtained from X-ray computed tomography [9–11]. H. A. Bale et al. implemented micro-Computed Tomography (μ-CT) to acquire the real geometry and analyze the evolution and distribution of thermal residual stresses in all-ceramic dental restorations. One advantage of μ-CT, unlike 3D laser scanners and cameras [12, 13] which only provide information about external surface of the model; is it provides a 3D model with internal features. This ability of micro-Computed Tomography makes it more suitable for FE models of multilayer geometries, particularly with inconstant thicknesses.

Finite element analysis generally produces a large data set of results. A real geometry model contains curved surfaces with details included which may not be necessary. Reducing minor details from the model that does not contribute to the results shortens the computational time and simplifies visualization. Most commercial FE software provides a graphical user interface (GUI) to interactively visualize the result. The FE result is a set of calculated physical properties like stresses or temperatures of a point cloud, wire frame, or faces with edges. The most common way to analyze FE results is to visualize using color shading on the part geometry.

M. Allahkarami • J.C. Hanan (✉)
School of Mechanical and Aerospace Engineering, Oklahoma State University, Tulsa, OK, USA
e-mail: jay.hanan@okstate.edu

L.S. Faraji
Department of Mechanical Engineering, The University of Tulsa, Tulsa, OK, USA

S. Quinn, X. Balandraud (eds.), *Residual Stress, Thermomechanics & Infrared Imaging, Hybrid Techniques and Inverse Problems, Volume 9*, Conference Proceedings of the Society for Experimental Mechanics Series, DOI 10.1007/978-3-319-42255-8_13

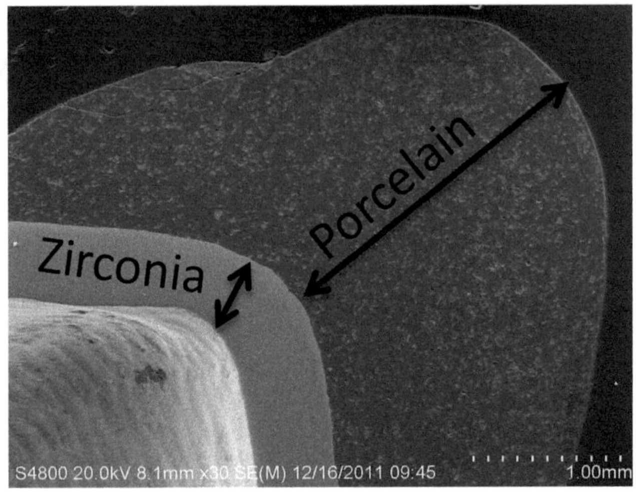

Fig. 13.1 SEM micrograph of a sectioned sample

This paper presents a method of correlating the result of finite element analysis to other geometry related physical properties. As an example, distribution of thermal residual stresses in all ceramic dental restorations was correlated to the local thickness of an outer layer of the bi-layer model [14].

Classically, interpreting mechanical simulation involves identifying stress concentration regions, or maximum and minimum stresses to compare designs or assure acceptable ranges. In the case of bi-layer dental restorations the aim of simulation is generally to design a crown with compressive residual stress for the veneer layer to prevent crack initiation [15, 16]. Generally measuring strength of interfacial bonding requires adapting a new measurement method [17]. Identifying the magnitude and type of residual stress at the interface of the core and veneer layer is another important design factor predictable by FE analysis.

13.2 Materials and Methods

An all-ceramic crown with 500 μm alumina core was fabricated by CNC machine from a CAD file with up to 2 mm porcelain veneer at the thickest locations. Figure 13.1 illustrates a low magnification scanning electron microscopy image (SEM) of the crown's cross section. The zirconia layer thickness is almost constant, while the veneer layer has variable thickness.

Micro-Computed Tomography (μ-CT) was performed using a Skyscan 1172 with a Tungsten (W) X-Ray source at 100 keV and a max power of 10 W. A 0.5 mm aluminum filter was used. At a resolution of 9.08 μm/pixel, 450 radiographs with averaging over eight frames were captured at 0.4° rotation steps.

13.3 Results and Discussion

13.3.1 Non Uniform Thickness Measurement From Micro-X-ray Tomography

The 3D reconstructed tomography and finite element surface meshing shown in Fig. 13.2a, b were carried out in 3D visualization software.[1] Tomography allows virtual cross sectioning to look at layer thickness without destroying the actual sample. A vertical cross section slice of the reconstructed (μ-CT data is presented in Fig. 13.2c. A code[2] was developed to extract the nodes that belong to the porcelain external surface node coordinates $f_p(x, y)$ and zirconia-porcelain interface surfaces $f_p(x, y)$.

[1] Amira 3.1.1 software by FEI Visualization Sciences Group

[2] Matlab 2015b by MathWorks

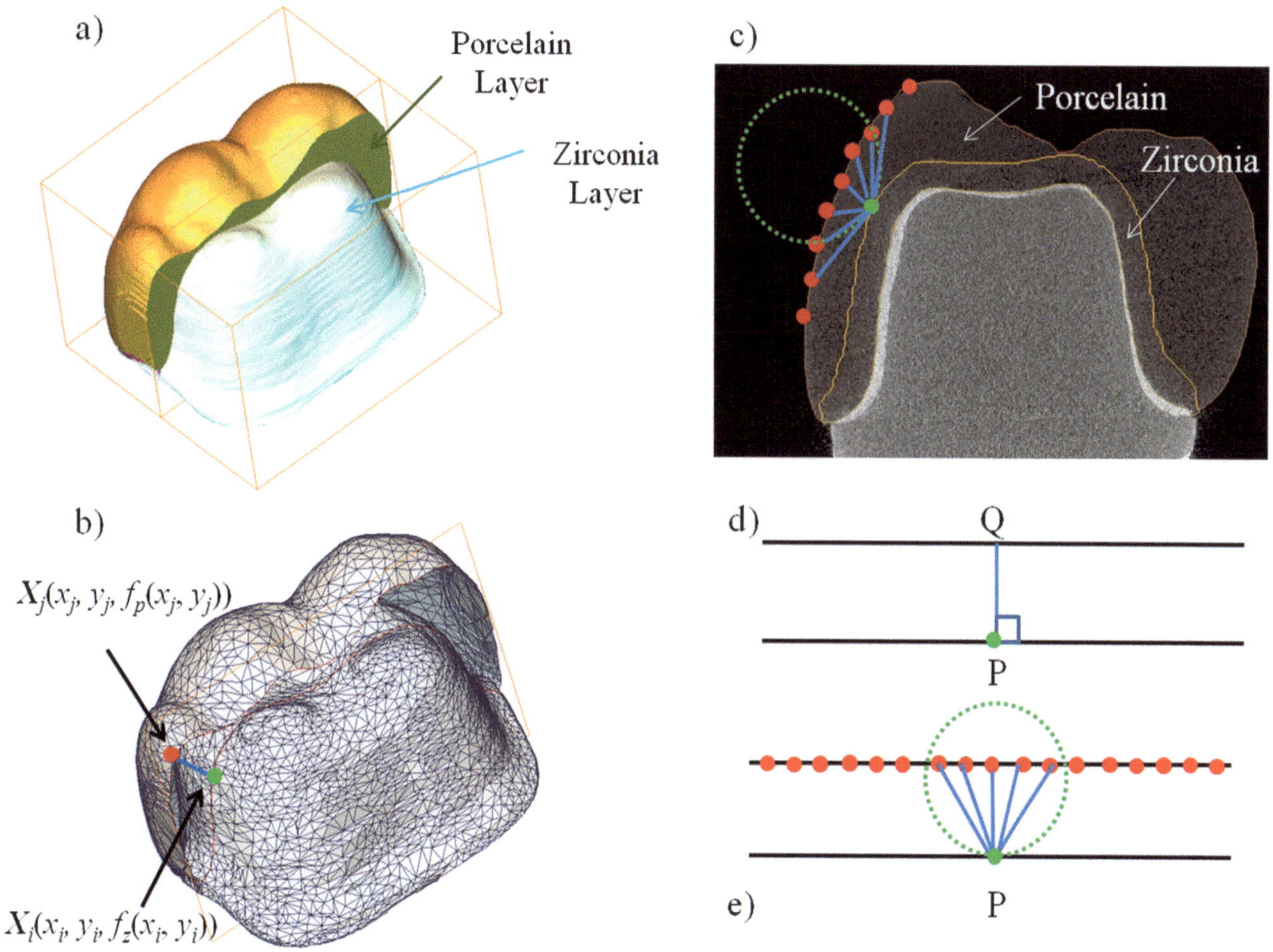

Fig. 13.2 (**a**) 3D graphical rendering of tomography from a bi-layer dental crown made of zirconia ceramic base covered by porcelain layer (**b**) 3D meshed surfaces (**c**) Minimum distance between two curved lines nodes (**d**) Distance between two straight lines (**e**) Minimum distance between the nodes on two straight lines

The $f_z(x, y)$ and $f_p(x, y)$ are two locally variant curved surfaces. In general, the distance between two curved surfaces is not well defied but it is possible to calculate the shortest surface-to-surface distance for the grid points laying on one of the two surfaces [14]. Let's consider the zirconia upper surface as a reference surface, and calculate the minimum distance to the porcelain surface at each point. We consider the shortest distance between each point on $f_z(x, y)$, the base surface, relative to the outside of the porcelain layer, $f_p(x, y)$ surface, a local thickness of the porcelain layer. The entire zirconia layer surface has been covered by porcelain and for each pair of "x, y" there are unique corresponding "$f_z(x, y)$ and $f_p(x, y)$ values. The two surface functions $f_z(x, y)$ and $f_p(x, y)$ always satisfy the condition: $f_z(x, y) \le f_\mathrm{p}(x, y)$.

Figure 13.2d shows that the distance between two parallel lines at a certain point, P, can be defined as the length of the normal vector that originates at P and ends at Q. Instead, if we grid the top line with very fine steps, the minimum distance between the nodes and point P in Fig. 13.2e is the same as distance PQ in Fig. 13.2d. The same analogy is extendable for the distance between two 3D curved surfaces at a certain point on one of the surfaces, as illustrated in Fig. 13.2c. The minimum distance for a point i on the interface surface that has a coordinate of $X_i = (x_i, y_i, f_z(x_i, y_i))$ is defined as,

$$d(x_i, y_i, z_i) = \min\left\{ \sqrt{(x_i - x_j)^2 + \left(y_i - y_j\right)^2 + \left(f_z(x_i, \mathrm{y_i}) - f_p\left(x_j, y_j\right)\right)^2}; \text{ for all possible } j \right\} \tag{13.1}$$

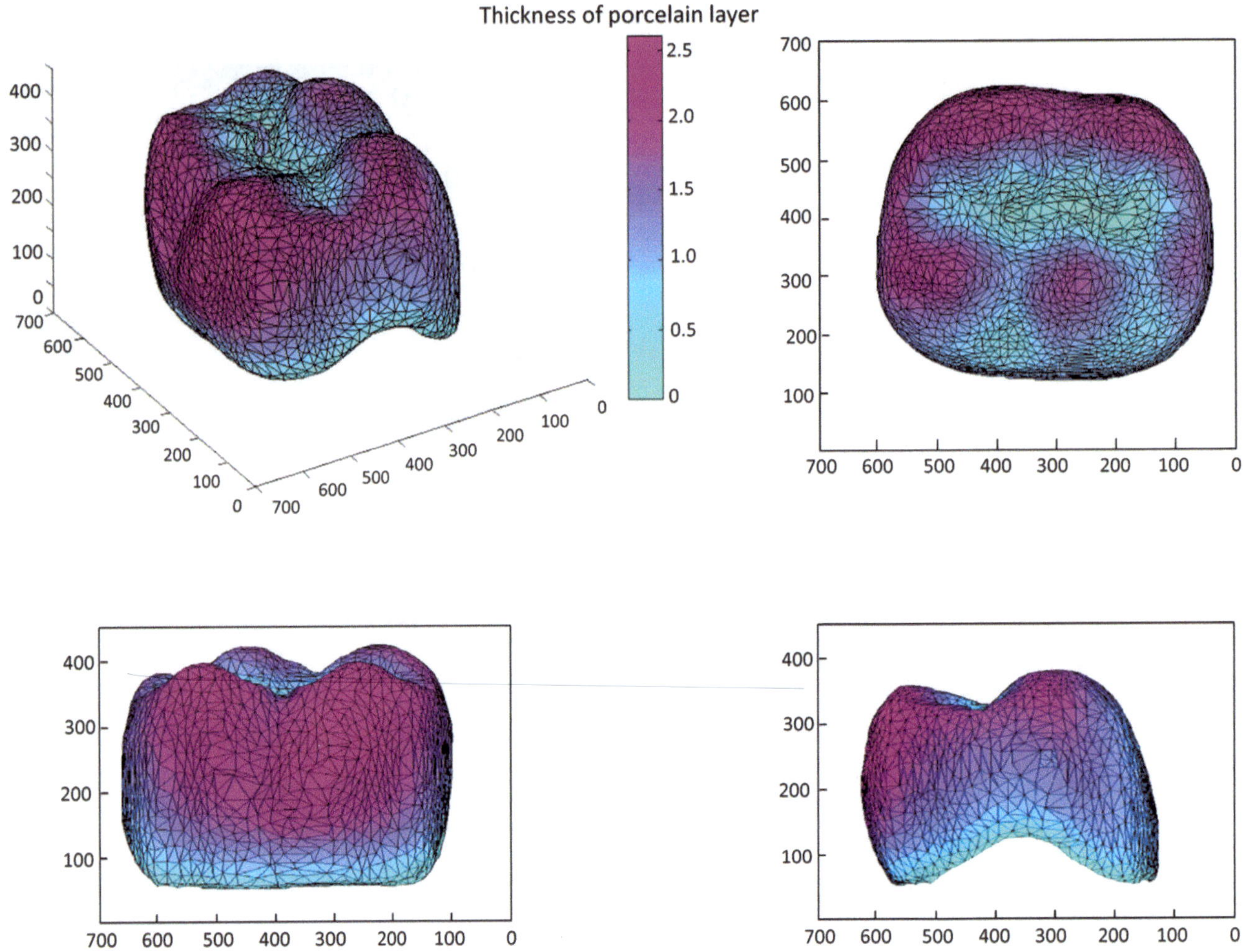

Fig. 13.3 Colored map of the minimum distance measurement

Where j is the index for a point j on the interface surface (zirconia surface) that has a coordinate of $X_j = \left(x_j, y_j, f_p\left(x_j, y_j\right)\right)$.

Using another code a colored 3D map of minimum distance measurement was created. Different rendering views of the results are shown in Fig. 13.3. Each triangle face has three nodes. The average of minimum distance for those three nodes was considered as the face minimum distance. Faces are colored based on their average minimum distance.

13.3.2 FE Model Development From Micro-Computed Tomography

A three-dimensional reconstruction software[3] was used for building tomographs from radiographs and obtaining two-dimensional slices saved in a 16-bit TIFF format. Based on the gray scale histogram, and global threshold, the veneer and core were automatically segmented. Automatic segmentation worked well for about 90 % of data set and rest manually corrected.[4] A volume mesh was generated for each of the segmented labels using tetrahedral elements generated after seed vertices provided by the surface model. The model was exported as an ".inp" file in Abaqus 6.8 format.

[3] NRecon 1.6.9.8 by Bruker microCT (formerly Skyscan Inc.)

[4] Data visualization and segmentation was carried out in Amira 3.1.1.

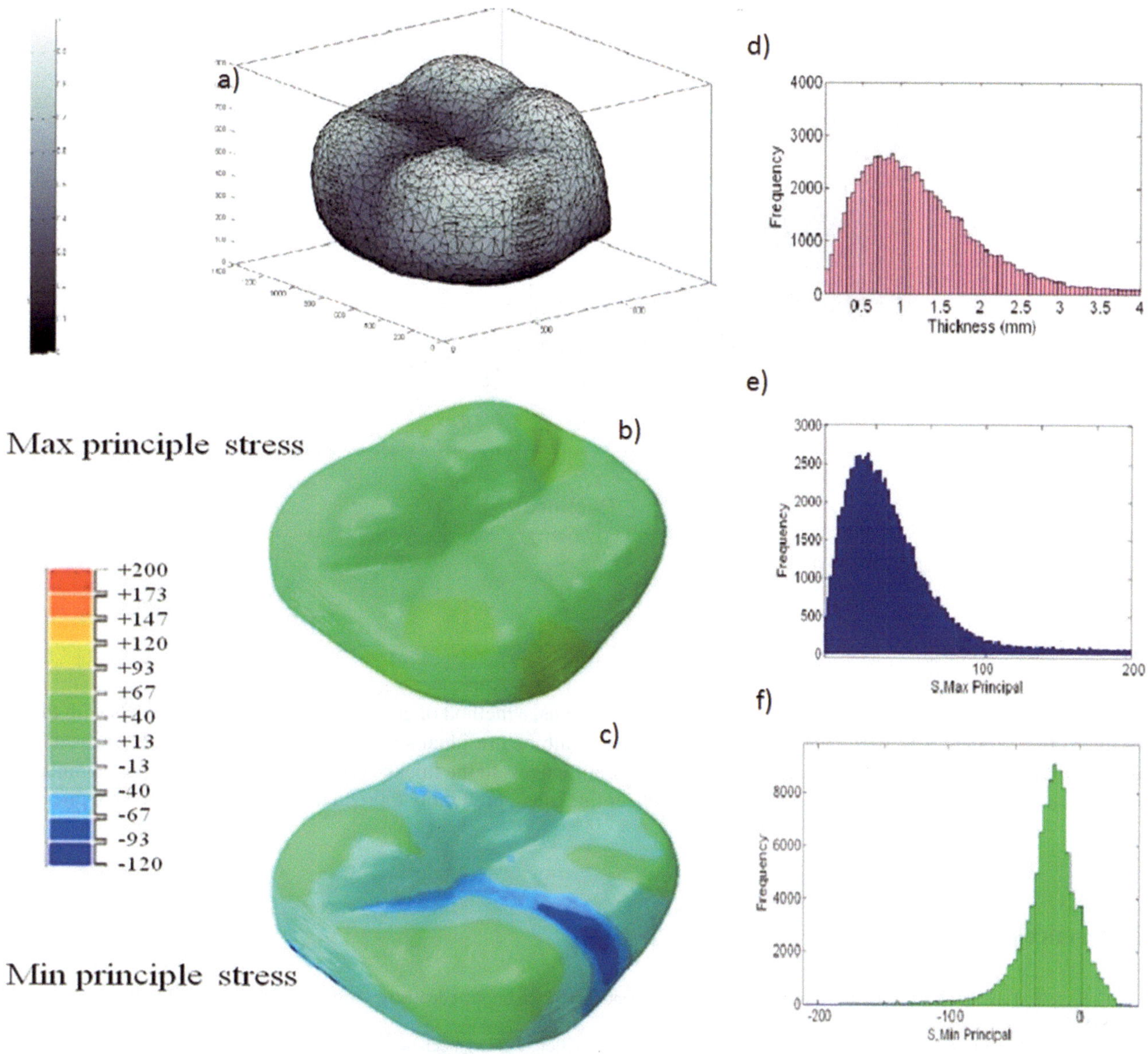

Fig. 13.4 (**a**) Surface of a crown colored with respect to the thickness underneath the veneer layer, (**b**) and (**c**) Maximum and minimum principal stresses, (**d**) Histogram of porcelain thickness distribution, (**e**) and (**f**) Histogram of the maximum and minimum principal stresses

Generally, due to mismatch in coefficients of thermal expansion there are residual stresses between the core (Zirconia) and veneer layers. In order to simulate the dental crown firing processing, a pre-defined temperature field of 1000 °C was applied and ramped down to 25 °C. The maximum and minimum principal stresses of the real crown geometry model from this simulation output are presented in Fig. 13.4b, c.

In Fig. 13.5 a scatter plot of the maximum principle stress is plotted vs. thickness showing a statistical correlation between the two variables. The results showed that at the surface of the crown there is on average a maximum principle stress of 100 ± 50 MPa. Higher magnitude stress levels of up to 200 MPa were observed in certain cusp regions in proximity with the occlusal surface which were associated with thinner veneer layers. In addition to regions having sharp anatomic features in the veneer layer which are prone to high residual stresses, in general, regions with stress concentration were observed at thin porcelain layer locations.

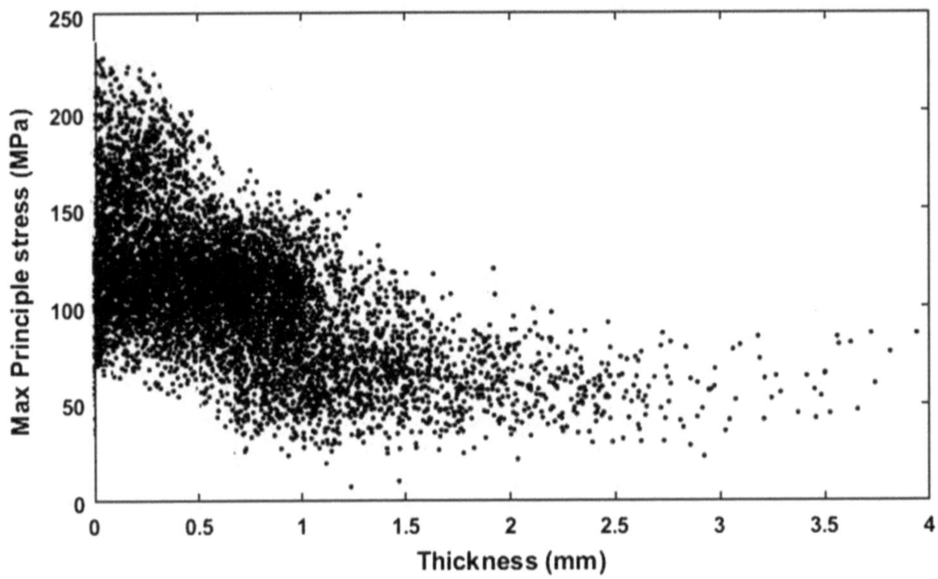

Fig. 13.5 Maximum principal stress vs thickness

13.4 Conclusions

In order to simulate residual stress in real geometry dental crowns, a method of generating finite element models with the use of micro-tomography was implemented. For the first time, combined residual stress simulation and thickness measurement was introduced as a method of correlating residual stress components to the thickness of the veneer layer. These simulations suggest development of crowns with more uniform veneer thickness, preferably more than 0.2 mm local porcelain thickness.

References

1. Geng, J.P., Tan, K.B.C., Liu, G.R.: Application of finite element analysis in implant dentistry: A review of the literature. J. Prosthet. Dent. **85**, 585–598 (2001)
2. Fervers, C.W.: Improved FEM simulation model for tire–soil interaction. J. Terramech. **41**, 87–100 (2004)
3. Gupta, S., Uday, V., Raghuwanshi, A.S., Chowkshey, S., Das, S.N., Suresh, S.: Simulation of blow molding using ansys polyflow. APCBEE Procedia **5**, 468–473 (2013)
4. Jayakumar, B., Allahkarami, M., Hanan, J.C.: Residual stress response to peening in metallic glass. Exp. Appl. Mech. **6**, 99–104 (2015)
5. Jayakumar, B., Allahkarami, M., Hanan, J.C.: Residual stress of individual aluminum grains from three dimensional X-ray diffraction. Exp. Appl. Mech. **6**, 123–130 (2015)
6. Bale, H.: Measurement and analysis of residual stresses in zirconia dental composites using micro X-ray dissertation, Ph.D. 3499704, Oklahoma State University, United States, Oklahoma, (2010)
7. Allahkarami, M., Hanan, J.C.: Finite element model generation based on stochastic analysis on AFM images. In: Imaging Methods for Novel Materials and Challenging Applications, vol. 3, pp. 417–421. Springer, New York (2013)
8. Allahkarami, M., Hanan, J.C., Bale, H.A.: Regeneration of surface roughness by the Langevin equation using stochastic analysis on AFM image of a carbon fiber. Appl. Surf. Sci. **257**(3), 857–860 (2010)
9. Amira 5 Software Users's Guide, Visage Imaging, 2009
10. Della Bona, M.B., Benetti, P., Duan, Y., Griggs, J.A.: Three-dimensional finite element modelling of all-ceramic restorations based on micro-CT. J. Dent. **41**, 412–419 (2013)
11. Bandla, S., Hanan, J.C.: Use of X-ray tomography for efficient package design. SPE-ANTEC Tech. Papers, 2248–2251, (2012)
12. Karasik, U.S.: 3D scanning technology as a standard archaeological tool for pottery analysis: Practice and theory. J. Archaeol. Sci. **35**, 1148–1168 (2008)
13. Gentilini, I., Shimada, K.: Predicting and evaluating the post-assembly shape of thin-walled components via 3D laser digitization and FEA simulation of the assembly process. Comput. Aided Des. **43**, 316–328 (2011)

14. Allahkarami, M., Bale, H.A., Hanan, J.C.: Characterization of nonuniform veneer layer thickness distribution on curved substrate zirconia ceramics using X-ray micro-tomography. Developments in Strategic Materials and Computational Design II, 153–163, (2011)
15. Allahkarami, M., Hanan, J.C.: Residual stress delaying phase transformation in Y-TZP bio-restorations. Phase Transit. **85**, 169–178 (2012)
16. Allahkarami, M., Hanan, J.C.: Mapping the tetragonal to monoclinic phase transformation in zirconia core dental crowns. Dent. Mater. **27**(12), 1279–1284 (2011)
17. Jayakumar, B., Allahkarami, M., Hanan, J.C.: Inter-cellular joining for amorphous honeycombs. Compos. Mater. Join. Technol. Compos. **7**, 25–33 (2013)

Chapter 14
Combining Hole-Drilling and Ring-Core Techniques

Antonio Baldi

Abstract The hole-drilling and ring-core techniques are two well-known approaches to residual stress measurement. They apparently are complementary: in the former, a small hole is drilled in the specimen and residual stresses are estimated from the displacement/strain field around the hole; in the latter a small ring is milled and residual stresses are computed from the displacement/strain field inside the ring. However, owing to the different stiffness of the constraining region, their sensitivity and depth range are somewhat different.

Although both techniques are implemented using strain gauges, the use of optical methods is nowadays largely accepted. As optical methods allow acquisition of data from both inside and outside the ring, in this work, we will try to exploit the different displacement release rates of the two regions to optimize residual stress measurement.

Keywords Residual stress • Hole drilling • Ring core • ESPI

14.1 Introduction

Hole-drilling and Ring-core are two well-known techniques for residual stress measurement. Although they are complementary techniques—the former estimates residual stress from the displacement field around a small hole, the latter pursues the same objective by observing the displacements inside a circular groove—they are usually seen as separate techniques because their experimental apparatus, experimental protocol and performance are different. Indeed, drilling a small hole is quite different from milling a circular ring (rotating speed of the spindle and cutting forces are completely different). Moreover, routing the wires of a hole-drilling rosette is significantly simpler than in the ring-core case because the latter requires either that the cables pass through the spindle in the plunge-milling case or a system to take them away from the spindle if orbital milling is used. Summing all the previous points, and considering that surface damage is much greater since the groove has to be large enough to mount a rosette i.e. a factor about six times larger than the hole, the ring core is used far less than hole drilling.[1] The same is true even when different measurement techniques are used. Indeed, it is quite easy to find dozens of articles using the optical experimental technique in combination with the hole drilling approach in literature, whereas only a few combine optical methods with the ring core approach, usually in very specific sectors and/or applications.

However, if we focus on optical methods, it is almost impossible to perform a ring-core test without performing a hole drilling experiment at the same time. Figure 14.1 (left) shows the fringe field acquired in a residual stress experiment combining the ESPI (Electronic Speckle Pattern Interferometry) technique and the ring-core approach. The ring-groove can easily be detected in the center of the image—material removal caused a complete loss of correlation of the phase fields before and after milling—but, most important, it is apparent that the phase field extends over almost the entire field of view,[2] i.e. at the same time we are imaging the region inside the groove (the area related to the ring-core technique) and a large region surrounding the groove, which can be viewed as a blind-hole drilling test. In fact, there is a small difference between the test shown and a "pure" blind-hole experiment: in the latter, the core is fully removed whereas in the current case, only a circular groove has been milled. Thus, the central core may contribute to the overall stiffness and affect the displacement field outside the groove. In the author's opinion this is quite improbable: it suffices to visualize the stress flow to infer that core contribution to the displacement field outside the groove, if any, is very small. In any case, the test shown in Fig. 14.1

[1] Note that there is no ring-core standard similar to [1].

[2] The top left corner is not correctly enlightened by the laser source, so it presents no fringes.

A. Baldi (✉)
Dipartimento di Ingegneria Meccanica, Chimica e dei Materiali, Università degli Studi di Cagliari, Via Marengo, 3, Cagliari 09123, Italy
e-mail: Antonio.baldi@dimcm.unica.it

S. Quinn, X. Balandraud (eds.), *Residual Stress, Thermomechanics & Infrared Imaging, Hybrid Techniques and Inverse Problems, Volume 9*, Conference Proceedings of the Society for Experimental Mechanics Series, DOI 10.1007/978-3-319-42255-8_14

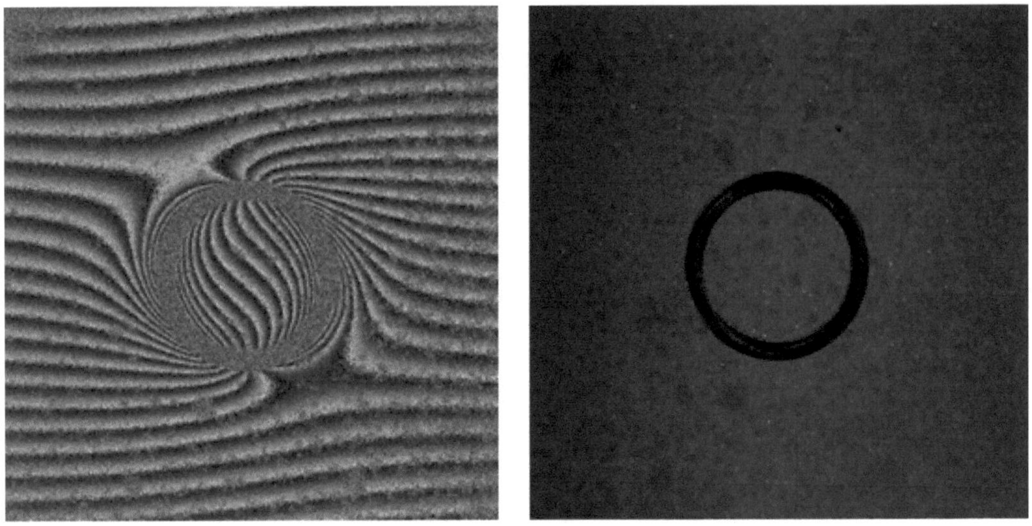

Fig. 14.1 A ESPI ring core experiment. *Left*: phase modulo 2π field (in plane sensitivity, $\lambda = 532$ nm); *right*: an image of the surface at the end of the experiment

corresponds to a blind-hole drilling case, which requires a numerical calibration procedure; by including the core in the Finite element mesh, the possible contribution will be automatically taken into account.

Looking at Fig. 14.1, the displacement fields inside and outside the slit appear completely different. Actually, from the theoretical viewpoint the two techniques are almost the same. According to Kirsch analysis [2], the displacement field around a hole after drilling can be written (in a cylindrical coordinate system) as [3]

$$\begin{aligned} u_r &= A(\sigma_x + \sigma_y) + B\left[(\sigma_x - \sigma_y)\cos 2\vartheta + 2\tau_{xy}\sin 2\vartheta\right] \\ u_\vartheta &= C\left[(\sigma_x - \sigma_y)\sin 2\vartheta - 2\tau_{xy}\cos 2\vartheta\right] \\ u_z &= F(\sigma_x + \sigma_y) + G\left[(\sigma_x - \sigma_y)\cos 2\vartheta + 2\tau_{xy}\sin 2\vartheta\right] \end{aligned} \tag{14.1}$$

where the A, B, C, F and G coefficients have to be computed by Finite Element Analysis in the blind hole case, while they are known in closed form in the hole-through case:

$$\begin{aligned} A &= \frac{r_0}{2E}(1+\upsilon)\varrho \\ B &= \frac{r_0}{2E}\left[4\varrho - (1+\upsilon)\varrho^3\right] \\ C &= -\frac{r_0}{2E}\left[2(1+\upsilon)\varrho + (1+\upsilon)\varrho^3\right] \\ G &= \frac{\upsilon t}{E}\varrho^2 \end{aligned} \tag{14.2}$$

ρ being the ratio of the hole radius to the distance from the hole center, t the thickness, E the Young modulus, ν the Poisson's ratio and F being null.

It can be shown that Eq. (14.1) are also valid in the ring core case, providing we use appropriate calibration coefficients:

$$A = (1-\nu)r/2E \quad B = C = (1+\nu)r/2E \tag{14.3}$$

but (14.3) is completely useless; indeed, both (14.2) and (14.3) assume a complete relaxation of strains, but while a hole-through is possible, in the ring core case this hypothesis corresponds to milling a full-depth groove to its complete removal. However, since calibration coefficients have to be computed numerically, this is unimportant. The main point is that Eq. (14.1) are in force both inside and outside the circular groove. Thus, including the core in the computation of residual stress components requires slight modifications to an existing code.

It is well known that with a residual stress level, the strain components measured with the ring-core technique are significantly larger than those measured using the hole-drilling approach. This can be explained by considering that the region surrounding the hole is constrained by the rest of the plate, whereas the displacements of the core are limited only by the bottom connection with the rest of the plate.[3] Thus, the strain release rate as a function of ring depth and the maximum measuring depth are different. This might suggest that it is possible to find an optimal combination of the two methods, using data from the core during the initial milling steps and moving progressively to data taken from the outside when the groove depth becomes significant.

In the following, we will explore this idea, first by analyzing the dependency of calibration coefficients on groove depth—most optical-methods-based algorithms compute residual stress components on the basis of displacements—then by trying the mixed technique in a practical computation.

14.2 Calibration Coefficients

As noted in the previous section, the stated relation between the ring-core and the hole-drilling techniques is based on the inspection of strains (i.e. it comes from the strain-gauge field). However, most of the approaches to residual stress estimation from optical data rely on displacements [3–7], because this is the variable measured by a large fraction of optical techniques (e.g. Electronic Speckle Pattern Interferometry, Digital Holographic Interferometry, Moiré Interferometry, Digital Image Correlation, ...). Thus, in this section we will inspect the dependency on depth of the coefficients *A–G* appearing in Eq. (14.1) for both the ring and hole sides. To this end, we will refer to the standard Integral Method [8, 9], using as paradigm the first loading step related to each depth increment (Fig. 14.2), i.e. the maximum absolute value of each drilling step.

Figures 14.3, 14.4 and 14.5 show the maximum absolute displacements related to the *A–G* coefficients. In particular *A* and *F* were computed by assuming $\sigma_x = \sigma_y = 1$, $\tau = 0$, whereas the coefficients *B*, *C* and *G* were estimated by assuming $\sigma_x = -\sigma_y = 1, \tau = 0$ (corresponding respectively to hydrostatic and pure shear loading conditions). To compute the graphs, a series of Finite Element simulations were performed assuming a unitary Young Modulus, a Poisson's ratio of 0.3 and a constant depth increment (from 0.1 mm down to 3 mm, by step 0.1 mm). Thus, to compute the full set of coefficients required for the Integral Method, 465 F.E. runs (i.e. $n\ (n + 1)/2$) were performed. As regards the geometry of the groove, the ring core radius is 3 mm whereas the hole radius (the external radius of the slit) is 4 mm. The external radius of the mesh is ten hole radii away and Finite Element computations were performed using an in-house-written code with semi-analytical elements, i.e. axial-symmetric elements expanded in Fourier's series.

Each graph displays two pairs of curves; the former is related to the boundary of the hole (ring) whereas the latter shows the same values estimated 1 mm away from the edge (i.e. at $r = 2$ mm for the ring and at $r = 5$ mm for the hole). The second set of curves is shown because, due to machining, the data extracted near the edge of the ring (hole) are quite often highly noisy and cannot be reliably used. Results far from the boundaries are usually consistent with maximum values, but magnitudes are usually significantly smaller (about $^1/_3$).

Examining the *A*-related graph (Fig. 14.3-left) the ring-related coefficient stabilizes before the hole-related one (as expected), but its absolute magnitude is always smaller (i.e. the opposite of reported behavior).

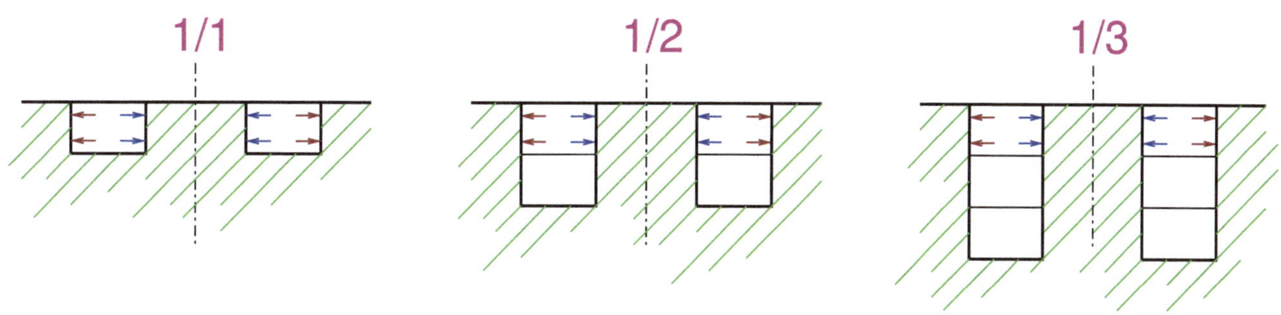

Fig. 14.2 Integral Method. Looking for the maximum coefficient for each drill step

[3] Note that we are imaging the top.

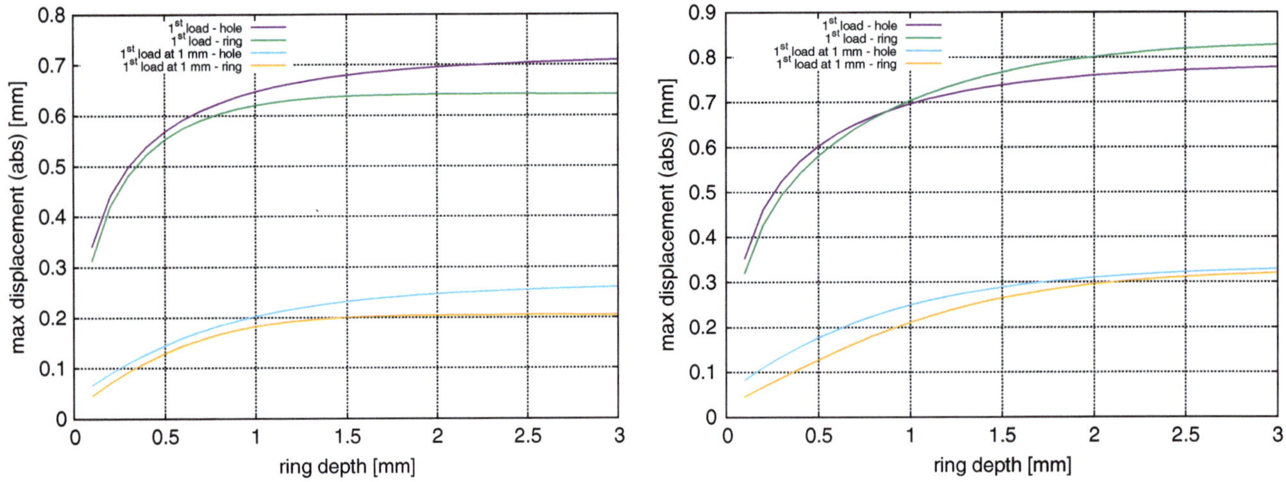

Fig. 14.3 A (*Left*) and B (*Right*) coefficients vs. drilling depth

Fig. 14.4 C coefficient vs. drilling depth

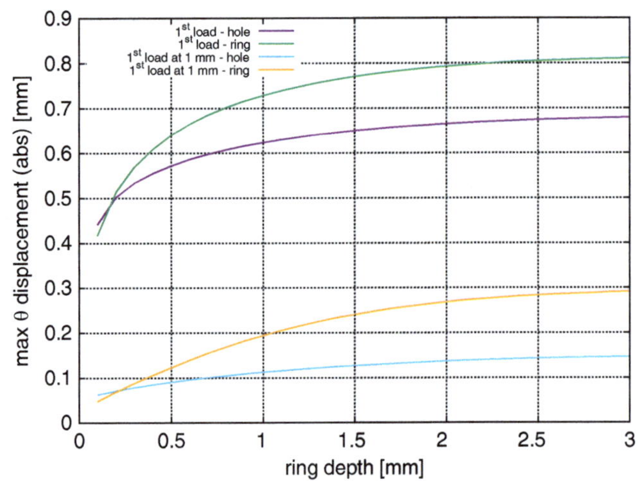

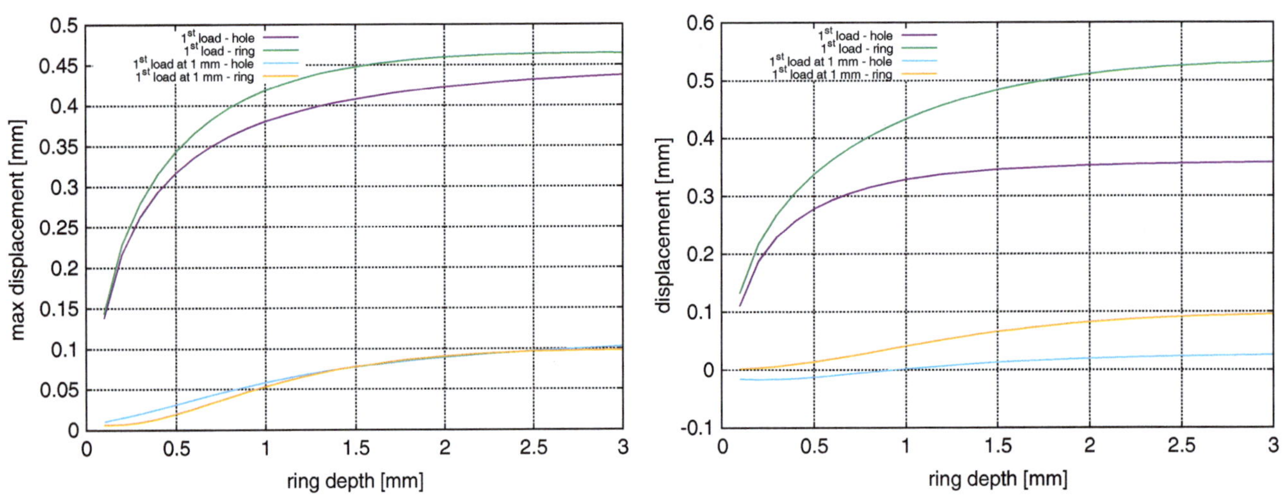

Fig. 14.5 F (*Left*) and G (*Right*) coefficients vs. drilling depth

On the contrary, the F coefficient (still related to hydrostatic loading, Fig. 14.5-left) performs exactly as described in literature: the ring-related coefficient is larger, but saturates before the hole-related one. It is to be noted that the difference in magnitude between ring- and hole-related coefficients is in any case small (about 5 %). This is true for A, B and F coefficients and can easily be verified experimentally (after removing rigid body components from Fig. 14.1, about the same number of fringes can be counted both in the core and the hole).

Finally, a significant difference between the ring- and hole-related coefficients can be observed for C and G. In both these cases the ring coefficients are larger; moreover, in the latter the hole-related curve exhibits saturation long before that of the ring. Considering that both the C and G coefficients are related to shear loading, we can conclude that when displacements are considered the two experimental techniques are almost the same from the viewpoint of the calibration coefficients, except in the case of high shear loading, where the ring-core approach appears to be somewhat better.

14.3 Numerical Tests

To test the relative performance of the hole-drilling, ring-core and combined algorithm, we have to know the expected residual stress components related to each depth increment. Thus, we opted to use synthetic data instead of experimental results. In the following sub-section, we will focus on input data generation, while in the successive one we will inspect the behavior of the various algorithms.

14.3.1 Reference Data Generation

Computing artificial phase fields for the algorithms of analysis is in principle quite simple: in the case of constant stress in the specimen, Eq. (14.1) allow for computation of displacement components given σ_x, σ_y and τ. The case of stress components variable with depth is only a bit more complex: if we assume a constant stress value for each depth increment (i.e. we follow the Integral Method concept), the observed displacements result from the sum of the contributions of the various instances of (14.1) where the stress components are related to the assumed value at the given depth and the calibration coefficients A–G have to be computed using F.E. analyses.

Moving from displacement components to phase data is straightforward: it is well known that optical interferometric methods are sensitive in a single direction only, thus it suffices to convert the cylindrical displacement data to Cartesian components and then project them in the sensitivity direction with a simple scalar product.[4] Phase data are then easily computed by scaling the projected displacements by the sensitivity factor.

Simulating the real acquisition chain is far more complex. Using the ESPI process as paradigm, the phase data result from the subtraction of two fields acquired before and after drilling. Each phase field is computed by a *phase shifting* algorithm involving several images (in this case five), thus final noise results from the complex interaction of various noise sources. Moreover, rigid body motion might occur between the two acquisitions. Thus, we opted to replicate all the computational steps. Since the intensity of each image of the *phase shifting* algorithm can be written as

$$I_i(x,y) = I_0(x,y) + \gamma(x,y)\cos\left[\varphi(x,y) + \Delta\varphi_i + \zeta(x,y)\right] \tag{14.4}$$

where $I_0(x, y)$ is the mean intensity, $\gamma(x, y)$ the modulation, $\varphi(x, y)$ the (known) phase and $\zeta(x, y)$ the acquisition noise, we opted to generate all the components of (4) before and after the virtual experiment.

Figure 14.6 shows the intensity modulation—a spherical function ranging from 48/256 to 208/256—and one of the intensity modulation maps required for image generation. Using them, the computed phase fields and the white noise matrices, we ended up with the phase modulo 2π fields. Figure 14.7 compares one of the theoretical phase fields (related to the third milling increment and assuming in-plane sensitivity in the x direction) with the corresponding "experimental" data. Note that, apart from the previously described process, a rigid body motion has been added to the phase field. The final input

[4] Note that if the object and/or viewing beams are not collimated, the sensitivity direction changes from point to point.

Fig. 14.6 *Left*: mean intensity, *right*: intensity modulation

Fig. 14.7 *Left*, theoretical phase field resulting from the imposed residual stress volumetric field. *Right*: the corresponding phase modulo 2π data

for the residual stress computational algorithm is then estimated from the phase modulo 2π data using a standard *phase unwrapping* algorithm [10].

14.3.2 Results

In the following, two different sets of results are shown. The former is related to theoretical input data while the latter refers to the "experimental" data, computed as described above. Note that similar tests were performed using various sensitivity directions, but results are substantially the same, thus only the first one, related to in-plane sensitivity in the x direction is reported.

Table 14.1 Results using noise-free data

Expected result			Hole drilling			Ring Core			Mixed		
σ_x [MPa]	σ_y [MPa]	τ [MPa]	σ_x [MPa]	σ_y [MPa]	τ [MPa]	σ_x [MPa]	σ_y [MPa]	τ [MPa]	σ_x [MPa]	σ_y [MPa]	τ [MPa]
100.	10.	25.	100.17	10.70	25.06	100.80	**57.27**	23.77	99.62	11.89	24.84
80.	30.	25.	79.88	29.21	24.98	77.99	**19.34**	26.77	80.45	29.59	25.19
60.	50.	25.	60.01	50.08	24.99	60.54	57.11	24.41	59.95	49.61	24.99
40.	70.	25.	39.99	70.00	25.00	40.52	**74.53**	24.60	40.01	70.00	25.00

Table 14.2 Results using noisy data

Expected result			Hole drilling			Ring Core			Mixed		
σ_x [MPa]	σ_y [MPa]	τ [MPa]	σ_x [MPa]	σ_y [MPa]	τ [MPa]	σ_x [MPa]	σ_y [MPa]	τ [MPa]	σ_x [MPa]	σ_y [MPa]	τ [MPa]
100.	10.	25.	99.23	9.46	24.92	100.63	**75.04**	23.34	98.67	12.18	24.61
80.	30.	25.	81.29	31.80	24.86	77.72	**17.74**	26.81	81.76	31.66	25.26
60.	50.	25.	59.74	51.40	25.62	60.63	51.80	25.91	59.69	48.40	25.29
40.	70.	25.	39.94	68.97	24.43	41.29	**81.71**	22.78	40.09	70.60	24.81

Tables 14.1 and 14.2 give an account of stress distribution in depth. Computations were performed using only the region outside the groove (labelled *Hole Drilling*), the region inside the circular slit (*Ring Core*) or both (*Mixed*). Apart from the set of pixels involved in the computation, the algorithm is the same for all tests (i.e. rigid body removal was performed even when analyzing theoretical data). Taking into account the reduced number of increments, no reference updating was performed (i.e. all phase fields result from the subtraction of the current phase from the first one). The best estimate for each quantity is underlined, whereas the bold font is used for grossly erroneous results.

Looking at the pattern of underlined/bold values in both tables, it is apparent that results are almost the same. A direct inspection shows that only the best estimate of σ_x in the first layer moves from the Hole-Drilling to the Ring-Core column, the former being worse than the latter by a tiny fraction of MPa; this is more or less expected because stress computation involves hundred of thousand of values, thus a robust estimation is expected. What is somewhat disappointing is the estimation of σ_y using only data from the core. Errors are substantial at low stress level and become progressively smaller. Considering that this behavior takes place even with theoretical data, it appears that rigid-body removal is significantly affecting results.[5]

14.4 Conclusions

In previous section we shown that the expected behavior—ring-core-based residual stress measurement more accurate in the initial milling steps, hole-drilling-based in the following—does not occur. This makes the use of ring core data of little interest: indeed, the number of pixels outside the hole is already huge, so adding a few more pixels does not significantly improve accuracy and/or robustness of the code. Moreover, the displacement data in the core area tend to be significantly noisier than those acquired in the outside region—the distance of all points of the core from the slit (the source of chips and dirt) is in any case less than a core radius—thus adding the core may be detrimental to the overall accuracy of the system.

[5] Note that while in the hole drilling case the stress-related signal is vanishing small far from the hole—it does make it simple to estimate rigid body components—in the ring core case there is no area unaffected by stress, thus rigid body removal is far more difficult, in particular when stress-related contributions are small.

References

1. ASTM E837-08e1. Standard test method for determining residual stresses by the hole-drilling strain-gage method. American Society for Testing and Materials, West Conshohocken, PA, 2008. doi:10.1520/E0837-08E01
2. Gustav, K.: Die Theorie der Elastizität und die Bedürfnisse der Festigkeitslehre. Z. Verein. D. Ing. **42**, 797–807 (1898)
3. Nelson, D.V., Makino, A., Fuchs, E.A.: The holographic-hole drilling method for residual stress determination. Opt. Lasers Eng. **27**, 3–23 (1997)
4. Baldi, A.: A new analytical approach for hole drilling residual stress analysis by full field method. J. Eng. Mater. Technol. **127**(2), 165–169 (2005)
5. Schajer, G.S., Steinzig, M.: Full-field calculation of hole drilling residual stresses from electronic speckle pattern interferometry data. Exp. Mech. **45**(6), 526–532 (2005)
6. Schajer, G.S., Winiarski, B., Withers, P.J.: Hole-drilling residual stress measurement with artifact correction using full-field DIC. Exp. Mech. **53**(2), 255–265 (2013)
7. Baldi, A.: Residual stress measurement using hole drilling and integrated digital image correlation techniques. Exp. Mech. **54**(3), 379–391 (2014). doi:10.1007/s11340-013-9814-6
8. Schajer, G.S.: Measurement of non-uniform residual stresses using the hole-drilling method. Part I—Stress calculation procedures. J. Eng. Mater. Technol. **110**(4), 338–343 (1988)
9. Schajer, G.S.: Measurement of non-uniform residual stresses using the hole-drilling method. part II—Practical application of the integral method. J. Eng. Mater. Technol **110**(4), 344–349 (1988)
10. Baldi, A.: Phase unwrapping by region growing. Appl. Optics **42**(14), 2498–2505 (2003)

Chapter 15
A Low-Cost Residual Stress Measuring Instrument

Antonio Baldi and Filippo Bertolino

Abstract The hole-drilling method is the approach most used for residual stress measurement and most of the commercially available instruments are based on this working principle. The idea is quite simple: drill a small hole on the surface of the components and measure the strain/displacement components resulting on the surface. The stress components can be estimated from these data with a reverse calibration process. The measurement of the displacement/strain field is usually performed using either strain gauges or interferometric optical methods, thus the cost of the instruments is significant.

Recently, it has been proposed to replace the interferometric method with an integrated Digital Image Correlation (i-DIC) approach: by using problem-specific displacement functions, all difficulties related to the low sensitivity of DIC are avoided and the measurement procedure simply requires acquisition of an image after each drilling step. This paper describes the development of a low-cost instrument based on this working principle: the imaging subsystem is based on a Raspberry-Pi camera module whereas step motors are controlled by an Arduino board. Finally, the frame is built using a 3D-printer.

Keywords Residual stress • Hole drilling • Integrated digital image correlation • Inverse method • Raspberry Pi

15.1 Introduction

Hole drilling is by far the technique most used for residual stress measurement. Its working principle is quite simple and consists of drilling a small hole on the surface of the specimen to look for surface strains. By assuming a plane stress state, it is possible to correlate the observed strains with the residual stress components on the surface of the specimen and, to some extent, with the stress distribution below the surface [1–3]. In the latter case, a staircase (step) function of depth is assumed for stress components and a numerical calibration is required; on the contrary, in the case of a hole through, a theoretical solution directly related to Kirsch's analysis of a plate in tension with a hole [4] can be derived.

According to Kirsch's analysis, the displacement field around a hole after drilling can be written (in a cylindrical coordinate system) as [5]

$$
\begin{aligned}
u_r &= A(\sigma_x + \sigma_y) + B\left[(\sigma_x - \sigma_y)\cos 2\vartheta + 2\tau_{xy}\sin 2\vartheta\right] \\
u_\vartheta &= C\left[(\sigma_x - \sigma_y)\sin 2\vartheta - 2\tau_{xy}\cos 2\vartheta\right] \\
u_z &= F(\sigma_x + \sigma_y) + G\left[(\sigma_x - \sigma_y)\cos 2\vartheta + 2\tau_{xy}\sin 2\vartheta\right]
\end{aligned}
\tag{15.1}
$$

where the A, B, C, F and G coefficients have to be computed by Finite Element Analysis in the blind hole case, whereas they are known in closed form in the hole through case:

A. Baldi (✉) • F. Bertolino
Dipartimento di Ingegneria Meccanica, Chimica e dei Materiali, Università degli Studi di Cagliari, Via Marengo, 3, Cagliari 09123, Italy
e-mail: Antonio.baldi@dimcm.unica.it

S. Quinn, X. Balandraud (eds.), *Residual Stress, Thermomechanics & Infrared Imaging, Hybrid Techniques and Inverse Problems, Volume 9*, Conference Proceedings of the Society for Experimental Mechanics Series, DOI 10.1007/978-3-319-42255-8_15

$$\begin{aligned} A &= \frac{r_0}{2E}(1+\upsilon)\varrho \\ B &= \frac{r_0}{2E}\left[4\varrho - (1+\upsilon)\varrho^3\right] \\ C &= -\frac{r_0}{2E}\left[2(1+\upsilon)\varrho + (1+\upsilon)\varrho^3\right] \\ G &= \frac{\upsilon t}{E}\varrho^2 \end{aligned} \tag{15.2}$$

ρ being the ratio of the hole radius to the distance from the hole center, t thickness, E the Young modulus, ν Poisson's ratio and F being null.

Based on (1) and (2), various solution techniques can be developed: the first, and still most common approach, relies on the measurement of radial strain at three different angular coordinates. Knowing the relative angular position, the system of the three instances of $\varepsilon_r(\vartheta_i) = s_i$, s_i being the ith measured strain and $i = 1\ldots3$, can be solved in closed form. Strains are usually measured using strain gauge rosettes, thus, appropriate calibration coefficients, taking into account the relatively large sampling area, have to be computed [1].

The use of strain gauges is not the only option. Indeed, the displacement field around the hole can be measured using several techniques; of these, optical methods are particularly interesting because they allow for non-contact measurement. Several approaches have been proposed using both interferometric [5–9] and non-interferometric optical techniques [10]. The first provides higher resolution, given the short wavelength of the laser light, and allow for a denser spatial sampling when compared with the others (mainly, Digital Image Correlation), but require costlier instruments.

Whatever the experimental technique, the hole drilling approach requires milling the hole and measuring the resulting displacement/strain around the hole. To this end, various ad-hoc instruments using either the strain gauge or the interferometric approach have been developed and are nowadays commercially available (e.g. [11–13]). All of them require high-precision components; this is particularly true for the interferometric (ESPI) system, but is generally true also for the strain-gauge-based systems. The latter class of instruments requires an accurate centering of the hole with respect to the rosette and a highly stable strain gauge amplifier; both points do not constitute a problem for the interferometric systems, because there is no rosette and the hole can easily be located by image analysis, but rigid body motion has to be limited in the sub-micrometer range and vibrations must be avoided. Thus, whatever the experimental technique adopted, the cost of the instrument will be significant. Note that, from the instrumentation viewpoint, DIC requirements are lower: a simple incoherent light source and a monochromatic camera suffice. However, both the sensitivity of the technique and the number of independent sampling points[1] are significantly lower; thus, standard DIC is not the optimal technique for residual stress measurement.

Recently, a new non-interferometric approach, able to solve most of the limitations of the DIC technique has been proposed for both isotropic [14] and orthotropic [15] materials. The basic idea is to integrate, inside the minimization functional of DIC, the displacement field related to hole drilling (i.e. (1)) instead of using general-purpose displacement functions. The resulting algorithm (iDIC) tries to adjust the controlling parameters of the functional, i.e. the stress components, in such a way as to obtain an optimal matching of the images of the hole acquired before and after drilling. On convergence, the estimated optimal stress values are the residual stress components. This is somewhat different from standard approaches, where stress identification requires two steps, i.e. a preliminary estimation of the displacement field followed by reverse calibration; moreover, since problem-specific shape functions are used, there is no need to partition the image into subsets and perform a series of local measurements. The net result is a significantly simpler and more robust algorithm. In particular, it is well known that standard DIC measurements are particularly critical near the hole boundary because of the presence of metal chips, of the potential contamination of the painted speckle field or, more simply, because part of the texture is no longer present. Normally, solving these problems requires an a-priori analysis of the displacement field and removal of the artifacts before the reverse calibration step [10], but they are no longer a complication for the iDIC algorithm because a single, large area around the hole is used for the analysis and small, local perturbations are unable to significantly affect the global behavior.

Using the iDIC approach, most of the experimental requirements are significantly reduced: rigid body motion can easily be compensated for and no decorrelation problem results from large translations. Moreover, since a single subset is used for

[1] The standard subset approach actually correlates the area around the measuring point with the corresponding region in the test image. If the successive area partially intersects the former one, the result will be partially correlated. Thus, to ensure statistical independence of results, the various subsets should not overlap.

correlation, image size need not be large. Thus, only the drilling phase has to be handled, obviously taking into account the requirement of retracting the driller to allow for acquisition of the image of the surface. With respect to the latter point, the diffusion of 3D printers has made available specialized hardware for motion control, thus it is possible to design a simple system for residual stress measurement, costing less than 500 €. In the following, we will focus on the interesting part of the instrument, namely the camera and the drilling subsystems, because the rest of the design (motion control and frame) are quite standard.

15.2 The Raspberry Pi Camera Module

The Raspberry Pi is a small-size computer board (somewhat larger than a credit-card, see Fig. 15.1). When connected to standard peripherals (keyboard, mouse, monitor or TV) it is capable of doing everything expect from a desktop computer; in particular, several languages—Python, Ruby, Perl—are available. Moreover, since it is able to execute either a Unix or the Windows operating system, code can be compiled on board using standard tools; finally, the Raspberry Pi is able to execute a webserver or a secure shell daemon, thus it can be interfaced over a network connection. What makes it so interesting is its cost (the Raspberry Pi model B+ shown in Fig. 15.1 costs about 21 €, whereas the Raspberry Pi 2 model B, using a quad core ARM Cortex-A7 processor, costs a few euros more; finally, the new Raspberry Pi 3 model B costs less than 40 €), the 40-pin GPIO connector (allowing for interfacing with external hardware) and, more important, the fact that the list of its peripherals includes both a standard and an infrared camera (Fig. 15.2).

Chip	Broadcom BCM2835 SoC
Core	ARM 11
CPU	700 MHz Low Power ARM1176JZFS Appl. Proc.
GPU	Dual Core VideoCore IV® Multimedia Co-Processor
Memory	512MB SDRAM
Operating System	Linux operating system; boots from Micro SD card
Dimensions	85 × 56 × 17 mm
Power supply	Micro USB socket 5V, 2A

Fig. 15.1 The Raspberry Pi model B+ and some of its specifications

Sensor	OmniVision OV5647
Sensor Resolution	2592 × 1944 (5 MPixel)
Sensor Image Area	3.76 × 2.74 mm
Pixel Size	1.4 × 1.4 μm
S/N Ratio	36 db
Dynamic Range	67 db @ 8× gain
Fixed Focus	1 m to infinity
Focal Length	3.6 ± 0.01 mm
Horizontal Field of View	53.5 ± 0.13 degree
Vertical Field of View	41.41± 0.11 degree

Fig. 15.2 The Raspberry Camera Module

In about 25 × 24 × 9 mm, the camera module allows for acquisitions of 5 MPixel images, thus, in principle we have more than enough data to perform Digital Image Correlation. However, some points have to be addressed; in particular, the low signal-to-noise ratio and the fixed focus may significantly affect the performance of DIC. Moreover, the sensor is equipped with a Bayer pattern, thus two of the RGB components of each pixel result from the interpolation of the data of the neighbor pixels. However, the camera driver allows for reading the raw data before entering the GPU post-processing pipeline (i.e. before auto-white balance, vignette compensation, smoothing, down-scaling and format encoding), thus, a suitable gray tone image can be obtained by averaging the four signals related to a Bayer cell. Indeed, there is no way to recover the filtered out intensity components (i.e. statistically independent RGB signals cannot be obtained), but the spectral contribution of each Bayer cell is always the same (given the BGGR Pattern used in the camera, we have 2 × Green, plus 1 × Red and 1 × Blue) thus allowing for comparison of the averaged value, assumed as the intensity value, obviously at the cost of accepting half the spatial resolution in both vertical and horizontal directions.

The SNR stated in the camera datasheet is quite low (36 db): indeed, by entering preview mode and observing a fixed scene, it is quite easy to detect flickering. However, since speed is not a constraint in our application, we can use an artificial image obtained by averaging several frames. To test this idea, we performed a simple test: we acquired 160 nominally identical images, thus, the noise-free intensity of each pixel can be quite reliably extracted from the average image. Knowing the "true" value of each pixel, it is quite simple to estimate the standard deviation of the intensity of each pixel. Figure 15.3 shows a portion of the Signal-to-Noise Ratio (SNR) map resulting from the analysis.

$$SNR_i = 20\log_{10}\left(\frac{\mu_i}{\sigma_i}\right) \tag{15.3}$$

where μ_i is the mean value of intensity of pixel i and σ_i is its standard deviation.

Maximum SNR has been mapped to white (i.e. 255) whereas minimum SNR has been mapped to dark gray (25). Saturated/defective pixels (i.e. pixels whose intensity does not change during the analysis) have been flagged using black (0). Results of Fig. 15.3 are mostly expected: due to the small pixel size, thermal noise probably constitutes a large fraction of the total. It being independent from photon count, the SNR results proportional to intensity (in other words, σ_i and μ_i are weakly correlated)

Figure 15.4 confirms the previous analysis. It shows the minimum, peak and mean SNR Vs gray tone. All of them increase with gray value. Note that the distribution of SNR related to each gray tone is not Gaussian and the mean values are significantly biased toward high values. Note also that the peak SNR is somewhat higher than the value given in the datasheet of the camera; this is expected, because each experimental value corresponds to four physical pixels (due to the binarization we performed to deal with the Bayer pattern).

Looking at Fig. 15.4, it is apparent that the performance of the camera is still unsatisfactory: even though the peak SNR becomes quite high for gray values above 200, the mean SNR stays well below 40 db. This means that only 6 bits are reliable and the others are noisy. To check the potential improvement, we re-performed the same analysis on images obtained by averaging 2, 4 and 8 successive frames. Figure 15.5 shows the results of the analysis: the impulse graph refers to the mean SNR of the original image shown as reference (it is worth remembering that due to the Bayer pattern this configuration cannot be used) whereas the continuous lines refer to increasing averaging levels.

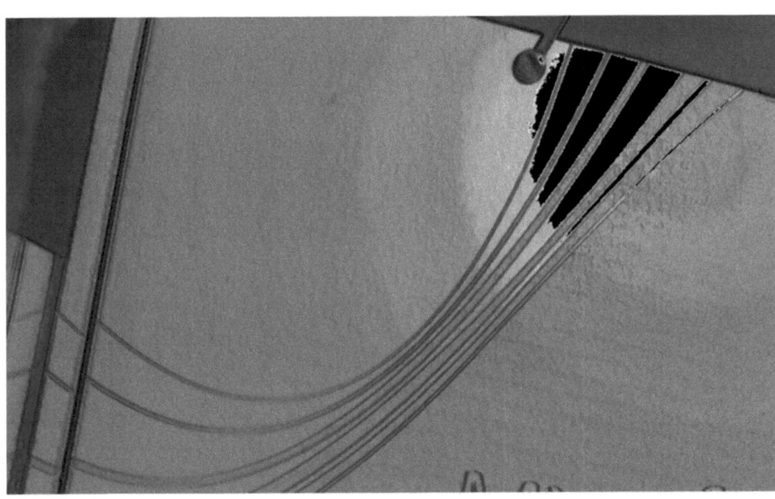

Fig. 15.3 Experimental SNR map (partial view). The black regions corresponds to saturated pixels

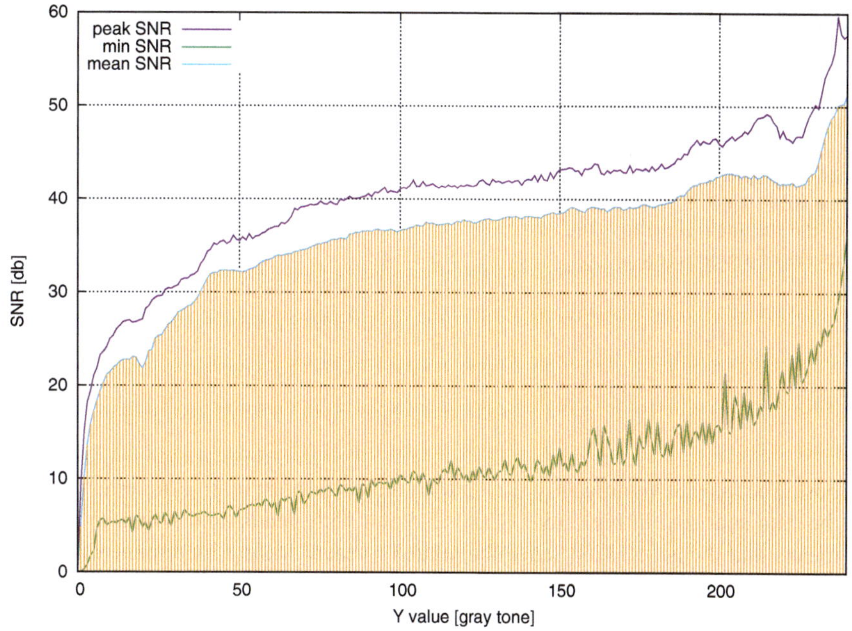

Fig. 15.4 Experimental SNR of the Raspberry Pi camera module. 2 × 2 binarization

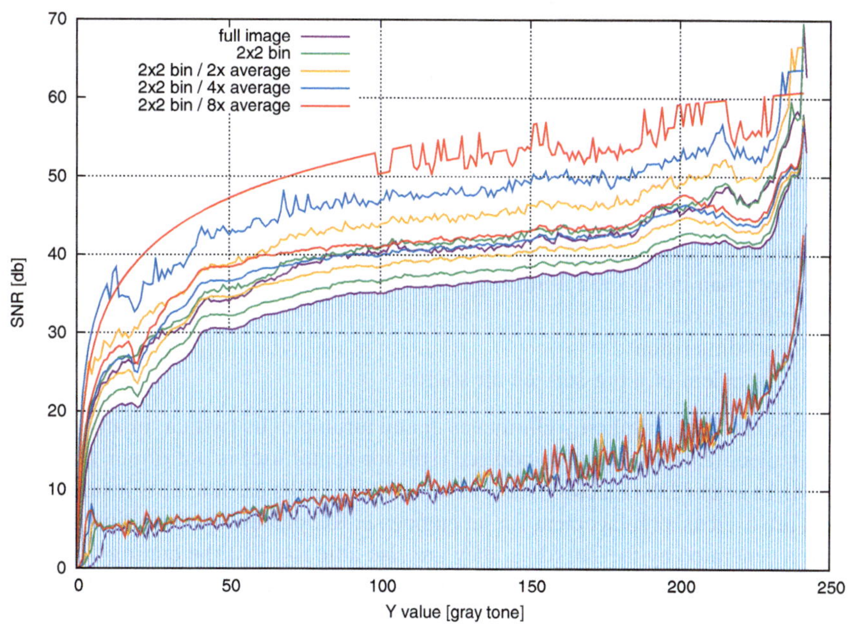

Fig. 15.5 Experimental SNR of the Raspberry Pi camera module using 2×, 4× and 8× averaging

Looking at Fig. 15.5, it is apparent that minimum, mean and peak SNRs behave differently with respect to the averaging process: the first is quite insensitive to the process and the various curves related to different averaging procedures are almost perfectly superimposed. On the contrary, the peak SNR shows a large improvement when moving from the simple 2 × 2 binarization to a binarization combined with image averaging (from 2× up to 8×). Finally, the mean SNR (probably the most interesting parameter) shows a clear saturation effect: if the improvement obtained when moving from the simple binarization to a 2× averaging is significant, the increment obtained with the successive step is moderate and the move from 4× to 8× offers few advantages. Thus, using a 2 × 2 binarization combined with 4× averaging appears to be the optimal solution. Note that in this configuration the peak SNR is above the 8 bit limit (48.1 db) for more than half the gray tone range, whereas the corresponding mean SNR is around 40 db, i.e. almost 7 bit.

A final note on timing: the OmniVision OV5647 chip allows for acquisition of 1080p sequences at 30 frames per second. However, in this operating mode it provides no access to low-level data, thus, the images have to be acquired using the still

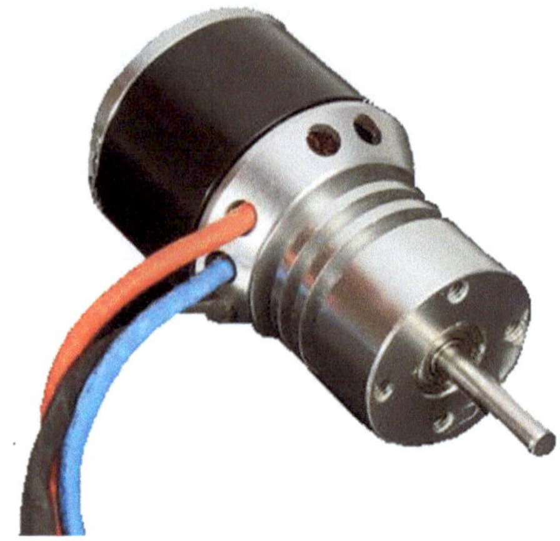

Fig. 15.6 An EDF outrunner engine. Note that an outrunner motor has stationary coils at the center, and the rotating magnet on the outside. This grants high torque and allows for very easy replacement of the shaft, which is connected to the rotating magnet by means of a grub screw

port. The latter is significantly slower than the video port: to give an order of magnitude, the Python script we used to acquire the data analyzed in Figs. 15.3, 15.4, 15.5 and 15.6, reads (and writes to the SD card) 1 full-frame-image per second. This means that the acquisition sequence related to each drilling increment will require about 4 s.

15.3 Drilling the Hole

The drilling subsystem is a critical point of the instrument. Given the design target (making the cost as low as possible), we opted not to use orbital milling because it would require adding an extra degree of freedom to the system. In any case, the driller has to satisfy a minimum set of requirements:

- A rotational speed of the spindle as high as possible (i.e. on the order of magnitude of tens of thousands of revolutions per minute);
- Ability to sustain the cutting forces; note that, because of our decision to use plunge milling, the axial force is somewhat larger than in the orbital milling case;
- Reduced cost;
- Small dimensions;
- Independent of a pressurized air source.

The last point implicitly eliminates dental turbine units (which, in any case, are too expensive). In the same way, the use of an electrical high-speed handpiece (still for dental use) is out of our (self-imposed) budget. The simplest solution to the problem is an EDF (Electric Ducted Fan) engine. EDFs are fan units inside a duct that produces more thrust than a propeller and are used as a replacement for jet engines in RC planes. By definition, this type of engine must be able to resist the axial force generated by the fan (thus, they are also able to resist the cutting force generated in our application); moreover, their rotational speed is quite high (common values range from 20,000 to 50,000 RPM); in the outrunner configuration they are able to deliver significant torque (i.e. a torque high enough to allow direct driving of the fan with no gearbox) and finally their cost is on the order of some tens of euros.

Figure 15.6 shows an EDF outrunner. To give an idea of its dimensions, the diameter of the mounting-hole circle is 18 mm, the total length is 42 mm and the outer diameter is 26 mm. Finally, the shaft diameter is 3 mm. Note that the shaft can easily be replaced (see Fig. 15.6) because it is nothing more than an axle supported by a couple of rolling bearings, thus, a simple option to install the mill is to remove the shaft and use the mill as the shaft of the EDF. However, this solution poses a limitation on the engine's rotational speed: due to their configuration, the rotational speed of outrunner engines is roughly proportional to the length of the engine (and obviously to the voltage supply). The total length of a 1 mm diameter mill is about 40 mm, thus, the maximum distance between the rolling bearings is about 30 mm (i.e. the engine shown cannot be used). A simple solution is to build an intermediate shaft, to be glued to the shank of the engine, allowing for mounting of a commercial collet (e.g. one of the collets in the Dremel *Quick Change Collet Nut Set*).

15.4 Conclusions

In the previous sections we have shown that a portable instrument for residual stress measuring, based on the hole-drilling technique and integrated Digital Image Correlation, can be assembled using relatively low-cost components. The iDIC approach makes the system tolerant to rigid body translation, but requires a relatively high-quality imaging system. This constraint cannot be satisfied directly, but our noise analysis shows that the target SNR can be obtained by averaging a relatively small number of successive frames.

An estimation of the cost of the instrument, based on the (probably non-exhaustive) sum of the cost of of the various components, amounts to less than four hundred euros. This value obviously does not include the design, construction and assembly time (we used a 3D printer for all the components of the frame) and completely neglects the development of the firmware. In any case, a user wanting to build a copy of the instrument (we plan to put the instrument in the public domain once all the validation tests have been finished) will need only to download the STL files, print the components and assemble the instrument.

References

1. ASTM E837-08e1. Standard test method for determining residual stresses by the hole-drilling strain-gage method. American Society for Testing and Materials, West Conshohocken, PA, 2008. doi:10.1520/E0837-08E01
2. Schajer, G.S.: Measurement of non-uniform residual stresses using the hole-drilling method. Part I—Stress calculation procedures. J. Eng. Mater. Technol. **110**(4), 338–343 (1988)
3. Schajer, G.S.: Measurement of non-uniform residual stresses using the hole-drilling method. part II—practical application of the integral method. J. Eng. Mater. Technol. **110**(4), 344–349 (1988)
4. Gustav, K.: Die Theorie der Elastizität und die Bedürfnisse der Festigkeitslehre. Z. Verein. D. Ing. **42**, 797–807 (1898)
5. Nelson, D.V., Makino, A., Fuchs, E.A.: The holographic-hole drilling method for residual stress determination. Opt. Lasers Eng. **27**, 3–23 (1997)
6. Schwarz, R.C., Kutt, L.M., Papazian, J.M.: Measurement of residual stress using interferometric moiré: A new insight. Exp. Mech. **40**(3), 271–281 (2000)
7. Steinzig, M., Ponslet, E.: Residual stress measurement using the hole drilling method and laser speckle interferometry: Part I. Exp. Tech. **27**(3), 43–46 (2003)
8. Schajer, G.S., Steinzig, M.: Full-field calculation of hole drilling residual stresses from electronic speckle pattern interferometry data. Exp. Mech. **45**(6), 526–532 (2005)
9. Baldi, A.: A new analytical approach for hole drilling residual stress analysis by full field method. J. Eng. Mater. Technol. **127**(2), 165–169 (2005)
10. Schajer, G.S., Winiarski, B., Withers, P.J.: Hole-drilling residual stress measurement with artifact correction using full-field DIC. Exp. Mech. **53**(2), 255–265 (2013)
11. Vishay. Milling Guide for Residual Stress Measurements. http://www.vishaypg.com/docs/11304/rs200.pdf
12. Sint Technology. Restan MTS3000, Automatic System for Residual Stress Measurement by Hole-Drilling. http://www.sintechnology.com/strain-gauge-measurements/RESTAN_MTS3000.html
13. Stresstech group. Prism—Residual stress measurement based on hole-drilling and ESPI. http://www.stresstechgroup.com/files/upload_pdf/604976/hd1000g%5FPrism%5Fstandard.pdf
14. Baldi, A.: Residual stress measurement using hole drilling and integrated digital image correlation techniques. Exp. Mech. **54**(3), 379–391 (2014). doi:10.1007/s11340-013-9814-6
15. Baldi, A.: Residual stress analysis of orthotropic materials using integrated digital image correlation. Exp. Mech. **54**(7), 1279–1292 (2014). doi:10.1007/s11340-014-9859-1

Chapter 16
Non-Destructive Internal Lattice Strain Measurement Using High Energy Synchrotron Radiation

Jun-Sang Park and John Okasinski

Abstract High energy synchrotron X-rays can penetrate large samples and real engineering components. Taking advantage of this capability, diffraction techniques using monochromatic X-rays have been widely used to measure the residual strains in engineering components. However, isolating a particular volume inside a large component and measuring the residual strain is a challenge when employing typical monochromatic X-ray techniques. In this work we describe a spiral slit system capable of isolating an interior volume in a polycrystalline sample and non-destructively measuring the lattice strains in the volume. An interference fit sample constructed from a Ni-based superalloy is used to demonstrate the capabilities of the system. We compare the strain results to those measured using a conical slit system, a more mature and established device. The results from several polycrystalline samples with non-cubic crystal symmetry are also presented.

Keywords Residual stress • Lattice strain • Synchrotron X-ray • Spiral slits • Conical slits

16.1 Introduction

Manufacturing processes can induce residual stress fields in mechanical components. The existence of residual stress in a component is particularly important for predicting the mechanical performance of a component. The ability to characterize the residual stress in a component is, therefore, of significant importance to create components that meet the intended design.

Residual stress characterization techniques can be broadly categorized as destructive and non-destructive. In the non-destructive category, diffraction-based techniques have been quite successful. Neutron scattering based techniques such as the time-of-flight (TOF) method are commonly used to characterize the residual stresses in large components made of high Z materials. With appropriate slitting mechanisms, the residual stresses inside a component can be characterized with spatial resolution on the order of millimeters. Synchrotron X-ray based techniques are also quite popular. The energy dispersive diffraction (EDD) method is sometimes utilized to characterize the residual stresses inside a component. Similar to the TOF method, appropriate slitting mechanisms and stages allow spatial resolution on the order of millimeters longitudinally (along the beam direction) and sub-millimeters perpendicular to the beam.

Residual stress characterization techniques utilizing angle dispersive diffraction (ADD) method is also frequently employed. In this case, the residual stresses can be characterized with micrometer spatial resolution in the direction perpendicular to the beam when appropriate X-ray optics are used. Furthermore, the availability of large area detectors means that many strain components can be measured simultaneously giving the users a more complete picture of the strain. However, in the case of ADD method, the ability to obtain the residual stress information in a volume of material located inside a component is not as simple. In this work, we compare two slitting mechanisms that achieves that task.

16.2 Background

Diffraction is governed by Bragg's Law:

$$\lambda = 2d_n \sin\theta_n \tag{16.1}$$

J.-S. Park (✉) • J. Okasinski
Advanced Photon Source, Argonne National Laboratory, 9700 S. Cass Avenue, Lemont, IL 60439, USA
e-mail: parkjs@aps.anl.gov

S. Quinn, X. Balandraud (eds.), *Residual Stress, Thermomechanics & Infrared Imaging, Hybrid Techniques and Inverse Problems, Volume 9*, Conference Proceedings of the Society for Experimental Mechanics Series, DOI 10.1007/978-3-319-42255-8_16

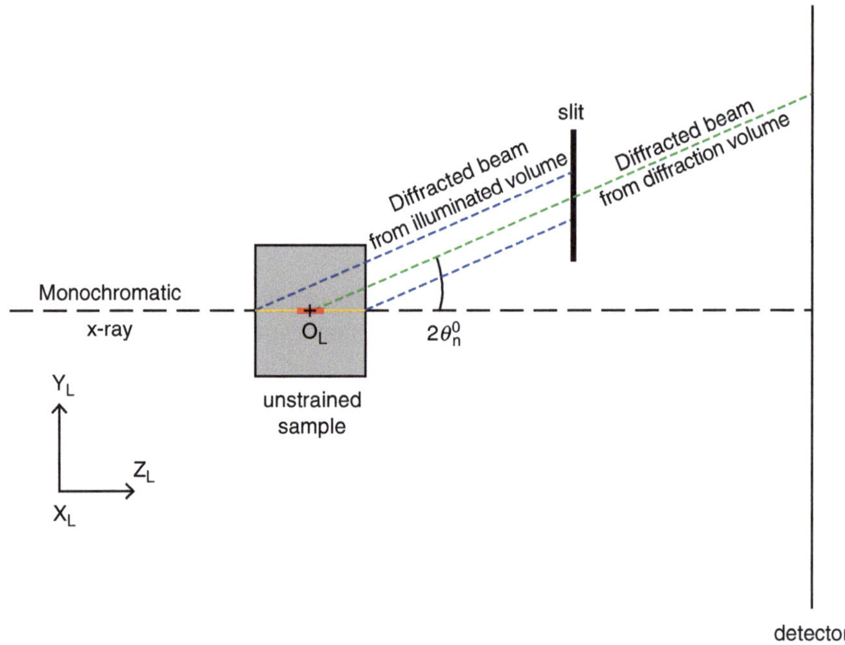

Fig. 16.1 A schematic of slit system to isolate an internal volume for the ADD method. A monochromatic x-ray beam illuminates a volume of material in a sample highlighted in *orange*. A slit is placed in the path of the diffracted beam emanating from the illuminated volume at a particular angle $2\theta_n^0$. Only the diffracted beam from the volume highlighted in *red* is allowed to pass through the slit and is recorded on the detector

In this equation, λ is the wavelength of the monochromatic X-ray, $d_\mathbf{n}$ is the plane spacing for a family of crystallographic planes {hkl} where its plane normal is denoted by **n**, and θ_n is the corresponding Bragg angle.

Several slitting mechanisms for the ADD method to obtain an isolated volume of material located along the X-ray beam have been proposed and demonstrated. For example, Nielsen et al. [1] have proposed a set of conical slits capable of isolating approximately 100–200 μm long volume inside the sample and measured the lattice strains in the volume non-destructively. Martins et al. [2] have proposed a set of spiral slits with similar longitudinal spatial resolution. A set of spider web slits has also been proposed [3].

In all these approaches, the working principle is similar. Figure 16.1 illustrates the geometry associated with the slit mechanism. In this figure, the slit is placed in the path of the diffracted beam. A monochromatic X-ray beam illuminates the entire sample in its path. For a crystalline material, if a slit is placed in the path of the diffracted beam at a reference or unstrained Bragg angle, $2\theta_n^0$, with respect to a point in space, O_L, the diffracted beam from a volume of unstrained material placed at O_L passes through the slit and is recorded on the detector. It is also the case that if a volume of strained material is placed in the path of the direct beam, the diffracted beam from the material can also pass through the slit and be recorded on the detector as long as the location of the volume is away from O_L.

In this article, a set of lattice strain measurements from an interference fit sample employing the conical and the spiral slits are presented. The lattice strain results are compared and contrasted. To highlight the advantage of the spiral slits, a set of lattice strain measurement on an interference fit sample manufactured from a titanium alloy is presented.

16.3 Experiments

The exact geometries and working principles of the conical slits and the spiral slits used in this work are described in [1] and [2] respectively. Figure 16.2 shows an example diffraction data recorded on the GE amorphous Si area with the conical slits system in place. Figure 16.3 shows an example diffraction data recorded on the area detector with the spiral slit system in place. While the conical slit system captures the full Debye-Scherrer rings with the exception of regions where the four spokes are holding the conical slit system together, only small sections of the Debye-Scherrer rings are recorded when the spiral slits are used.

The geometry of the interference fit sample used in this work is described in detail by Park et al. [4]. Figure 16.4 shows the schematic of the interference fit sample. The sample is composed of a hub manufactured from a nickel superalloy and a shaft manufactured from a tool steel bar stock. The outer diameter of the hub is 24.000 mm and the inner diameters are 12.580 mm and 12.700 mm. The thickness of the hub is 1.250 mm. The diameter of the shaft is 12.7 mm. Because of the tapered inner hole, we anticipate a stress gradient through the thickness of the hub.

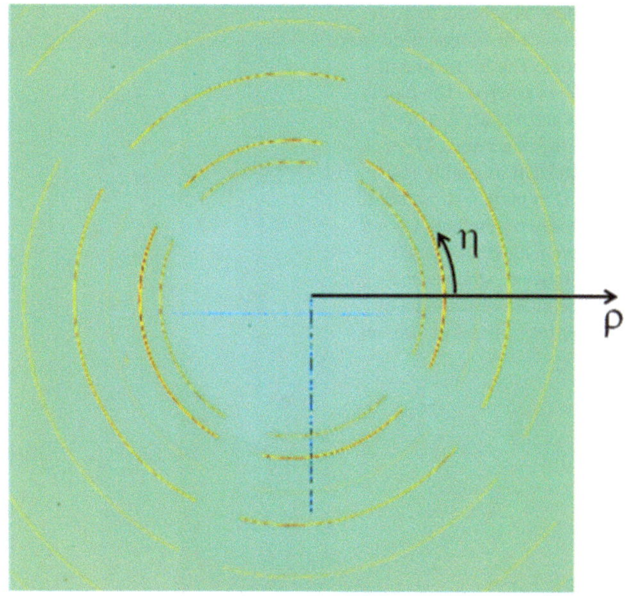

Fig. 16.2 A diffraction image recorded on the area detector with the conical slit system in place. Full Debye-Scherrer rings are recorded on the detector with the exception of the four spokes used to hold the conical slits system intact

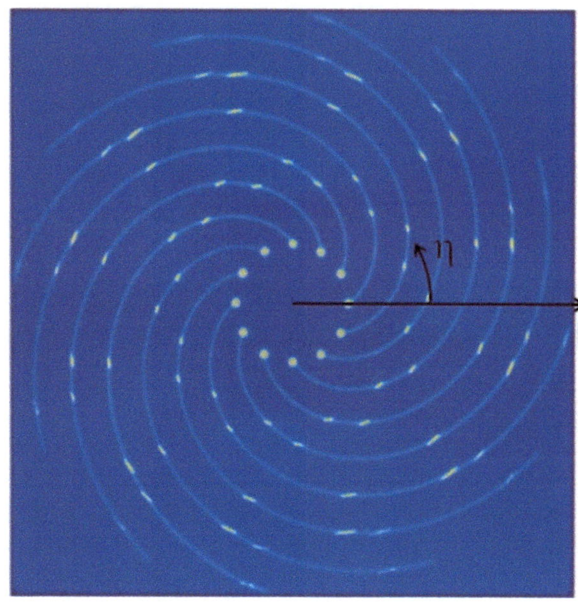

Fig. 16.3 A diffraction image recorded on the area detector with the spiral slit system in place. There are 12 spiral slits and only partial Debye-Scherrer rings are recorded on the detector

The second interference fit sample manufactured from a titanium alloy is described in Park et al. [5]. In this sample, the inner hole of the hub is not tapered; its inner diameter is 12.636 mm and the outer diameter is 31.750 mm. The thickness of the hub is 1.000 mm. Because there is no taper in the inner hole of the hub, we do not anticipate a stress gradient through the thickness of the hub.

The diffraction data are reduced as described in [2, 6]. To summarize, the detector is calibrated with the CeO_2 calibration powder data. The diffraction images are binned to produce spectra. The peaks of interest in the spectra are fitted using a pseudo-Voigt function and their peak positions are extracted. In the case of the data set collected using the conical slits, the peak positions as well as the location of the sample with respect to O_L are used to determine the location of the diffraction volume along **z** and the associated strain. In the case of the data set collected using the spiral slits, the radial and the azimuthal positions (ρ and η) are used to compute the location of the diffraction volume along z and the associated strain. In this work, we focus on the nominal xx- and the yy- components of strain measured along the **y-z** plane on the hub.

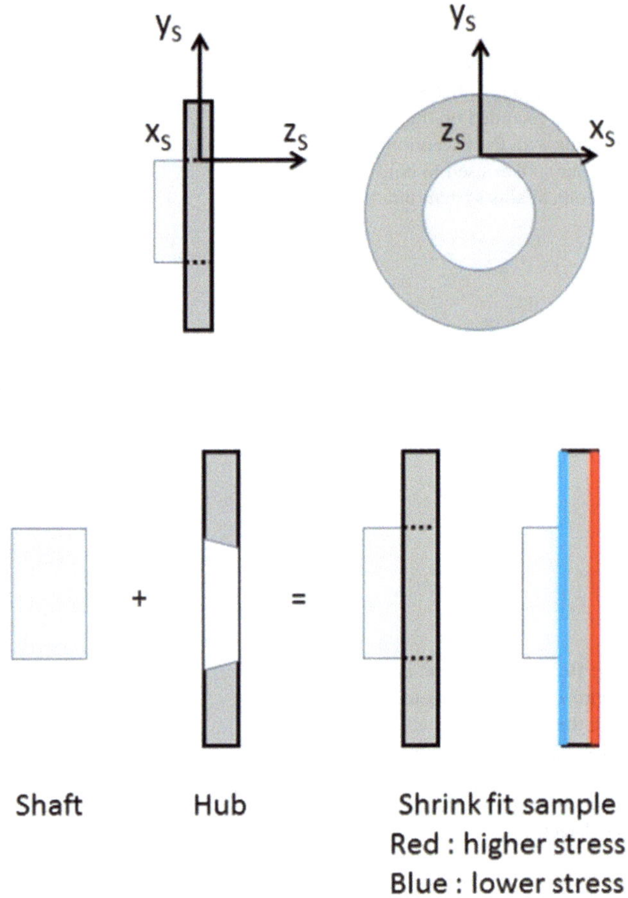

Fig. 16.4 A schematic of the interference fit sample used in this work. The inner hole of the hub is tapered. The hub is heated and a shaft is placed inside the inner hole. When the hub/shaft assembly is allowed to cool, a stress field is imposed on the hub. Larger interference induces larger magnitudes of stresses and smaller interference induces smaller magnitudes of stresses

16.4 Results and Discussion

Figure 16.5 shows the xx- and yy- components of strain measured along the **y-z** plane of the interference fit sample manufactured from nickel superalloy. While there are other reflections available, the {200} reflection was used to measure the strain. As anticipated, the xx-component of strain is tensile and the yy-component of strain is compressive in the **y-z** plane in both cases. The magnitudes of strain are larger when the interference is larger (at z = 0.61 mm) and are smaller when the interference is smaller (at z = −0.61 mm).

The strains from the conical slits are in general larger in magnitude. This may be attributed to larger azimuthal averaging necessary when the spiral slits are used to measure the strains. When the conical slits are used, the sample is moved along the beam and diffraction images are recorded. Through this motion, the longitudinal strain distribution is measured. Referring to Fig. 16.1, the sample is moved about O_L. During this motion, the diffracted beams from individual diffraction volumes in the illuminated volume are recorded on the detector. The peak positions from the diffracted beams are used to determine the strains and the location of the diffraction volume inside the hub. As long as the Debye-Scherrer rings show powder-like qualities (no diffraction spots), a small azimuthal bin can be used to integrate the images and to carry out the subsequent analysis. On the other hand, when the spiral slits are used, the location of the peak along a spiral slit is related to the longitudinal location of the diffraction volume; the sample does not have to be translated along O_L. The consequence of this is that while a great deal of time is saved because the sample is not translated about O_L, larger azimuthal bins maybe necessary.

The real strength of the spiral slits is the ability to deal with non-cubic crystal symmetry. The conical slits used in this work is designed for cubic crystal symmetry. To investigate a polycrystalline material with non-cubic crystal symmetry, one can (1) build a dedicated set of conical slits for the non-cubic crystal symmetry or (2) tune the wavelength of the monochromatic X-ray to allow a small subset of reflections to go through the cubic conical slits. However, both options

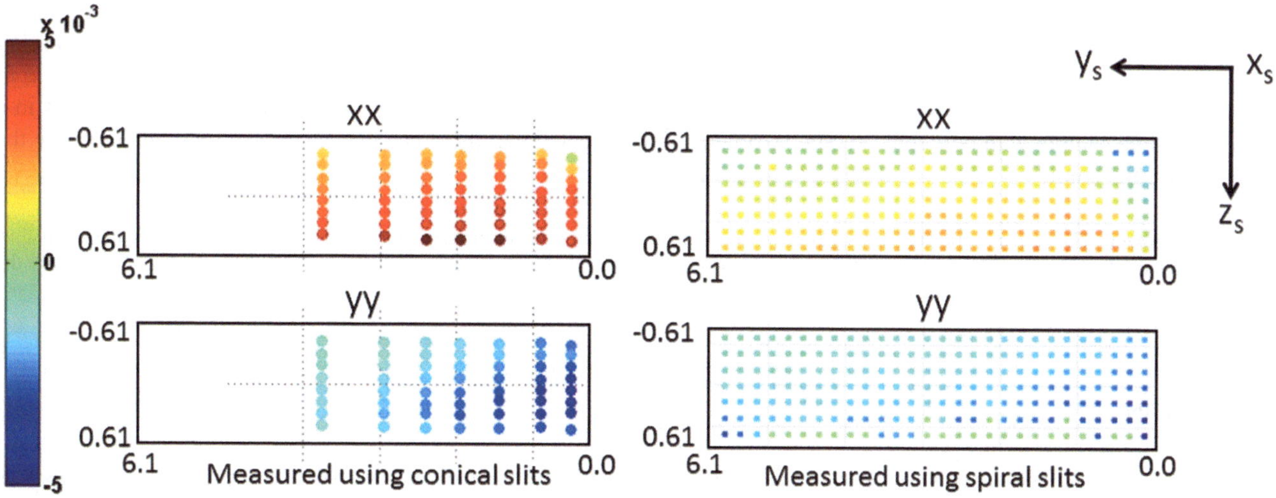

Fig. 16.5 The xx- and yy- components of strain measured along the **y-z** plane of the nickel superalloy interference fit sample. The *black boxes* outline the hub dimension in millimeters in the **y-z** plane. The locations of the colored dots inside the *black boxes* indicate where measurements were made and the colors of the dots show the magnitude of the strain component. Along y, y = 0 mm correspond to the inner most position in the hub and y = 6.1 mm correspond to the outer most position in the hub

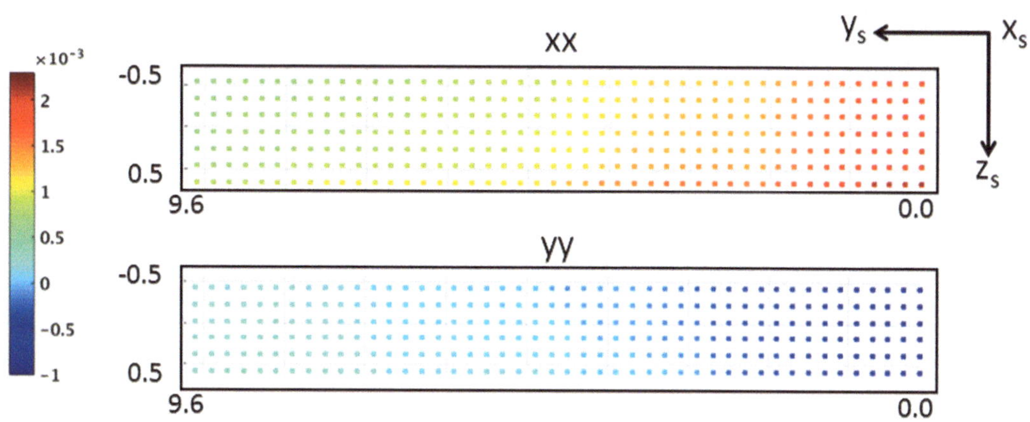

Fig. 16.6 The xx- and yy- components of strain in the hub of the titanium interference fit sample. At y = 0.0 mm, the hub and the shaft are in contact. The strain gradient along the thickness of the hub is measured using the spiral slits

are not optimal. With the spiral slit system, any crystal symmetry can be investigated; with appropriate rotation of either the slit or the sample, the full Debye-Scherrer rings can be measured.

Figure 16.6 shows the xx- and yy- components of strain in the hub of the interference fit sample manufactured from the titanium alloy. The {10.2} reflection was used to measure the strains. As anticipated, the xx-component of strain is tensile and the yy-component of strain is compressive. The magnitudes of the two components are the largest near the shaft and are the smallest at the outermost position on the hub. There is also a significant strain gradient (5×10^{-4} difference between the highs and the lows along the thickness of the hub) through the thickness of the hub. This was not anticipated but with the spiral slits, we were able to quantify it.

16.5 Conclusions

High energy synchrotron X-rays are capable of penetrating a large structural component to measure the strain in the component. This strain is typically the average strain in the entire volume illuminated by the X-ray in an angle dispersive diffraction setup. With a set of slits placed in the path of the diffracted X-rays, we can also measure the strains in individual volumes along the beam path. In this work, an interference fit sample manufactured from a nickel superalloy is used to benchmark the conical and the spiral slit systems. We compared the strain fields in an interference fit sample measured using the conical and the spiral slit systems. The measured strains fields compare favorably. We also used an interference fit manufactured from a titanium alloy to demonstrate the spiral slit system's capability to measure the internal strains in a polycrystalline sample with non-cubic crystal symmetry. It is revealed that there is a measurable strain gradient inside the interference fit sample.

Acknowledgment The authors appreciate Dr. V. Honkimäki at the European Synchrotron Radiation Facility, France for loaning the spiral slits. This research used resources of the Advanced Photon Source, a U.S. Department of Energy (DOE) Office of Science User Facility operated for the DOE Office of Science by Argonne National Laboratory under Contract No. DE-AC02-06CH11357.

References

1. Nielsen, S., Wolf, A., Poulsen, H., Ohler, M., Lienert, U., Owen, R.: A conical slit for three-dimensional XRD mapping. J. Synchrotron Rad. **7**, 103–109 (2000)
2. Martins, R., Honkimaki, V.: Depth resolved investigation of friction stir welds made from AA2024 / AA2024 and AA2024 / AA6082 using a spiral slit and high energy synchrotron radiation. Mater. Sci. Forum **490–491**, 424–429 (2005)
3. Sinsheimer, B. Bouet, S. Ghose, E. Dooryhee and R. Conley.: X-ray powder diffraction with spiderweb slits (poster), 2014
4. Park, J.-S., Lienert, U., Dawson, P.R., Miller, M.P.: Quantifying three-dimensional residual stress distributions using spatially-resolved diffraction measurements and finite element based data reduction. Exp. Mech. **53**(9), 1491–1507 (2013)
5. Park, J.-S., Ray, A., Dawson, P., Lienert, U., Miller, M.: Determination of residual stress in a microtextured alpha titanium using high energy sychrotron x-ray. Journal of Strain Analysis for Engineering Design, no. accepted
6. Miller, M.P., Bernier, J.V., Park, J.-S., Kazimirov, A.: Experimental measurement of lattice strain pole figures using synchrotron x rays. Rev. Sci. Instrum. **76**(11), 113903–113911 (2005)

Chapter 17
Discussion on X-Ray and HDM Residual Stress Measurements

C. Barile, C. Casavola, and V. Moramarco

Abstract The stress field remaining in some materials without application of external sources of stress is known as residual stress. These residual stresses are produced in almost all manufacturing processes or may occur during the life of structures. They have a fundamental role in welded joints because they affect the way to design structures (e.g. the safety coefficients), their fatigue life and their corrosion resistance. Quantify, as well as possible, the residual stress field is one of main issues for mechanical engineers. To this purpose, in the last decades, several techniques have been developed. Hole Drilling Method (HDM) and X-Ray Diffractometry (XRD) are two of the most diffused and the only standardized techniques to measure the stress field in depth on welded structures. Although both methods declare to make an accurate measurement of the stresses, few comparisons of these techniques applied on the same structure are described in scientific literature.

The aim of the present work is to compare the residual stress measurement between HDM and XRD on Ti-6Al-4 V (Grade 5) laser butt-welding joints to understand if a significant difference exists between the results obtained by the two methods or they may be considered comparable.

Keywords Residual stress • Temperature field • Hole drilling method (HDM) • X-Ray diffractometry (XRD) • Laser butt-welding joints

17.1 Introduction

Residual stresses are present in a material even in absence of an external load as a consequence of manufacturing processes. Evaluation of residual stresses is an important task in view of the fact that they can introduce dimensional distortions and they have an essential role in welded joints because they affect the way to design structures (e.g. the safety coefficients), their fatigue behavior and their corrosion resistance [1, 2].

Nowadays, several techniques have been developed to measure residual stress into the materials [3]. However, just some of these are quantitative and non-destructive. X-Ray diffractometry (XRD) is quantitative and non-destructive experimental technique to measure the residual stress on the surface of metallic components. XRD methods can provide very accurate results, even if they require expensive equipment and a relatively long procedure for the residual stress depth profiling [4]. In this technique, the measure of the variation of the diffractometric angle of crystallographic lattice related with is deformation permit to calculate the residual stress state of the material.

On the other hand, the Hole Drilling Method is the most general approach used to measure residual stress in depth profiles in materials [5]. It consists in executing a very small hole, tolerable or repairable, and then in measuring the relieved strain by using both a strain gage rosette and the electronic speckle pattern interferometry in an area close to the hole itself. Three major aspects are characteristics of the HDM approach and in the last years they greatly develop [6, 7], providing significant advances of the technique. They are the execution of the drilled hole [8, 9], the measurement of the resulting deformations around the hole and the computation of the corresponding residual stresses by using different mathematical algorithm. A large number of factors contribute to determine the final quality of the measurement such as the drilling rotation speed [10–12], the hole shape [13–18], the number of increment steps [19, 20] and the cutter milling diameter [12].

In particular, in this paper XRD technique and HDM combined with strain gage rosette are applied on five titanium grade 5 (Ti-6Al-4 V) laser butt-welding joints plates [21]. Ti6Al4V is a biocompatible material that can be used in different applications such as chemical, aerospace, naval and biomedical engineering. Titanium and its alloys are materials of great

C. Barile • C. Casavola (✉) • V. Moramarco
Dipartimento di Meccanica, Matematica e Management (DMMM), Politecnico di Bari, Viale Japigia 182, 70126 Bari, Italy
e-mail: katia.casavola@poliba.it

S. Quinn, X. Balandraud (eds.), *Residual Stress, Thermomechanics & Infrared Imaging, Hybrid Techniques and Inverse Problems, Volume 9*, Conference Proceedings of the Society for Experimental Mechanics Series, DOI 10.1007/978-3-319-42255-8_17

interest in engineering field because of their excellent mechanical properties. They have high resistance to corrosion and fatigue, good mechanical properties at high temperatures, low density. Aircraft industry, in particular, appears very interested in a wider adoption of this material that would allow reducing the total mass of the airplane by reducing, consequently, the fuel consumption and the greenhouse emissions. Welding technology is widely used to assemble metal parts because of high productivity and flexibility [22]. One of the issue associated with welding Ti-based alloys results from their poor thermal conductivity, which leads to the creation of a steep thermal gradient across the weld region, and hence results in high residual stresses. Ti-based alloys are also susceptible to the formation of porosity and inclusions (e.g. hard alpha) when welded in the presence of hydrogen, oxygen, nitrogen, or carbon contamination, which results in reduced mechanical properties. No many results are reported in scientific literature about the comparison of different techniques in measuring these kind of stresses [23]. For this reason the aim of this work is to apply both HDM and XRD approaches and to compare the measurement the residual stress measurement in order to understand if a significant difference exists between the results obtained by the two methods or they may be considered comparable [24, 25].

17.2 Material and Methods

The laser welding process has been carried out on five laser butt-welded plates in Ti-6Al-4 V (Titanium Grade 5) 200 mm wide, having two different thickness 3 and 5 mm (Table 17.1).

Their chemical composition appear in Table 17.2. Thermo-physical properties and mechanical characteristics at room temperature (20°C) have been reported in Table 17.3. The Table 17.4 sums up the parameters used in the laser welding process.

It was defined the longitudinal axis or y-axis parallel to the axis of welding, then it was defined transverse axis or x-axis the one perpendicular to the axis of welding. Figure 17.1 shows the scheme of the performed measurements, as well as the direction of the principal direction.

The residual stress measurements were performed using two different techniques: the Hole Drilling Method and X-Ray Diffraction. For each plate six points were defined, three for each side respect to the welding cord. Each side was measured using one technique. The two methods were performed separately. Measured points are spaced 50 and 50 mm from the edge. Point B is located on the central axis of the plate. The XRD measurements were made only on the surface.

Table 17.1 Plate sizes and characteristics

Plate name	Length [mm]	Width [mm]	Thickness [mm]	Welding	Type of joint
Plate 1	200	200	5	Laser	Butt
Plate 2	200	200	5	Laser	Butt
Plate 3	300	200	5	Laser	Butt
Plate 4	200	200	3	Laser	Butt
Plate 5	300	200	3	Laser	Butt

Table 17.2 Chemical composition Ti-6Al-4 V (Grade 5)

Al [%]	Fe [%]	Ti [%]	V [%]	Other [%]
5.5–6.75	<=0.4	87.7–91	3.5–4.5	2.65–3.3

Table 17.3 Physical and mechanical characteristics

Parameter	value
Density [kg/m^3]	4430
Melting temperature [°C]	1604
Thermal conductivity [W/m-K]	6.70
Thermal expansion coefficient [1/K]	$8.60 \cdot 10^{-6}$
Modulus of Elasticity, E [GPa]	113.8
Yield Tensile Strength, σ_y [MPa]	880
Ultimate Tensile Strength, σ_u [MPa]	950
Vickers Hardness (HV)	349

Table 17.4 Process parameters of laser welding process

Plate name	Laser power [W]	Laser beam speed [mm/s]
Plate 1		
Plate 2		
Plate 3	9000	60
Plate 4		
Plate 5	6000	60

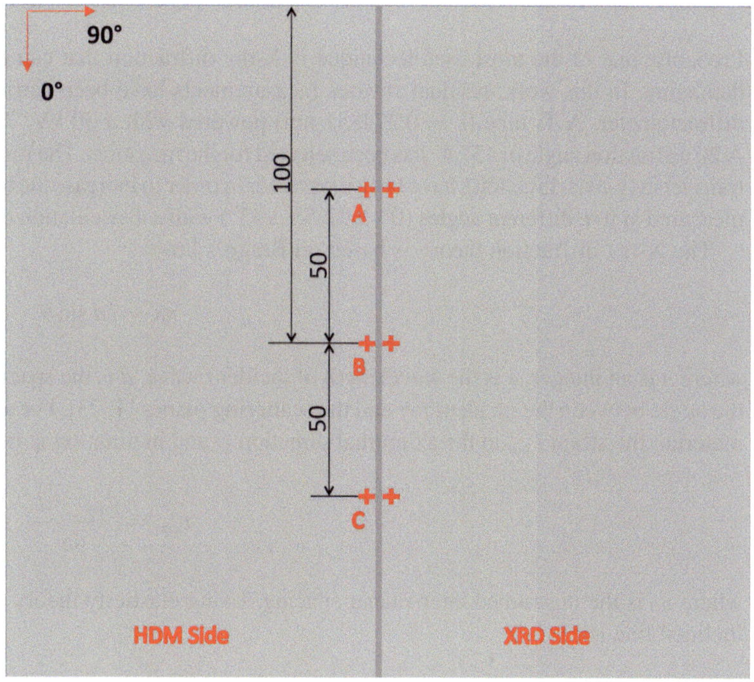

Fig. 17.1 Plate geometry and measurement points

17.2.1 Hole Drilling Method

The most widely used technique for measuring residual stress is the HDM combined with strain-gage rosette for stress relaxation. The principle involves the introduction of a small hole at the location where residual stresses are measured. Due to the hole drilling residual stresses are relieved and the corresponding strains on the surface are measured using a special three- (or six-) elements strain gage rosette bonded around the hole. From these measured strains, residual stresses are calculated using appropriate calibration constants referred to the particular type of strain gauge rosette used. The procedure for residual stress measurement using HDM is described in ASTM standards with designation E837-13 [26]. In most practical cases, the residual stresses are not uniform with depth. The incremental hole drilling method is an improvement of the basic hole drilling method since it executes the drilling in a series of small steps, which improves the versatility of the method and enables stress profiles and gradients to be measured. A high-speed air turbine and carbide cutters are used to drill the hole, without introducing any further machining stresses and thereby modifying the existing stress system.

The strain data at pre-determined depths are accurately acquired. Different stress calculation algorithms are used to evaluate the residual stress field starting from the measured strains. The most known techniques used for measuring non-uniform stress are *integral methods* and *hdm*, both of them are implemented in the commercial system used to perform measurements. In any case, a precise and severe measurement practice is required to obtain good quality holes and strain data and to properly use the analysis method in order to reach an accurate assessment of the residual stress field.

The measurements of residual stresses by hole-drilling were performed using the Restan system of SINT Technology. This system carries out measurement according to the ASTM standard. The equipment includes a high-speed air turbine (up to 400,000 rpm) which drives an inverted drill-bit cone with diameter of 1.6 mm, carbide tungsten TiAlN coated and a two axes microscope for correct centring of the milling cutter on the strain gage rosette. System was used to properly execute the hole and for measuring the effective diameter of the hole, after the drilling execution. Each hole was characterized by a diameter of about 1.8 mm and a depth of 1.5 mm reached in 30 steps distributed according to a polynomial function.

This allows increasing the number of steps in the proximity of the surface and decreasing it as increasing depth. The strain gauge deformation were recorded after each drilling step by using the acquisition system Quantum X HBM. H*dm* calculation method was used to estimate the variation of the residual stresses through the thickness; it was based on ten calculation steps distributed according to a linear function.

17.2.2 X-Ray Diffraction Method

Probably, one of the most used technique is X-ray diffraction that can measure the residual stress of the structures without damaging. In this work, residual stresses measurements have been carried out using an Xstress 3000 G3R Stresstech X-ray diffractometer. A Ti tube ($\lambda = 0.274851$ nm) powered with a 30 kV, 7 mA current and a 3 mm collimator has been used. A 2θ diffraction angle of 137.4° has been selected for the measures. The residual stresses, both longitudinal (x-axis direction) and transverse (y-axis direction), have been measured. In order to increase the quality of the measure, the stress in each point has been measured at five different angles (0°, ±22.5°, ±45°) with a φ oscillation of ±3°. The detection distance has been set at 50 mm.

The X-ray diffraction theory is based on Bragg's law

$$n\lambda = 2d\sin\theta \tag{17.1}$$

where n is an integer, λ is the wavelength of incident wave, d is the spacing between the planes in the atomic lattice, and θ is the angle between the incident ray and the scattering planes [1, 23]. For a shear stress-free biaxial stress field in an untextured material, the strain $\varepsilon_{\varphi\psi}$ in the azimuthal direction φ and inclination ψ from the plane normal is determined by the equation:

$$\varepsilon_{\varphi\psi} = \frac{d_{\varphi\psi} - d_0}{d_0} \tag{17.2}$$

where d_0 is the unstrained inter-planar spacing. Using elasticity theory for an isotropic solid shows that the strain along an inclined line is:

$$\varepsilon_{\varphi\psi} = \frac{1-\nu}{E}\left(\sigma_1\cos^2\varphi + \sigma_2\sin^2\varphi\right)\sin^2\psi - \frac{\nu}{E}(\sigma_1 + \sigma_2) \tag{17.3}$$

If we consider the strains in terms of inter-planar spacing, and use the strains to evaluate the stresses, then it can be shown that

$$\sigma_\varphi = \frac{E}{(1+\nu)\sin^2\psi}\left(\frac{d_\psi - d_n}{d_n}\right) \tag{17.4}$$

where d_n is the inter-planar spacing measured in normal direction to the plan. This equation allows calculating the stress in any chosen direction from the inter-planar spacing determined from two measurements, made in a plane normal to the surface and containing the direction of the stress to be measured. For stress determination the $\sin^2\psi$ method was used. A number of XRD measurements are made at different ψ tilts. Than, the inter-planar spacing is measured and plotted as a curve. The stress can be calculated from the gradient of the line and with basic knowledge of the elastic properties of the material

$$\sigma_\varphi = \left(\frac{E}{1+\nu}\right)m \tag{17.5}$$

where m is the gradient of the d vs. $\sin^2\psi$ curve.

To evaluate the effects of the paint coating on the measure of the residual stress tension by X-ray diffractometer the intensity of the peak and value of the angle of diffraction (2θ) with and without coating will be studied.

17.3 Results

Figures 17.2 and 17.3 show calculated residual stress respectively along the longitudinal and the transversal directions with respect to the weld cord. Plotted residual stresses were calculated by *hdm* method. Stresses are plotted versus the hole depth. Some "unexpected" values were measured at first and at last drilling steps probably due to both the surface preparation for strain gage bonding and to ill conditioned matrix calculation. Experimental data show a more regular residual stress trend at central depths.

Few slight differences in stress trends could be due to different welding station used for the plates, but in any case no great differences are shown between results of 3 and 5 mm plates.

Figures 17.4 and 17.5 show the comparison between the principal residual stresses and their orientation measured by XRD and the second step of HDM. In this way, it results ensured that approximately the same depth of measurement is considered for both techniques.

In general, principal directions for both techniques show a good agreement, finding that the principal stresses are approximately parallel to the welding cordon. Principal stresses, instead, show more comparable results for 3 mm plates, while in the 5 mm plates some differences have been founds. These differences could be due to the fact that, for the thick plates, the superficial stress state are less representative of the internal state of stress.

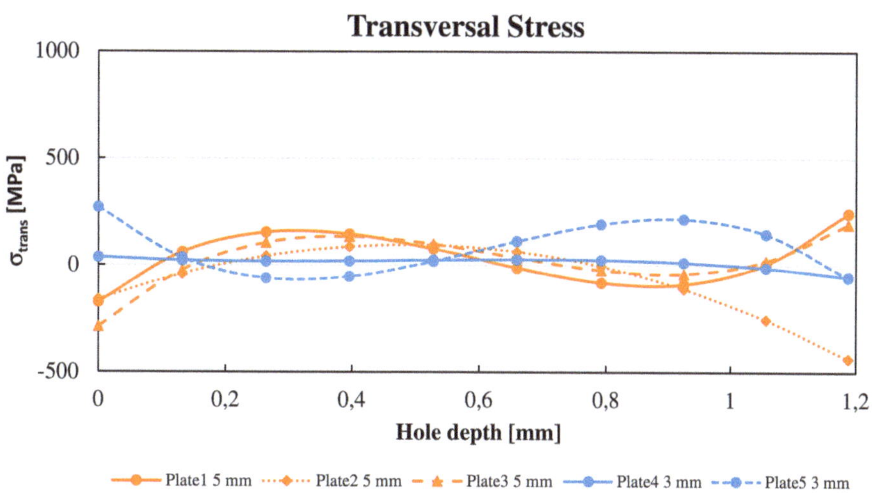

Fig. 17.2 Stress measured transversal to the weld cord

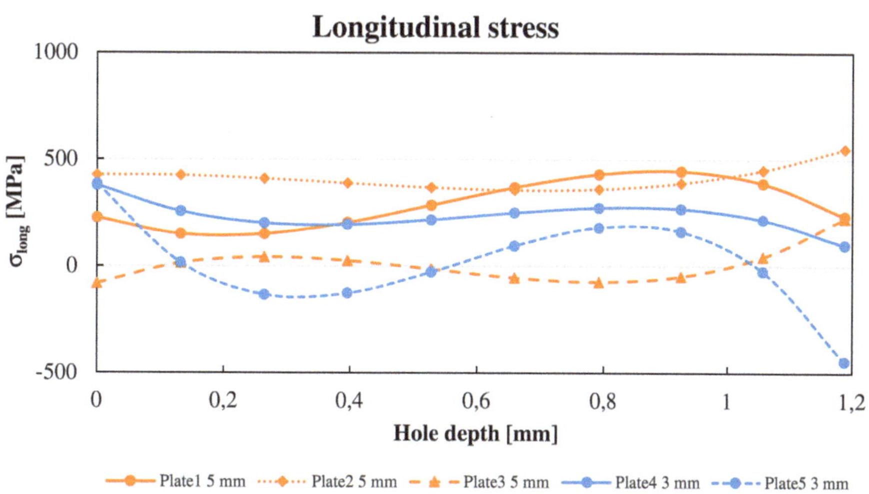

Fig. 17.3 Stress measured longitudinal to the weld cord

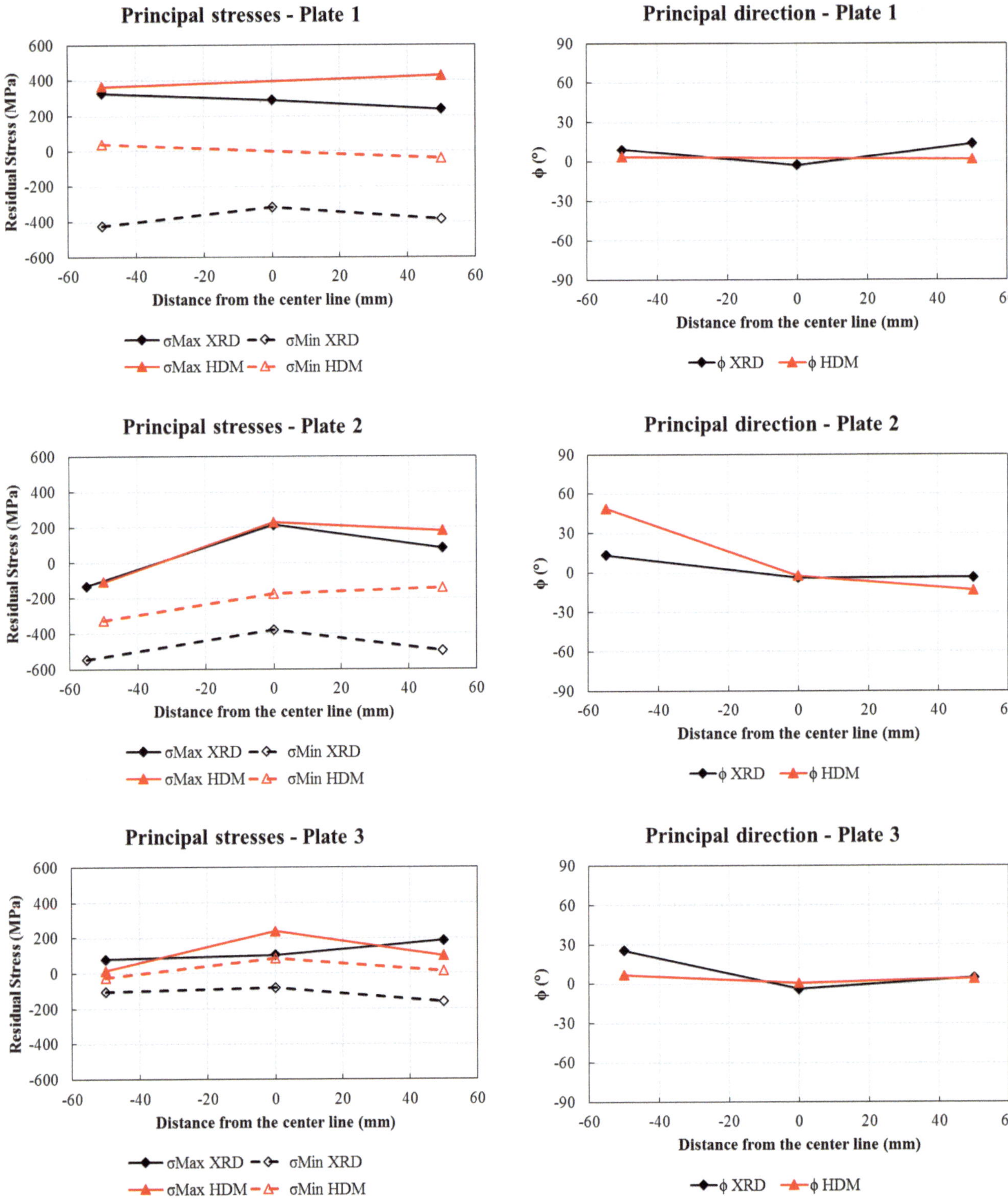

Fig. 17.4 HDM vs. XRD principal stress and principal stress orientation compared in 5 mm plates

Anyway it should be take into account that residual stress measured by means of HDM gives the average stress related to the area (the hole diameter range between 1.8 and 2.0 mm) where the material is drilled, for XRD the surface analysed is shorter than in previous case. This characteristic is fundamental in laser welding since the heat-affected zone typically is very limited and this justify the differences between the two approaches. Anyway, a more reliable comparison could be reached by applying in depth XRD measurements using chemical attack.

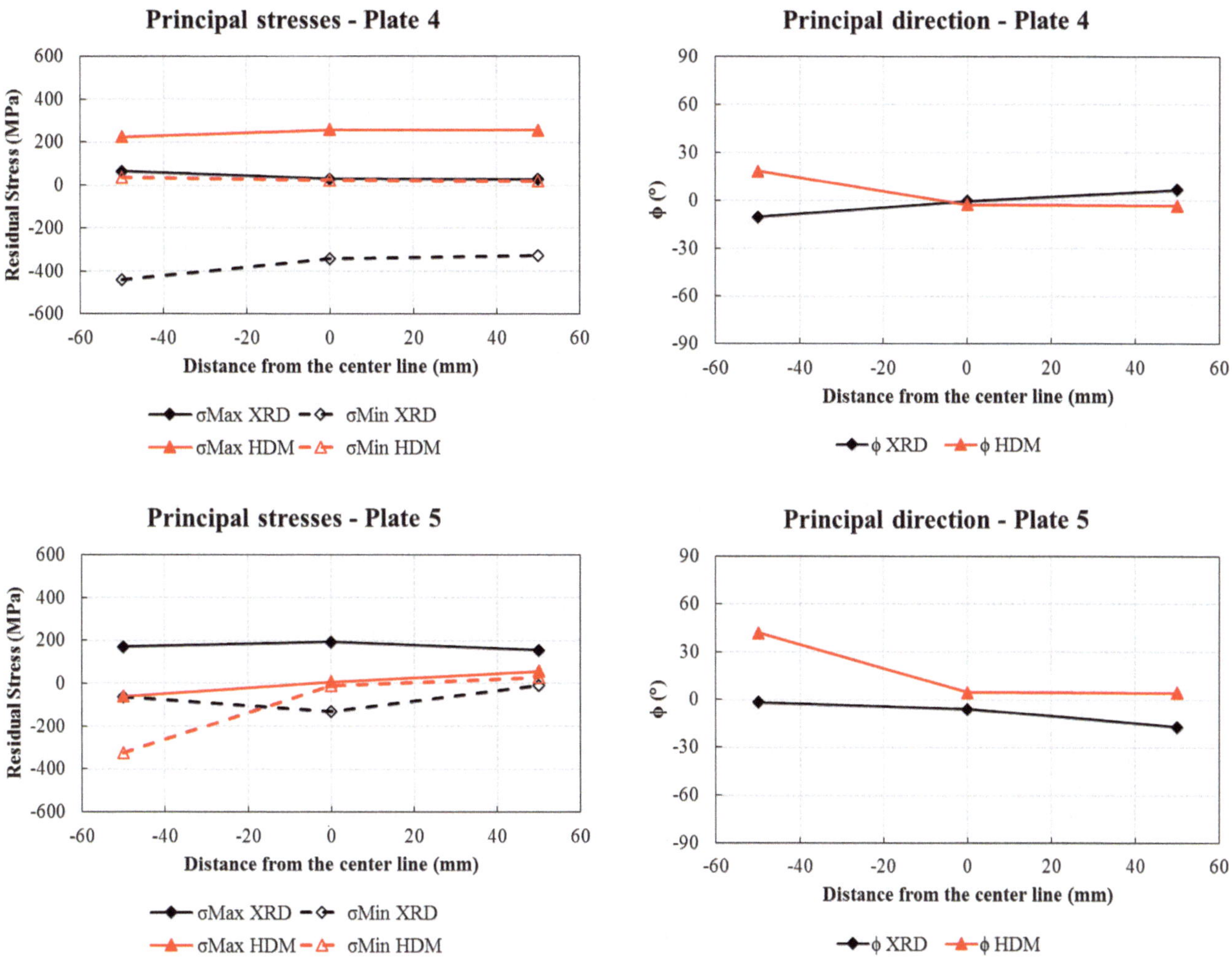

Fig. 17.5 HDM vs. XRD principal stress and principal stress orientation compared in 3 mm plates

17.4 Conclusion

Titanium grade 5 is a material widely used in different industrial fields as chemical, aerospace, naval and biomedical engineering. Recently, welding technology are improving, allowing to assemble titanium parts. Nevertheless, its poor thermal conductivity leads to the generation of a steep thermal gradient across the weld region resulting in high residual stresses. In this work residual stress measurements were carried out by means of HDM and XRD: the last one gives a more localized value of stress than the first approach. Measured points are located along the weld cord. Analysis of measured residual stresses suggests some considerations: HDM measurements don't show large differences between 3 mm and 5 mm plate thickness, XRD measurements confirmed the principal stress orientation obtained in HDM, few differences are shown in stress trend between the two techniques but they could be justified considering the analysis area of both approaches.

Future works could give more reliable comparison between HDM and XRD, if considering the in-depth measurements also for XRD by using a chemical attack.

References

1. Kudryavtsev, Y.F.: Residual stress. In: Sharpe Jr., W.N. (ed.) Springer Handbook of Experimental Solid Mechanics, pp. 371–388. Springer, Boston (2008)
2. Withers, P., Bhadeshia, H.: Residual stress part 2—Nature and origins. Mater. Sci. Technol. **17**, 366–375 (2001)

3. Rossini, N.S., Dassisti, M., Benyounis, K.Y., Olabi, A.G.: Methods of measuring residual stresses in components. Mater. Des. **35**, 572–588 (2012)
4. Casavola, C., Cazzato, A., Moramarco, V., Pappalettere, C.: Influence of the clamps configuration on residual stresses field in friction stir welding process. J. Strain Anal. Eng. Des. **50**, 232–242 (2015). 201
5. Measurements Group: Measurement of Residual Stresses by Hole-Drilling Strain Gage Method, Tech Note TN-503-6. Vishay Measurements Group, Raleigh (2001)
6. Schajer, G.S.: Hole-Drilling Residual Stress Measurements at 75: Origins, Advances, Opportunities. Exp. Mech. **50**(2), 245–253 (2010)
7. Barile, C., Casavola, C., Pappalettera, G., Pappalettere, C.: Remarks on residual stress measurement by hole-drilling and electronic speckle pattern interferometry. Scientific World Journal **2014**, 487149 (2014). 7
8. Barile, C., Casavola, C., Pappalettera, G., Pappalettere, C.: Overview of the effects of process parameters on the accuracy in residual stress measurements by using HD and ESPI. Conference Proceedings of the Society for Experimental Mechanics Series, 9, pp. 113–118
9. Barile, C., Casavola, C., Pappalettera, G., Pappalettere, C.: Considerations on the choice of experimental parameters in residual stress measurements by hole-drilling and ESPI. Frattura ed Integrità Strutturale **30**, 211–219 (2014)
10. Flaman, M.T.: Brief investigation of induced drilling stresses in the center-hole method of residual-stress measurement. Exp. Mech. **22**(1), 26–30 (1982)
11. Barile, C., Casavola, C., Pappalettera, G., Pappalettere, C., Tursi, F.: Drilling speed effects on accuracy of HD residual stress measurements. Conference Proceedings of the Society for Experimental Mechanics Series, Vol. 8, pp. 119–125, 2014
12. Upshaw, D., Steinzig, M., Rasty, J.: Influence of drilling parameters on the accuracy of hole-drilling residual stress measurements. Conference Proceedings of the Society for Experimental Mechanics Series, Vol. 8, pp. 95–109, 2011
13. Kiel, S.: Experimental determination of residual stresses with the ring-core method and an on-line measuring system. Exp. Tech. **16**(5), 17–24 (1992)
14. DeWald, A.T., Hill, M.R.: Improved data reduction for the deep-hole method of residual stress measurement. J. Strain Anal. **38**(1), 65–78 (2003)
15. Cheng, W., Finnie, I.: Residual Stress Measurement and the Slitting Method. Springer, New York (2007)
16. An, Y., Schajer, G.S.: Residual stress determination using cross- slitting and dual-axis ESPI. Exp. Mech. **50**(2), 169–177 (2010)
17. Prime, M.B.: Cross-sectional mapping of residual stresses by measuring the surface contour after a cut. J. Eng. Mater. Technol. **123**(2), 162–168 (2001)
18. Barile, C., Casavola, C., Pappalettera, G., Pappalettere, C.: Consideration on temperature fields and internal radius of analysis in hdm + espi residual stress measurements. 13th IMEKO TC15 Youth Symposium on Experimental Solid Mechanics, pp. 11–14, 2014
19. Vangi, D.: Data management for the evaluation of residual stresses by the incremental hole drilling method. J. Eng. Mater. Technol. **116**(4), 561–566 (1994)
20. Grant, P., Lord, J., Whitehead, P., Fry, T.: The application of fine increment hole drilling for measuring machining-induced residual stresses. Appl. Mech. Mater. **3–4**, 105–110 (2005)
21. Barile, C., Casavola, C., Pappalettera, G., Pappalettere, C.: Analysis of the effects of process parameters in residual stress measurements on Titanium plates by HDM/ESPI. Meas. **48**(1), 220–227 (2014)
22. Denga, D., Kiyoshima, S.: Numerical simulation of residual stresses induced by laser beam welding in a SUS316 stainless steel pipe with considering initial residual stress influences. Nucl. Eng. Des. **240**, 688–696 (2010)
23. Stone, H.J., Withers, P.J., Holden, T.M., Roberts, S.M., Reed, R.C.: Comparison of three different techniques for measuring the residual stresses in an electron beam-welded plate of WASPALOY. Metall. Mater. Trans. A **30**, 1797–1808 (1999)
24. Noyan, I.C., Cohen, J.B.: Residual Stress—Measurement by Diffraction and Interpretation. Springer, New York (1987)
25. Fontanari, V., Frendo, F., Bortolamedi, T., Scardi, P.: Comparison of the hole-drilling and X-ray diffraction methods for measuring the residual stresses in shot-peened aluminium alloys. J. Strain Anal. Eng. Des. **40**, 199–209 (2005)
26. ASTM E837-13a: Standard test method for determining residual stresses by the hole-drilling strain-gage method, ASTM International, West Conshohocken, (2013)

Chapter 18
Reducing Full-Field Identification Cost by Using Quasi-Newton Methods

J. Neggers, F. Mathieu, S. Roux, and F. Hild

Abstract The sensitivity fields required for identification methods such as Finite Element Method Updating (FEMU) or Integrated Digital Image Correlation (IDIC) are usually expensive to compute and require considerable amounts of memory storage. This research evaluates the application of Proper Orthogonal Decomposition (POD) to provide efficient means to reduce the weight to express these fields. It is shown that, as little as four orthogonal modes are required to adequately express the sensitivity fields needed to identify the parameters of a Voce isotropic hardening model.

Keywords Identification • Inverse methods • Model reduction • Digital image correlation • Integrated DIC

18.1 Introduction

Currently, there is an abundance of advanced material models that allow for the accurate prediction of critical nonlinear phenomena such as plasticity, viscosity, damage, fracture. However, the application of such models is often limited to academic cases, mostly, because of challenges in the identification of the model parameters. With the rapid increase in experimental data richness, mostly due to the improvement of digital cameras is making these challenges easier to address. Techniques such as Digital Image Correlation (DIC) allow for the measurement of full-field displacements fields. However, for an identification procedure, the displacement field is merely an intermediate quantity. Frequently, the constitutive model at hand is implemented in a Finite Element Method (FEM) framework. Consequently, most identification techniques confront the measured fields with their numerical counterparts (e.g., FEMU [1–3] or Integrated-DIC [4, 5]).

In practice, the identification process is treated as an optimization problem that is solved by iterative minimization of the difference between the two fields. Such an optimization algorithm (e.g., Gauss-Newton) typically consider Jacobian and Hessian operators that are usually approximated with finite differences. This in turn requires that the FEM simulation is performed numerously to obtain each sensitivity field. Identifying advanced material models often involves intricate geometrical behavior calling for large FE meshes, which are computationally inefficient.

The presented work details the benefits of applying Proper Orthogonal Decomposition (POD) to the sensitivity fields needed for Integrated-DIC and FEMU. The innate reduction to the data weight and computational efficiency are the first objective. However, the singular modes required to perform POD allow for retrospection of the original sensitivity fields to provide insight into the required quality of the FE model, both in space and time discretizations.

18.2 Methods

The following discussion will focus on Integrated-DIC. However, the presented reduction methods equally apply to FEMU. For identification procedures, the definition of where the measurement stops and where the data processing starts is fuzzy and depends on what is considered as the object of interest. Within the Integrated-DIC framework, the unknown material parameters (or geometrical parameters) are defined as the quantities of interest. Following this line of thought, the experimental data are directly, i.e. without translation, cast into a cost function

J. Neggers (✉) • F. Mathieu • S. Roux • F. Hild
LMT Cachan, ENS Cachan/CNRS/Unversité Paris Saclay, 61 avenue du Président Wilson, 94235 Cachan Cedex, France
e-mail: jan.neggers@lmt.ens-cachan.fr

S. Quinn, X. Balandraud (eds.), *Residual Stress, Thermomechanics & Infrared Imaging, Hybrid Techniques and Inverse Problems, Volume 9*, Conference Proceedings of the Society for Experimental Mechanics Series, DOI 10.1007/978-3-319-42255-8_18

$$\chi(\mathbf{p}) = \alpha_u \int_\tau \int_{roi} [G_0(\mathbf{x}) - G(\mathbf{x},t,\mathbf{p})]^2 \, d\mathbf{x}\, dt + \alpha_f \int_\tau \int_{roi} \left[f_{exp}(t) - f_{sim}(t,\mathbf{p})\right]^2 d\mathbf{x}\, dt \tag{18.1}$$

where G_0 is an image in the reference configuration and G is the series of images acquired at all consecutive time instances during the experiment. f_{exp} and f_{sim} are the measured and simulated forces, respectively. The cost function is integrated over the considered region of interest in space (roi) and time (τ). The parameters are listed in the n-dimensional vector $\mathbf{p}$, which represents the unknowns. The function χ is similar to the brightness conservation equation typically found in DIC. The first difference is the time-integration to allow for the identification of the parameters using all time steps at once (time integration exists in DIC, see [6], but is not standard). The second difference is the additional term involving the forces, which is required to include force data vital for the identification of stiffness or stress parameters. The α_u and α_f factors are included to weigh both terms with respect to each others. They are based on estimations of the image and force acquisition uncertainties respectively, for details see [5].

The parameters are obtained by minimizing Eq. (18.1), which is typically achieved by iterative linearizations and corrections, for instance using a Gauss-Newton algorithm. Assuming an initial guess for the parameters, the update increment $[\delta p]$ is obtained by solving a linear system

$$[\delta p] = [H]^{-1}[J] \tag{18.2}$$

where $[H]$ is the Hessian matrix and $[J]$ the Jacobian matrix, both containing an image and a force part

$$[H] = \alpha_u[H_u] + \alpha_f[H_f] \tag{18.3}$$

$$[J] = \alpha_u[J_u] + \alpha_f[J_f] \tag{18.4}$$

The components of each part are obtained by linearization of Eq. (18.1)

$$[H_u]_{ij} = \sum_t \sum_{\mathbf{x}} ([\boldsymbol{\varphi}_u]_i \cdot [\nabla G_0][\nabla G_0] \cdot [\boldsymbol{\varphi}_u]_j), \quad [J_u]_i = \sum_t \sum_{\mathbf{x}} ([\boldsymbol{\varphi}_u]_i \cdot [\nabla G_0][G_0 - G]), \tag{18.5}$$

$$[H_f]_{ij} = \sum_t \left([\varphi_f]_i [\boldsymbol{\varphi}_f]_j\right), \qquad [J_f]_i = \sum_t \left([\varphi_f]_i [f_{exp} - f_{sim}]\right), \tag{18.6}$$

where the summation over t and $\mathbf{x}$ represents the quadrature over all time steps and all pixels. For the force Hessian and Jacobian the integration over space is omitted assuming a single force measurement, which only varies over time. The matrices $[\boldsymbol{\varphi}_u]$ and $[\varphi_f]$ are the sensitivity fields and are defined as the derivative of the displacement (and force) towards each parameter. This derivative is often approximated by finite differences

$$[\varphi_u]_i = \frac{\partial[\mathbf{u}]}{\partial p_i} \approx \frac{[\mathbf{u}] - [\mathbf{u}]_i}{\varepsilon p_i} \tag{18.7}$$

$$[\varphi_f]_i = \frac{\partial[f_{sim}]}{\partial p_i} \approx \frac{[f_{sim}] - [f_{sim}]_i}{\varepsilon p_i} \tag{18.8}$$

where $[\mathbf{u}]$ and $[f_{sim}]$ are the displacement field and force from a simulation using the reference set of parameters, while $[\mathbf{u}]_i$ and $[f_{sim}]_i$ are obtained from a simulation with one parameter perturbed by a scale factor $\varepsilon = 0.01$.

Obtaining these sensitivity fields requires running $n + 1$ simulations, which can easily be a significant portion of the identification algorithm. Moreover, these sensitivity fields require evaluations at the pixel level for each time step, thereby inducing matrices of considerable size for memory storage (or memory paging very detrimental to computation time). On the other hand the Hessian matrix is usually modest in size, with only $n \times n$ components, with n being the number of unknown material parameters. This contrast in size between the sensitivity fields and the Hessian is the consequence of the Integrated-DIC procedure where the maximum data set is applied to directly obtain the minimum number of unknowns.

The direct transformation from the data rich pixel space to the parameter space is important for robust and accurate identification. However, it is also impractical due to the sheer size of the sensitivity fields. Moreover, the FE simulation applied to approximate the sensitivity fields has a limited spatial resolution. Therefore, it is desirable to split the transformation into two parts, first from the pixel space to the FE nodal space, then from the nodal space to the parameter space

$$[\varphi_u]_i = [\mathbf{S}]_i[N] \tag{18.9}$$

where [**S**] contains the nodal values of the sensitivity fields for all time steps and all parameters, while [*N*] is a matrix containing the FE shape functions expressed at the pixel level. Assuming that the number of nodes is less than the number of pixels, [**S**] is smaller than $[\boldsymbol{\varphi}_u]$. The FE shape functions only depend on the reference geometry, which usually is constant for all time steps and all identification iterations, therefore [*N*] only has to be computed once. This decomposition is natural if an FE simulation is used within the identification process, and usually provides a significant reduction to the weight of the sensitivity fields. It merely defines an intermediate space (i.e., the mesh) on which to express the quantity between the data space defined by the experiment and the space defined by the unknown material parameters. However, the computational effort required to compute, process and store [**S**] in the memory is still considerable. The mesh size requirements to solve the case at hand may be more stringent than strictly required to accurately describe [**S**]. Therefore, in this work, the POD method is applied to [**S**] to reduce the order of the sensitivity fields by proper decomposition

$$\mathbf{S}_i(\mathbf{x}, t) \approx \sum_j a_j \mathbf{U}_{ij}(\mathbf{x}) V_j(t) \tag{18.10}$$

The sensitivity fields are decomposed into a series of spatial fields $\mathbf{U}_{ij}$, which depend on the parameters p_i, and temporal functions V_j, with a_j the corresponding principal values. The functions **U** and *V* are obtained by singular value decomposition of the matrix [**S**]. This matrix is arranged with one row for each nodal value and each vector component for each parameter, while the columns represent the time steps. Each combination of $\mathbf{U}_{ij}$ and V_j expresses a principal mode of the sensitivity fields, ordered by significance by their singular values a_j. The approximation quality depends on the number of modes applied in this factorization. The goal is to reduce the weight of [**S**] and thus only a limited number of modes will be retained. The number of modes required depends on the kinematic complexity of the case at hand, but in practice this will be less then ten.

The model reduction introduces a new data space, intermediate between the FE nodal space and the material parameter space. In total four discrete spaces have been introduced, namely, (1) the pixel space or data acquisition space, (2) the FE mesh or numerical space, (3) the singular modes or POD space, and (4) the material parameter space. Assuming that the constitutive model is chosen and the experiment is performed, the first and the last space are chosen. In any translation between two spaces, potentially information is lost, especially when the second space is poorer than the first. However, the proposed intermediate spaces are chosen exactly to optimize the efficiency while maximizing the quality of the retained information. The proposed identification method is iterative, and it is always possible to use more singular modes or a finer mesh when closer to convergence. The goal of the proposed model reduction is to expedite the iterative process.

18.3 Test Case

To illustrate the reduction in data weight of the proposed reduction method, a real test case is discussed in which the elasto-plastic material parameters of a Voce isotropic hardening model (Fig. 18.1a [7]) are identified from a tensile experiment on an Aluminum alloy (AA2219) sample (Fig. 18.1b). The sample is loaded cyclically, where the maximum displacement for the first cycles is incrementally increased (Fig. 18.1c–d). The displacements given in Fig. 18.1c are those on the left and right edges of the mesh (as shown in Fig. 18.1b), which are measured by global DIC using the same mesh.

The four parameters schematically shown in Fig. 18.1a are considered the unknown parameters (i.e., $p = [E, \sigma_0, n, \sigma_\infty]^T$). The corresponding four sensitivity fields are shown in Fig. 18.2. It should be noted that these sensitivity fields are defined over space and time, and they are vector fields with components for the displacements u_x, u_y and the force f. Figure 18.2 only shows two cross-sections of for each parameter, one in space at $t = 390$ s and one in time at $x = 16.9$ mm. When these fields are applied for identification purposes, the identification residual is successfully minimized and the obtained parameters are $E = 69$ GPa, $\sigma_y = 375$ MPa, $n = 37$ [-] and $\sigma_\infty = 510$ MPa.

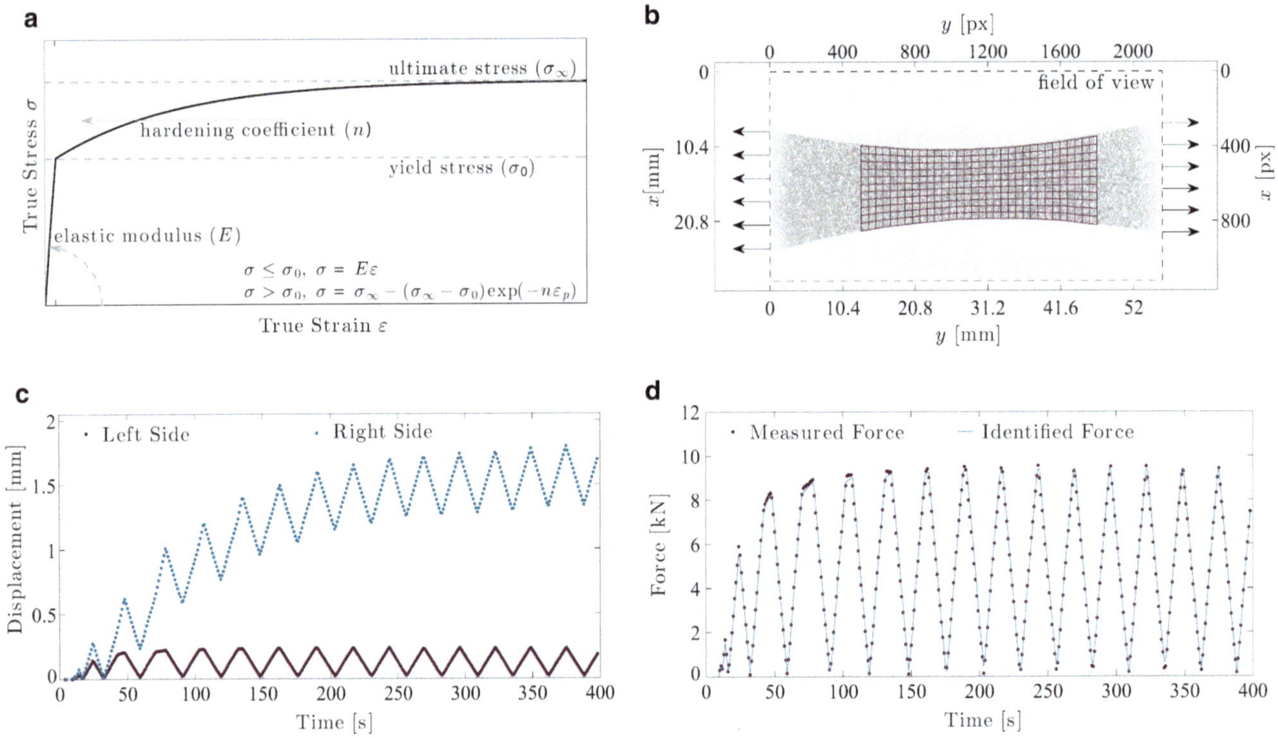

Fig. 18.1 The constitutive model (**a**) and the experiment performed for the identification ((**b**),(**c**),(**d**)) of Voce's parameters for AA2219. (**a**) Voce elasto-plastic material model, (**b**) image of the sample with the FE mesh, (**c**) applied displacement to each side of the mesh, (**d**) measured and identified force

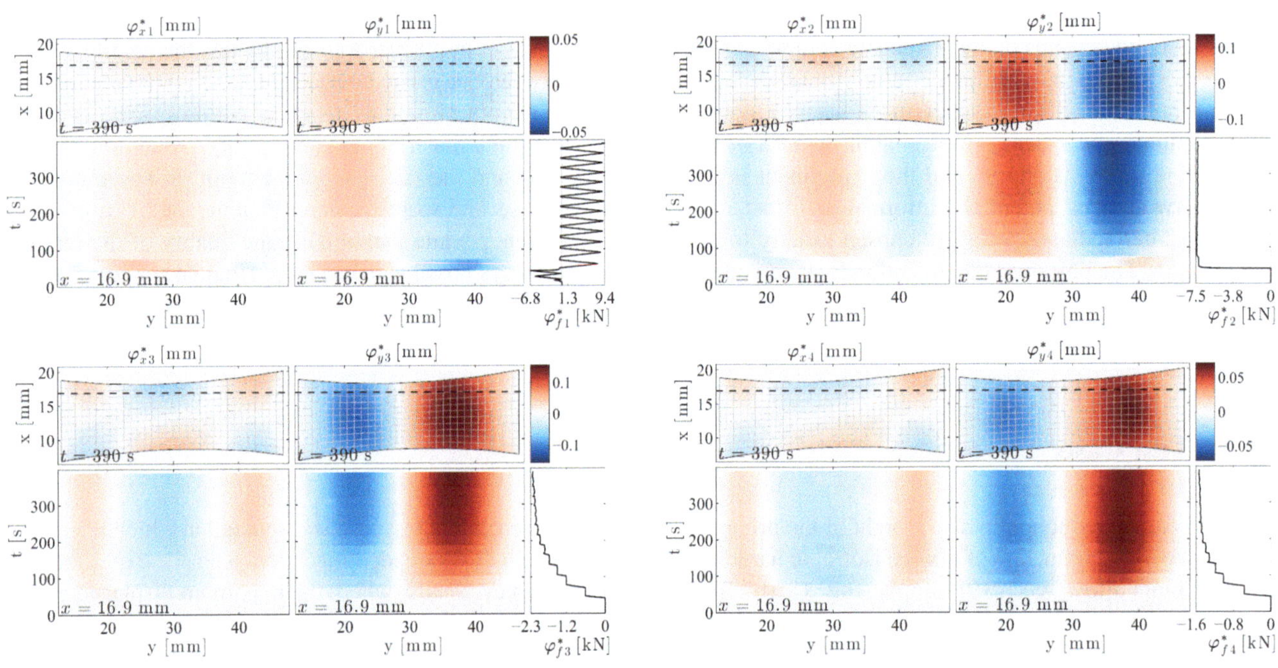

Fig. 18.2 Sensitivity fields for all four Voce material parameters

The original sensitivity fields, expressed at the pixel level, contain approximately 2×10^9 values, when written at the nodes, the number of values is reduced to 1.2×10^6, and when four singular modes are considered the required data footprint is reduced to 14×10^3. Figure 18.3 shows that the difference between nodal sensitivity fields, and their representative fields reconstructed from a limited number of modes decreases rapidly. Consequently, using only four modes allows for a rich

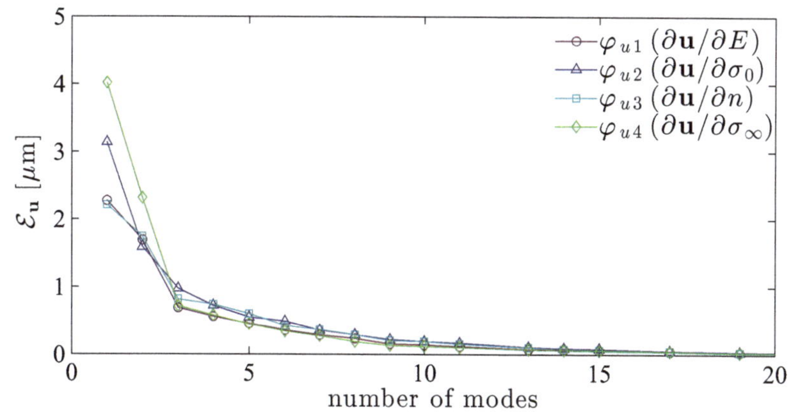

Fig. 18.3 The truncation error for reducing the sensitivity fields as a function of the number of singular modes

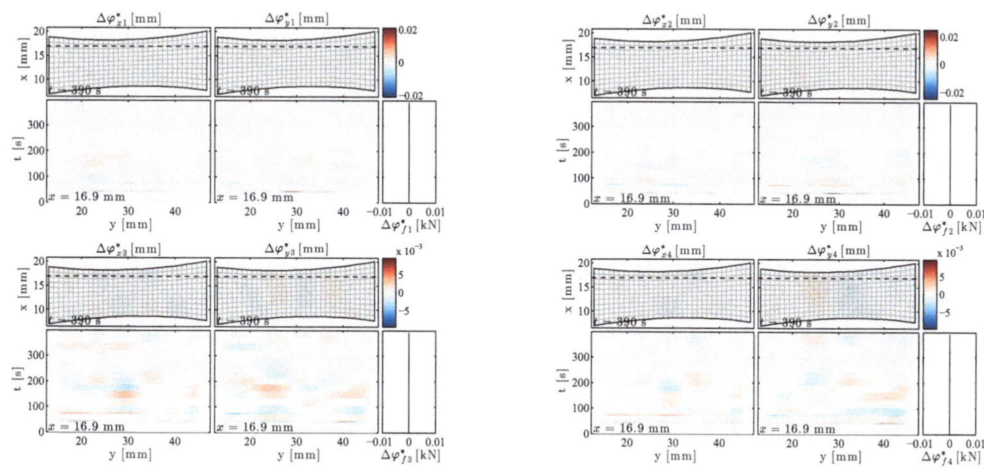

Fig. 18.4 The truncation error made by using four singular modes

enough expression of the sensitivity fields to converge close to the solution. The difference between the nodal sensitivity fields and the reduced fields is shown in Fig. 18.4. There is zero difference for the force sensitivity because the force is not included in the reduction. This is a choice justified by the fact that the force measurements were already light in their data representation.

18.4 Discussion and Conclusions

The sensitivity fields required for identification methods such as FEMU and Integrated-DIC typically have a large data footprint. Moreover, these fields are typically required very often, usually with only small variations with respect to each other. Consequently, the computational efficiency will benefit from methods that express these fields more efficiently. The Proper Orthogonal Decomposition method has proven to be highly applicable and powerful to achieve this goal. The sensitivity fields of an elasto-plastic identification routine could be expressed efficiently with only four orthogonal modes.

Any reduction in data leads to a loss of accuracy. However, the proposed model reduction method is chosen exactly to maximize the efficiency while maximizing the information content in the retained data. Importantly, the proposed method allows a gradual increase in the approximation quality by including more modes. Therefore, a strategy where the number of included modes is increased when approaching convergence is feasible and may prove to be advantageous.

Moreover, the orthogonal modes can also be used as tools, to identify where a high spatial or temporal discretization quality is required. Therefore, these modes can be applied to guide auto-meshing routines or time-stepping routines to create FE models that only have a refined mesh where needed for identification purposes.

References

1. Avril, S., Bonnet, M., Bretelle, A.S., Grédiac, M., Hild, F., Ienny, P., Latourte, F., Lemosse, D., Pagano, S., Pagnacco, E., et al.: Overview of identification methods of mechanical parameters based on full-field measurements. Exp. Mech. **48**(4), 381–402 (2008)
2. Mottershead, J.E., Link, M., Friswell, M.I.: The sensitivity method in finite element model updating: a tutorial. Mech. Syst. Signal Proc. **25**(7), 2275–2296 (2011)
3. Passieux, J.C., Bugarin, F., David, C., Périé, J.N., Robert, L.: Multiscale displacement field measurement using digital image correlation: application to the identification of elastic properties. Exp. Mech. doi:10.1007/s11340-014-9872-4 (2014)
4. Réthoré, J., Muhibullah, X., Elguedj, T., Coret, M., Chaudet, P., Combescure, A.: Robust identification of elasto-plastic constitutive law parameters from digital images using 3D kinematics. Int. J. Solids Struct. **50**(1), 73–85 (2013)
5. Mathieu, F., Leclerc, H., Hild, F., Roux, S.: Estimation of elastoplastic parameters via weighted FEMU and integrated-DIC. Exp. Mech. doi:10.1007/s11340-014-9888-9 (2014)
6. Besnard, G., Leclerc, H., Hild, F., Roux, S., Swiergiel, N.: Analysis of image series through global digital image correlation. J. Strain Anal. Eng. Des. **47**(4), 214–228 (2012)
7. Voce, E.: A practical strain hardening function. Metallurgia **51**, 219 (1955)

Chapter 19
Parameter Identification of Nonlinear Viscoelastic Material Model Using Finite Element-Based Inverse Analysis

Salah U. Hamim and Raman P. Singh

Abstract This study focuses on identifying the parameters of a nonlinear viscoelastic model from Berkovich nanoindentation experiment of an epoxy polymer using finite element-based inverse analysis approach. Instead of traditional approach of online optimization of model parameters, where finite element computation is placed inside of the optimization algorithm, this study utilizes a surrogate or meta-modeling approach. The surrogate model, which is based on Proper Orthogonal Decomposition (POD) and Radial Basis Function (RBF), is trained with finite element load–displacement data obtained by varying the different model parameters in a parameter space. Once trained POD–RBF based surrogate model is used to approximate the nanoindentation simulation data inside a multi-objective Genetic Algorithm. Current efforts are focused to validate identified parameter set of nonlinear viscoelastic model for different experimental conditions (e.g. maximum load, loading/unloading rate).

Keywords Taguchi orthogonal array • Nonlinear viscoelastic model • Finite element analysis • Radial basis function • Proper orthogonal decomposition

19.1 Introduction

Polymer materials have found applications in a wide variety of industries in the last few decades e.g. automotive, aerospace, packaging, and microelectronics. Unlike most materials polymer exhibit time-dependent mechanical response. Due to the inherent viscoelastic or viscoplastic behavior, understanding long-term mechanical response of these materials has been a challenge.

In addition to that, these materials are often used in micro- or nano-scale applications, e.g. thin films. Conventional testing methods, which can only provide the macro-scale mechanical behavior, are not suitable in characterizing nano- or micro-scale behavior of these materials [1, 2]. If a material system is non-homogeneous, such as ultraviolet irradiated polymer surface or nanofiller reinforced polymer, macro-scale test data fails to reflect the localized changes in a material [3]. In these situations nanoindentation or depth sensing indentation (DSI) can provide nano-scale mechanical behavior due to its high spatial resolution [4].

However, relating nanoindentation load–displacement data to mechanical properties requires suitable analytical or numerical methods [5, 6]. For materials exhibiting simple elastic or elastoplastic behavior, use of nanoindentation has been widely reported [7–10]. On the contrary, for materials exhibiting time-dependent mechanical behavior, the application of nanoindentation is still a challenge [11].

In this study, model parameters of a nonlinear viscoelastic model has been identified using finite element-based inverse analysis and a global optimization technique known as genetic algorithm (GA). Calibrating a complex mechanical constitutive relationship with the help of genetic algorithm is computationally expensive when finite element analysis is placed within the optimization algorithm. To circumvent this issue, a surrogate model trained with finite element data is used within the optimization loop. The surrogate model has been created using Proper Orthogonal Decomposition (POD) and Radial Basis Function (RBF) technique. The number of training points for the surrogate model has been reduced by utilizing the sensitivity of individual model parameters.

S.U. Hamim (✉) • R.P. Singh
School of Mechanical and Aerospace Engineering, Oklahoma State University, Stillwater, OK 74075, USA
e-mail: raman.singh@okstate.edu

S. Quinn, X. Balandraud (eds.), *Residual Stress, Thermomechanics & Infrared Imaging, Hybrid Techniques and Inverse Problems, Volume 9*, Conference Proceedings of the Society for Experimental Mechanics Series, DOI 10.1007/978-3-319-42255-8_19

19.2 Experimental Details

19.2.1 Materials

An epoxy polymer, named EPON 862, was selected for carrying out nanoindentation experiments. EPON 862 is a diglycidyl ether of bisphenol F (DGEBF). The curing agent used for this resin system was a moderately reactive, low viscosity aliphatic amine curing agent, called Epikure 3274. Both of these chemicals were supplied by Miller-Stephenson Chemical Company, Inc., Dunbury, Connecticut.

Epoxy and hardener was mixed at 100:40 weight ratio and hand-mixed using a glass-rod for 5–10 min. The mixture was then degassed for around 10–20 min to remove any entrapped air bubbles. The mixture was then poured into an aluminum mold and cured at room temperature for 24 h and subsequently post-cured at 121 °C for 6 h. The final sample was cut from the prepared epoxy plate using a bandsaw. Sample surface preparation was carried out by polishing using standard metallographic techniques.

19.2.2 Nanoindentation

Nanoindentation experiments were conducted on an MTS Nanoindenter XP (Agilent Technologies, Santa Clara, CA, USA) using a load-controlled scheme with a Berkovich tip. The maximum load was set to be 1.0 mN for the experiments. A triangular loading profile was chosen with 30, 45, 60, and 240 s durations. The durations were kept constant for both the loading and unloading segments.

Before conducting the actual experiments the Berkovich tip was calibrated using a fused silica reference material. Also, the acceptable thermal drift rate was chosen to be 0.15 nm/s. After ensuring that the thermal drift rate was stable and below the target drift rate nanoindentation experiments were carried out.

19.2.3 Material Model

In this study, a spring–dashpot model developed by Marin and Pao [12] was used. In linear case this model is generally called four-parameter Burgers model [13] and it is formed by a serial connection of a Maxwell element to a Voigt element. For an increased relaxation spectrum, the viscoelastic response can be modeled by increasing the number of Voigt elements.

The nonlinear characteristic is introduced when the dashpot constants (m_s and m_t) take values other than unity. In the three-dimensional model, the total strains are calculated as the summation of the elastic (ε^e), transient creep (ε^t), and steady creep strains (ε^s) [14]. In this study, the nonlinear creep deformation is assumed to be incompressible. Under these assumptions, the three-dimensional nonlinearly viscoelastic law can be expressed as:

$$\varepsilon_{ij}^{e} = \frac{1+\nu}{E}\sigma_{ij} - \frac{\nu}{E}\sigma_{kk}\delta_{ij} \tag{19.1}$$

$$\dot{\varepsilon}_{ij}^{s} = C_s J_2^{m_s}(t) s_{ij}(t) \tag{19.2}$$

$$\dot{\varepsilon}_{ij}^{t} + \frac{\varepsilon_{ij}^{t}}{t_\varepsilon} = \frac{C_t}{t_\varepsilon} J_2^{m_t}(t) s_{ij}(t) \tag{19.3}$$

where E, ν are the Young's modulus and Poisson ratio, respectively; J_2 is the second invariant of the deviator stress tensor s; C_s, C_t, m_s, m_t, t_ε are the nonlinear material parameters. σ is the Cauchy stress tensor; i, j are the indices ranging among 1–3. δ_{ij} is the Kronecker delta which used in the context of summation convention with the well-known property $\delta_{ij} = 1$ when i = j and $\delta_{ij} = 0$ otherwise. Small deformations are assumed in the formulation. When more than one Voigt element is included in the model, the total strain components can be given as the sum of elastic, steady creep, and transient creep components for all Voigt elements,

$$\varepsilon_{ij} = \varepsilon_{ij}^{e} + \varepsilon_{ij}^{s} + \varepsilon_{ij}^{t} = \varepsilon_{ij}^{e} + \varepsilon_{ij}^{s} + \sum_{i=1}^{n} \varepsilon_{ij}^{t_i} \tag{19.4}$$

where n is the number of Voigt elements. Equations (19.2) and (19.3) can also be written in integral form:

$$\varepsilon_{ij}^{s} = C_s \int_0^t J_2^{m_s}(t') s_{ij}(t') dt' \tag{19.5}$$

$$\varepsilon_{ij}^{t} = \frac{C_t}{t_\varepsilon} e^{(-t/t_\varepsilon)} \int_0^t J_2^{m_t}(t') s_{ij}(t') e^{(-t'/t_\varepsilon)} dt' \tag{19.6}$$

An UMAT was developed in order to implement the nonlinear Burgers model. UMAT requires the tangent stiffness matrix of the material model for finite element calculations. For implementation of the nonlinear Burgers viscoelastic model, the UMAT required temporal discretization. This was done following the procedure implemented by Kucuk et al. [15, 16]. A simple, stable integration operator for these equations is the central difference operator:

$$\dot{f}_{t+\frac{1}{2}\Delta t} = \frac{\Delta f}{\Delta t}, f_{t+\frac{1}{2}\Delta t} = f_t + \frac{\Delta f}{2} \tag{19.7}$$

where f is a function, f_t is its value at the beginning of the increment, Δf is the change in the function over the increment, and Δt is the time increment.

Tangent stiffness matrix $\delta \Delta \sigma / \delta \Delta \varepsilon$ of the constitutive model, with $\Delta \sigma$ being the stress increments and $\Delta \varepsilon$ the strain increments, can be derived by applying central difference operator to the rate-dependent constitutive equations (Eqs. (19.1)–(19.3)). The compliance matrices obtained for these three equations are given below–

$$C = \begin{bmatrix} 1/E & -\nu/E & -\nu/E & 0 & 0 & 0 \\ & 1/E & -\nu/E & 0 & 0 & 0 \\ & & 1/E & 0 & 0 & 0 \\ & & & \frac{1+\nu}{E} & 0 & 0 \\ & & & & \frac{1+\nu}{E} & 0 \\ & & & & & \frac{1+\nu}{E} \end{bmatrix}_{symmetric} \tag{19.8}$$

$$C = \Delta t C_s J_2^{m_s}(t) \begin{bmatrix} 1/3 & 0 & 0 & 0 & 0 & 0 \\ & 1/3 & 0 & 0 & 0 & 0 \\ & & 1/3 & 0 & 0 & 0 \\ & & & 1/2 & 0 & 0 \\ & & & & 1/2 & 0 \\ & & & & & 1/2 \end{bmatrix}_{symmetric} \tag{19.9}$$

$$C = \frac{\Delta t}{2t_\varepsilon + \Delta t} C_t J_2^{m_t}(t) \begin{bmatrix} 2/3 & 0 & 0 & 0 & 0 & 0 \\ & 2/3 & 0 & 0 & 0 & 0 \\ & & 2/3 & 0 & 0 & 0 \\ & & & 1 & 0 & 0 \\ & & & & 1 & 0 \\ & & & & & 1 \end{bmatrix}_{symmetric} \tag{19.10}$$

From Eq. (19.4), the total compliance is now

$$\frac{\delta \Delta \varepsilon_{ij}}{\delta \Delta \sigma_{kl}} = \frac{\delta \Delta \varepsilon_{ij}^e}{\delta \Delta \sigma_{kl}} + \frac{\delta \Delta \varepsilon_{ij}^s}{\delta \Delta \sigma_{kl}} + \frac{\delta \Delta \varepsilon_{ij}^t}{\delta \Delta \sigma_{kl}} \tag{19.11}$$

By investigating the total compliance matrix, system tangent stiffness matrix (Jacobian matrix) $\frac{\delta \Delta \sigma_{ij}}{\delta \Delta \varepsilon_{kl}}$ can be obtained from Eq. (19.11). It should be noted that the Jacobian matrix in Eq. (19.11) accounts only for the elastic deformation and creep deformation caused by load or stress increment. The rest of the creep strain is developed over the time period during the time increment and controlled by the applied stress. An artificial stress increment is introduced to include this creep strain in the system equation. This part of creep strain can be extracted as

$$\Delta \varepsilon' = \Delta t C_t J_2^{m_t}(t) s_{ij}(t) + \frac{1}{2t_\varepsilon + \Delta t}\left(2\Delta t C_t J_2^{m_t}(t) s_{ij}(t) - 2\Delta t \varepsilon^t\right) \tag{19.12}$$

A stress increment $\Delta \sigma' = C \Delta \varepsilon'$ is then added into the system equation to account for the creep strain in Eq. (19.12), with C being the Jacobian stiffness matrix calculated from Eq. (19.11).

19.2.4 Finite Element Modeling

The 3D finite element model of nanoindentation experiment was constructed using commercial finite element package ABAQUS (Dassault Systemes, Providence, RI). The Berkovich indenter was modeled as discrete rigid body, while the sample was modeled as deformable body.

To ensure accuracy of the simulation results, the sample was modeled with finer mesh near the contact area, where the stress and strain generated was much higher. The contact between the indenter and the sample was defined as surface-to-surface contact, where the indenter was designated as *master surface* and the sample was as designated as *slave surface*. The contact was assumed to have sliding friction with a friction coefficient of 0.25. The element types for the sample was chosen from the eight-node brick element family (C3D8). Material behavior of the sample was defined in the model using a subroutine (UMAT). Figure 19.1 shows the schematic of the finite element nanoindentation experiment model.

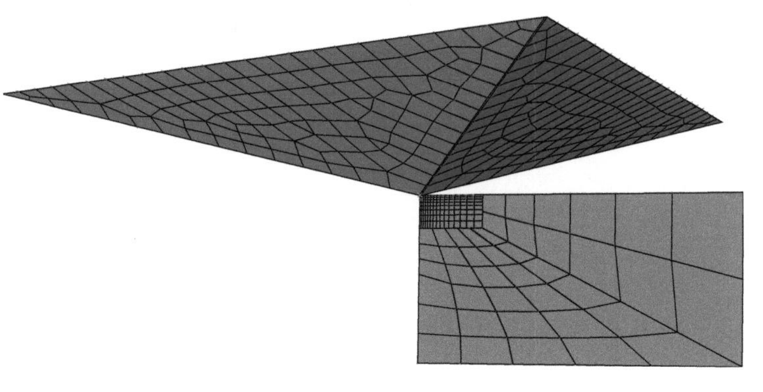

Fig. 19.1 Schematic of finite element model of Berkovich nanoindentation

Table 19.1 Levels of nonlinear Burgers model parameters

Parameters	Level 1	Level 2	Level 3
E	3	3.25	3.5
C_s	0.02	0.06	0.1
m_s	0.15	0.25	0.35
C_t	0.15	0.25	0.35
m_t	0.2	0.5	0.8
t_ε	0.1	0.25	0.4

19.2.5 Design of Experiments for Sensitivity Analysis

Before generating finite element simulation data by varying the model parameters, a sensitivity study of the parameters was conducted. This information helped to reduce the number of finite element simulation that was used to train a POD–RBF based surrogate model [17].

The nonlinear Burgers model that was chosen to represent the behavior of the epoxy has seven independent parameters. These parameters are E, ν, C_s, m_s, C_t, m_t, and t_ε. It is already known that a nanoindentation load–displacement response is not highly influenced by Poisson's ratio, ν [18–20]. Therefore, in order to keep the number of independent parameters to a minimum, ν was given a constant value of 0.34, and was not included in the sensitivity analysis scheme.

Sensitivity analysis was carried out using Analysis of Variance (ANOVA) technique. The data required for ANOVA was generated using the Taguchi Design of Experiments (DOE) method. In this study, the six nonlinear model parameters were varied in three equidistant levels. A statistical software, Minitab (Minitab Inc., State College, PA, USA) was used to design the experiments. For six parameters, where each parameters were varied in three levels, Taguchi L_{27} orthogonal array design was appropriate. Table 19.1 shows the levels of the six individual parameters of the nonlinear Burgers model. The experimental design required a total of 27 individual computer experiments. In these 27 experiments, except for material model parameters every other parameters (e.g. load, loading–unloading time, boundary conditions) were kept same.

Each of these 27 computer simulations resulted in data in terms of indenter displacement. The resulting value of error function, δ was calculated using the Eq. (19.13). This was then utilized in ANOVA to determine the effect of each parameters on the error function.

$$\delta = \frac{1}{n}\sum\left[\left(h_{exp}^{i} - h_{sim}^{i}\right)^{2}\right] \tag{19.13}$$

In Eq. (19.13), $i = 1, 2, 3, \ldots, n$, and n is the number of data points in a single nanoindentation simulation or experiment.

19.2.6 POD–RBF Surrogate Model

The POD theory, also known as PCA, was developed to approximate a function over some domain of interest based on the known relationships between the input and the output [21–23]. This study followed the POD–RBF procedure outlined by Buljak [24] and Rogers et al. [25]. POD–RBF method requires creating snapshots (input–output relationships of the system) from which the surrogate model could be established. Each of the data that provides a one-to-one relationship between the input and the output is called a snapshot. The more snapshots or training data points that could be utilized to generate the surrogate model the better the approximation becomes.

However, the computational burden associated with generating large number of snapshots becomes the limiting factor in obtaining very high-fidelity predictions from the surrogate model. Sensitivity analysis could be utilized to reduce the number of snapshots without sacrificing approximation error [17]. Hence, in this study, a similar approach was adopted to reduce the computation burden of training the surrogate model for nonlinear Burgers model.

Once the appropriate number of levels for different parameters were selected using information from sensitivity analysis, a full factorial approach was taken to generate the input parameter sets. These parameter sets were combined to produce the input matrix, **P**. Finite element simulation experiments were carried out for every individual parameter sets and their corresponding indenter displacement data was assimilated in the snapshot matrix, **U**, which can be defined as–

$$U = \begin{bmatrix} u_1^1 & u_1^2 & \cdots & u_1^M \\ u_2^1 & u_2^2 & \cdots & u_2^M \\ \vdots & \vdots & \vdots & \vdots \\ u_N^1 & u_N^2 & \cdots & u_N^M \end{bmatrix} \tag{19.14}$$

In this study, four different experimental conditions were utilized for which the training data would be generated. In these experimental conditions, the maximum load was kept constant at 1 mN, while the strain rate was varied from 1/30 s^{-1} to 1/240 s^{-1}. One surrogate model was created for each of those experimental conditions using finite element data. The approximations from each surrogate model was compared against their own experimental indenter displacement data to form the objective or error function.

A Multiquadratic RBF was chosen for this study. Hamim and Singh reported that the value of the shape parameter (c_j) did not influence the POD–MQ RBF surrogate model's performance significantly [17]. So, for this study it was chosen 0.5.

19.2.7 Genetic Algorithm

A multi-objective genetic algorithm-based optimization procedure was used to identify the parameters of the nonlinear Burgers model. The procedure was implemented using MATLAB's (Mathworks Inc., Natick, MA, USA) global optimization toolbox.

An initial population of 200 was randomly created with a uniform distribution (*Double vector* population type). Scores of the first and all subsequent generations were determined by evaluating the fitness function that was submitted to the program via MATLAB script. *Selection* of the worthy candidates for being the next generation parent were carried out via tournament of size 2. Eighty percent of the next generation population was produced via *crossover*, while the remainder of the was created through *mutation*. Gaussian mutation was selected, where a random number from a Gaussian distribution centered on zero was added to each vector entry of an individual. *Scale* and *Shrink* parameters were set to 1 for this study.

In this study, the crossover function was chosen to be *intermediate*. Ratio = 1 was used for creating next generation children. Forward migration direction was chosen. This meant individuals from n^{th} subpopulation would replace individuals from $(n+1)^{th}$ subpopulation and so on. The migration fraction and interval were chosen to be 0.2 and 20, respectively. Total number of generations for the optimization algorithm was chosen to 100× number of parameters, i.e. 100 × 6 = 600 for this study. The fitness (error) function tolerance was chosen to be $1e^{-4}$.

19.3 Results and Discussion

19.3.1 Sensitivity Analysis

Table 19.2 shows the result of sensitivity analysis carried out using Taguchi-based design of experiments. The data of 27 experiments carried out according to L_{27} orthogonal array was used to get information about the sensitivity of output towards individual parameters.

Table 19.2 Analysis of Variance (ANOVA) for different parameters

Source	DF	Adj SS	Adj MS	F-Value	% Contribution
E	2	5597,885,401	2798,942,700	3630.33	11.20
C_s	2	16,004,929,654	8002,464,827	10,379.48	32.01
m_s	2	22,166,899,947	11,083,449,973	14,375.63	44.34
C_t	2	6207,522,908	3103,761,454	4025.69	12.42
m_t	2	14,961,103	7480,552	9.70	0.03
t_e	2	92,652	46,326	0.06	0.00
Error	14	10,793,843	770,989		
Total	26	50,003,085,508			

The '% Contribution' data, which is a measure of variation contributed by individual parameters towards the output, shows that except for t_ε all other parameters contributed towards the overall variation of output. However, the contribution was significantly influenced by the 'steady state' parameters (C_s and m_s).

19.3.2 Surrogate Model Training and Inverse Analysis

The findings from the sensitivity analysis was taken into account to revise the number of levels for each nonlinear Burgers model parameter. As discussed, t_ε showed no influence over the output of the nanoindentation simulations. Thus, in order to reduce computational expense, t_ε was given a constant value.

The two parameters that were the most influential of the remaining five, C_s and m_s, were varied at four levels. Meanwhile, moderately influential two parameters, E and C_t, were varied at three levels, while m_t was varied in two levels.

Table 19.3 shows the corresponding levels for each parameters that were selected based on the sensitivity analysis. In a full factorial basis, a total of $3 \times 4 \times 4 \times 3 \times 2 \times 1 = 288$ finite element simulations were carried out in order to generate the surrogate model for every single experimental conditions. In each of these simulations, 100 load–displacement data points were used to represent the nanoindentation plot. Since there were four individual experimental conditions to represent, a total of four surrogate models were developed. The snapshot matrix used to generate each of these surrogate model had dimensions of 100×288.

After the POD model reduction process was carried out and the RBF coefficients were calculated, the POD–RBF surrogate model was ready to approximate nanoindentation data within the specified parametric space (Table 19.1). An objective function was written in MATLAB where each surrogate model's output was compared against the corresponding experimental data. This objective function was used within the MATLAB Global Optimization Toolbox to run multi-objective genetic algorithm-based global optimization. The optimization algorithm was set to run in parallel mode until it met the stopping criteria. Table 19.4 shows the result from the global optimization algorithm.

The optimized set of parameters were the numerical best fits depending on the objective function that produces the numerical difference between the predicted and experimental data. Figure 19.2 shows the comparison of predicted and experimental data for all four experimental cases. These were the experimental conditions that were closely followed in creating finite element models and were used to train the predictive or surrogate model. From Fig. 19.2 it can be seen that all four surrogate model outputs were very close to the corresponding experimental data. This demonstrated the fact that the multi-objective genetic algorithm-based optimization procedure was successful in finding a common minima taking the constraints in to consideration.

Although, the surrogate model prediction's were mostly close with the experimental data few inconsistencies were observed. For example, the final unloading portion data for the loading–unloading time $t = 30$ s did not match very well. Similar behavior was observed for $t = 45$ s, even though qualitatively the difference between prediction and experiment diminished. For higher loading–unloading time, e.g. $t = 60$ s and 240 s, the difference was noticeably very small.

Table 19.5 shows quantitative variation between the various plots in Fig. 19.2. Variations between the plots has been represented in terms of RMSE, R^2, Avg. Error, and % Error. As it can be seen, quantitative discrepancies for different plots

Table 19.3 Parametric space of nonlinear Burgers parameters for surrogate training

Parameters	No of points in space	Parametric value space
E	3	3, 3.25, 3.5
C_s	4	0.02, 0.045, 0.07, 0.1
m_s	4	0.35
C_t	3	0.15, 0.25, 0.35
m_t	2	0.2, 0.8
t_ε	1	0.25

Table 19.4 Optimized nonlinear Burgers model parameters

Parameters	E	ν	C_s	m_s	C_t	m_t	t_ε
Optimized	3.28	0.34	0.09	0.20	0.24	0.47	0.25

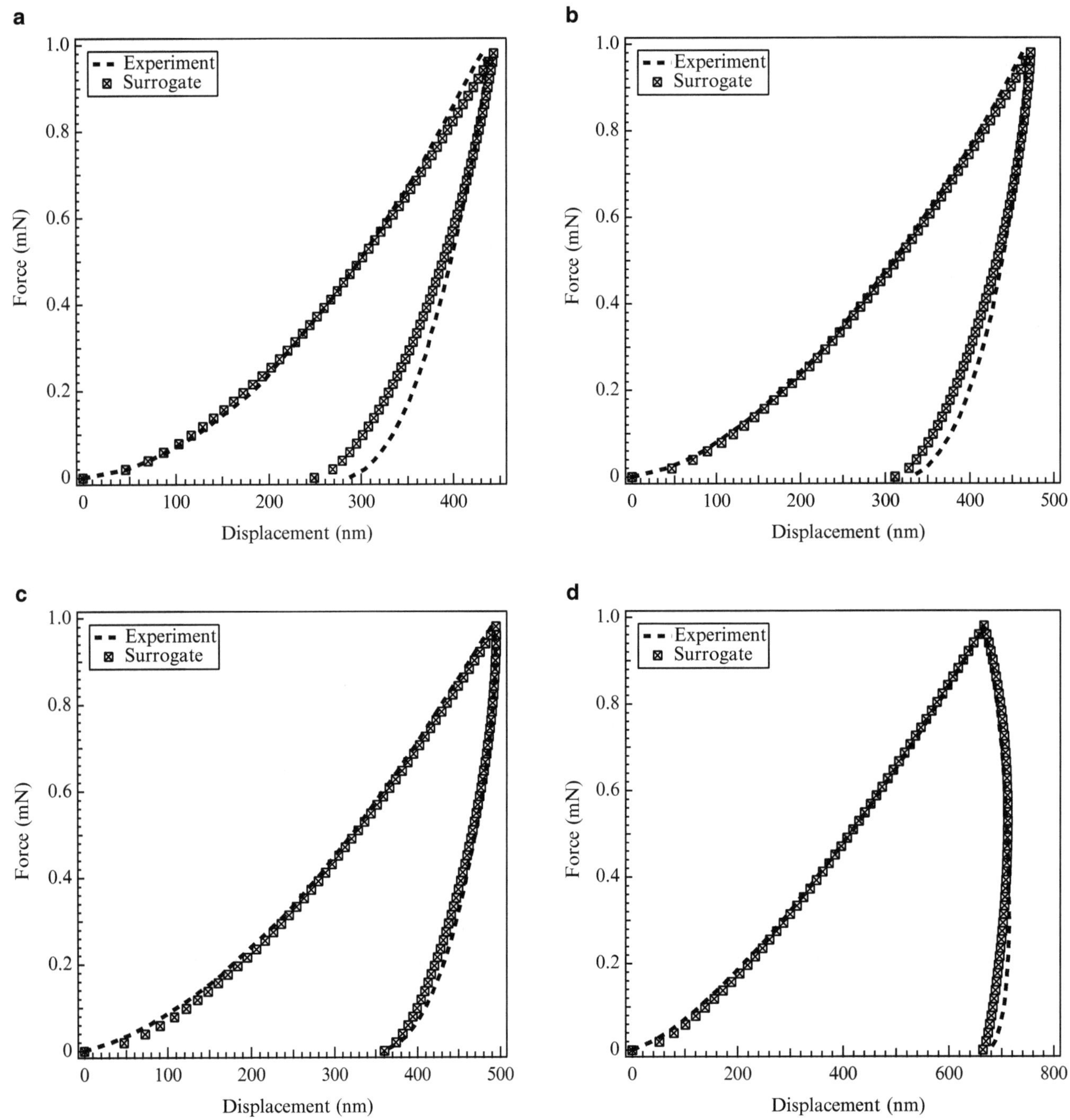

Fig. 19.2 Experiment vs. surrogate model for calibrated nonlinear Burgers model parameters. (**a**) Loading–unloading time = 30 s, (**b**) loading–unloading time = 45 s, (**c**) loading–unloading time = 60 s, (**d**) loading–unloading time = 240 s

Table 19.5 Variation between different plots

Conditions		RMSE	R^2	Avg. Err. (nm)	% Error
P_{max} = 1. 0 mN	t = 30 s	13.23	0.9821	9.11	2.81
	t = 45 s	9.17	0.9867	7.11	2.34
	t = 60 s	6.72	0.9891	5.70	2.81
	t = 240 s	9.01	0.9893	7.07	2.33

P_{max} = maximum load, t = loading–unloading time

of Fig. 19.2 were found to be comparable with each other. Another observation that could be made was that Fig. 19.2b and d both showed almost same quantitative variation. Although, Fig. 19.2d's match looked slightly better than Fig. 19.2b if perceived visually.

19.4 Conclusion

In this study, the model parameters for a nonlinear viscoelastic model has been identified using finite element based inverse analysis. A genetic algorithm-based optimization procedure was utilized where the load–displacement data was approximated using a surrogate model. A sensitivity analysis based on Taguchi method and ANOVA analysis was carried out to reduce the number of training points for the surrogate model. The POD–RBF based surrogate model was capable of producing good approximation for the nanoindentation experiment. The identified parameter set produced good match between experimental and surrogate-based load–displacement data.

Ascertaining that the material model parameter set that has been extracted from the inverse analysis procedure is indeed the global parameter set that would satisfy all possible material response is a challenge. In order to deal with this challenge, material responses from other experiments, such as compression, tension, or flexural tests could be included in the process. For some materials carrying out the aforementioned tests may not be feasible, e.g. thin films, coatings, biological cells. In those cases improving the confidence in the optimized parameter set could be established by obtaining material response data from multiple nanoindentation experiments, such as changing the cone angle for a pyramidal indenter tip, or using spherical tips with different radii.

Another way of finding additional constraints for the numerical analysis would be use additional experimental data from the same nanoindentation experiment. For example, if imprint geometry or residual depth profile data could be harnessed from a nanoindentation experiment and used in the objective function, the probability of finding the unique model parameter set increases.

An investigation is currently ongoing to validate the identified parameter set for different experimental conditions.

Acknowledgements We gratefully acknowledge that this work is funded in part or fully by a grant through the Oklahoma Nanotechnology Applications Project (ONAP) (Grant no. O9-20) and NASA Experimental Program to Stimulate Competitive Research (EPSCOR) (Grant no. NNXO9AP68A).

References

1. Hamim, S.U.A.: Variation of mechanical properties due to hygrothermal ageing and permanent changes upon redrying in clay/epoxy nanocomposites. ProQuest Dissertations and Theses, p. 49 (2011)
2. Hamim, S.U., Singh, R.P.: Effect of hygrothermal aging on the mechanical properties of fluorinated and nonfluorinated clay-epoxy nanocomposites. Int. Sch. Res. Not. **2014**, 1–13 (2014)
3. McKee, C., Last, J., Russell, P., Murphy, C.: Indentation versus tensile measurements of Young's modulus for soft biological tissues. Tissue Eng. B Rev. **17**(3), 155–164 (2011)
4. Hamim, S.U., Mishra, K., Singh, R.P.: Effect of UV exposure on mechanical properties of POSS reinforced epoxy nanocomposites. In: 2014 Annual Conference on Experimental and Applied Mechanics (2014)
5. Zhang, J., Michalenko, M.M., Kuhl, E., Ovaert, T.C.: Characterization of indentation response and stiffness reduction of bone using a continuum damage model. J. Mech. Behav. Biomed. Mater. **3**(2), 189–202 (2010)
6. Engels, P., Begau, C., Gupta, S., Schmaling, B., Ma, A., Hartmaier, A.: Multiscale modeling of nanoindentation: from atomistic to continuum models. Nanomechanical Analysis of High Performance Materials, pp. 285–322. Springer, Netherlands (2013)
7. Doerner, M., Nix, W.: A method for interpreting the data from depth-sensing indentation instruments. J. Mater. Res. **1**(04), 601–609 (1986)
8. Oliver, W., Pharr, G.: An improved technique for determining hardness and elastic modulus using load and displacement sensing indentation experiments. J. Mater. Res. **7**, 1564–1583 (1992)
9. Li, X., Bhushan, B.: A review of nanoindentation continuous stiffness measurement technique and its applications. Mater. Charact. **48**(1), 11–36 (2002)
10. Fischer-Cripps, A.: Nanoindentation. Springer, New York (2004)
11. Ngan, A.: Nanomechanical characterization of soft materials. In: Tiwari, A. (ed.) Nanomechanical Analysis of High Performance Materials. Springer, Netherlands (2014)
12. Marin, J., Pao, Y.: An analytical theory of the creep deformation of materials. J. Appl. Mech. **20**, 245–252 (1953)
13. Richter, H., Misawa, E., Lucca, D., Lu, H.: Modeling nonlinear behavior in a piezoelectric actuator. Precis. Eng. **25**(2), 128–137 (2001)
14. Shames, I., Cozzarelli, F.: Elastic and Inelastic Stress Analysis. Taylor and Francis, Washington (1997)

15. Kucuk, Y.: Simulation of non-linear viscoelastic behavior of cross-linked mesoporous silica aerogels by depth-sensing indentation. Indian J. Eng. Mater. Sci. **19**(4), 260–268 (2012)
16. Kucuk, Y., Mollamahmutoglu, C., Wang, Y., Lu, H.: nonlinearly viscoelastic nanoindentation of PMMA under a spherical tip. Exp. Mech. **53** (5), 731–742 (2012)
17. Hamim, S.U., Singh, R.P.: Taguchi-based design of experiments in training POD-RBF surrogate model for inverse material modeling using nanoindentation. Inverse Prob. Sci. Eng. (2016). doi:0.1080/17415977.2016.1161036
18. Magnenet, V., Giraud, A., Homand, F.: Parameter sensitivity analysis for a Drücker–Prager model following from numerical simulations of indentation tests. Comput. Mater. Sci. **44**(2), 385–391 (2008)
19. Ma, Y., Zhang, Y., Yu, H.-F., Zhang, X.-Y., Shu, X.-F., Tang, B.: Plastic characterization of metals by combining nanoindentation test and finite element simulation. Trans. Nonferrous Metals Soc. China **23**(8), 2368–2373 (2013)
20. Clément, P., Meille, S., Chevalier, J., Olagnon, C.: Mechanical characterization of highly porous inorganic solids materials by instrumented micro-indentation. Acta Mater. **61**(18), 6649–6660 (2013)
21. Chatterjee, A.: An introduction to the proper orthogonal decomposition. Curr. Sci. **78**(7), 808–817 (2000)
22. Liang, Y.C., Lee, H.P., Lim, S.P., Lin, W.Z., Lee, K.H., Wu, C.G.: Proper orthogonal decomposition and its applications - Part I: theory. J. Sound Vib. **252**(3), 527–544 (2002)
23. Ly, H., Tran, H.: Modeling and control of physical processes using proper orthogonal decomposition. Math. Comput. Model. **33**(1–3), 223–236 (2001)
24. Buljak, V.: Inverse Analyses with Model Reduction. Computational Fluid and Solid Mechanics. Springer, Berlin/Heidelberg (2012)
25. Rogers, C., Kassab, A., Divo, E., Ostrowski, Z., Bialecki, R.: An inverse pod-rbf network approach to parameter estimation in mechanics. Inverse Prob. Sci. Eng. **20**, 749–767 (2012)

Chapter 20
Stiffness Heterogeneity of Multiply Paperboard Examined with VFM

Anton Hagman, J.M. Considine, and Mikael Nygårds

Abstract Mechanical heterogeneity of a multiply paperboard was characterized in uniaxial tension using DIC and VFM. The specimen was divided into three subregions based on axial strain magnitude. VFM analysis showed that the subregions had stiffnesses and Poisson's ratio's that varied in a monotonically decreasing fashion, but with the stiffness differences between subregions increasing with applied tensile stress. An Equilibrium Gap analysis showed improved local equilibrium when comparing a homogeneous analysis with the subregion analysis. Although only a single specimen was examined, results suggest that high stiffness regions provide only marginal improvement of mechanical behavior. The analysis also showed that even though the subregions themselves were non-contiguous, their mechanical behavior was similar.

Keywords VFM • DIC • Stiffness • Paperboard • Heterogeneity

20.1 Introduction

Characterization of paperboard mechanical heterogeneity will help provide papermakers relevant information for control during production. Currently, variation of mechanical properties may not be manifested until converting operations or consumer use. The objective of this work is to demonstrate the ability of different Virtual Fields Method (VFM) formulations for characterization of paperboard stiffness heterogeneity.

Current state-of-the-art is to measure mechanical heterogeneity on a macroscale by performing many tests. This work takes a different approach by characterizing heterogeneity on local scale O(cm). Successful characterization could lead to reduced fiber requirements to meet property specifications and reduced drying expenses, which represent the two most costly components in the papermaking process.

Strain and deformation heterogeneity in paper and paperboard have been observed by many researchers. Strains in tensile specimens have been found to vary from near zero to almost three times the mean macroscopic strain in the specimen [1, 2]. Poor formation (high local grammage variation) increases strain heterogeneity [3, 4] of uniaxial tensile specimens. Korteoja et al. [5] found that a direct relationship between formation and tensile strength is unlikely, due to competing influences of non-uniform intra-fiber bonding, varying fiber density (along fiber length), uneven planar distribution of fines (very short fibers, < 1 mm long), uneven z-direction distribution of fillers, and non-oriented drying strains. In subsequent work [6], they performed 4000+ tensile tests and determined that $max(E)/mean(E)$ ranged from 1.2 to 3.1 and $min(E)/mean(E)$ ranged from 0.1 to 0.7, where E is initial elastic modulus. Ostoja-Starzewski and Castro [7] cross-correlated formation images with the strain measured during biaxial tests of two paper materials and concluded that the representative volume element, a parameter used to describe heterogeneous materials, was on the order of 10 times the floc size. Yamauchi and Murakami used thermography to examine well-formed and poorly formed papers [8]; the poorly formed papers had non-uniform temperature maps quite early during tensile testing.

This work uses different VFM [9] analyses on previously published work [10] where digital image correlation (DIC) was used to examined a five-ply paperboard in uniaxial tension. DIC provides spatially dense data with excellent strain resolution, both of which are important in characterization of heterogeneous materials. VFM is fast, is simple to program,

A. Hagman
Graduate Student, KTH and BiMaC Innovation, Teknikringen 8, 100-44 Stockholm, Sweden

J.M. Considine (✉)
Materials Research Engineer, USDA Forest Service, Forest Products Laboratory,
One Gifford Pinchot Drive, Madison WI 53726, USA
e-mail: jconsidine@wisc.edu

M. Nygårds
Research Manager, Innventia and BiMaC Innovation, Box 5604, 114-86 Stockholm, Sweden

S. Quinn, X. Balandraud (eds.), *Residual Stress, Thermomechanics & Infrared Imaging, Hybrid Techniques and Inverse Problems, Volume 9*, Conference Proceedings of the Society for Experimental Mechanics Series, DOI 10.1007/978-3-319-42255-8_20

Table 20.1 DIC system components and parameters

Technique	Stereo image correlation
Cameras	VDS Vosskühler (Osnabrück, Germany) Model CCD-1300QB
Lens	Schneider Kreuznach (Hauppauge, NY) 2.8/50, 16 mm
Sensor/digitization	1280 × 1024, 8-bit
Camera noise (% of range)	0.5 %
Lighting	Direct white light
Lens/Imaging distance	5 mm, 70 cm to specimen, 30 cm between cameras
Pixel to mm conversion	1 pixel = 0.09 mm
Software	Aramis 6.1
ROI	112 mm × 89 mm
Subset, step	15, 13
Shape function, accuracy, residual	Bilinear, 0.04 pixels, 20 gray levels
Displacement resolution	6.8 μm
Strain window	3 × 3
Strain resolution	190 $\mu\varepsilon$

is capable of multiple parameter assessment from a single test, and requires no FEM code. VFM has been successfully employed to determine anisotropic plate bending stiffnesses [11] and through-thickness rigidities of thick composite tubes [12].

20.2 Material and Testing

Material was board 'B' from previous work [10] and was a five-ply paperboard, thickness 0.35 mm, grammage 220 g/m^2, and density 628 kg/m^3. Although tests in both principal material directions were reported in that earlier work, only a 2-direction test was analyzed here.

A single tensile test was examined on a specimen 50 mm wide × 100 mm long that was tested using line clamps. Test was performed on an Instron® (Instron, Norwood, MA, USA) test machine at a grip elongation of 2.5 mm/s. A speckle pattern was laser printed on the specimen and was examined using an Aramis® (GOM mbH, Braunschweig Germany) stereo DIC system. Details of the DIC analysis are listed in Table 20.1. The test environment was controlled at 23 °C, 50 % RH.

20.3 Virtual Fields Method

VFM was used to identify stiffness behavior during tensile testing. Pierron and Grédiac [9] provide a thorough development of VFM; its major points are abbreviated here. For a plane stress problem, the Principle of Virtual Work can be written as

$$\int_S (\sigma_1 \varepsilon_1^* + \sigma_2 \varepsilon_2^* + \sigma_6 \varepsilon_6^*) dS = \int_{L_f} \overline{T}_i u_i^* dl \qquad (20.1)$$

where S is the area of 2-D domain, σ_i are stresses within S, u_i^* are kinematically admissible virtual displacements, ε_i^* are virtual strains associated with u_i^*, $\overline{T}_i$ are tractions applied on boundary of S, and L_f is the portion of S over which $\overline{T}_i$ are applied.

Assuming a homogeneous linear elastic isotropic material, the constitutive equation, using contracted index notation, is given by

$$\begin{Bmatrix} \sigma_1 \\ \sigma_2 \\ \sigma_6 \end{Bmatrix} = \begin{bmatrix} Q & \nu Q_{12} & 0 \\ \nu Q & Q & 0 \\ 0 & 0 & Q\left(\frac{1-\nu}{2}\right) \end{bmatrix} \begin{Bmatrix} \varepsilon_1 \\ \varepsilon_2 \\ \varepsilon_6 \end{Bmatrix} \tag{20.2}$$

Substituting Eq. (20.2) into Eq. (20.1) gives

$$\int_S Q\left[\left(\varepsilon_1\varepsilon_1^* + \varepsilon_2\varepsilon_2^* + \frac{1}{2}\varepsilon_6\varepsilon_6^*\right) + \nu\left(\varepsilon_1\varepsilon_2^* + \varepsilon_2\varepsilon_1^* - \frac{1}{2}\varepsilon_6\varepsilon_6^*\right)\right]dS = \int_{L_f} \overline{T}_1 u_1^* dl \tag{20.3}$$

Equation (20.3) contains two unknowns, Q, ν, information that is measured in DIC, ε_1, ε_2, ε_6, and information provided by a load cell, $\overline{T}_1$. The virtual displacement, u_1^*, and virtual strains, ε_1^*, ε_2^*, ε_6^*, are terms that are user defined and whose only limitation is kinematic admissibility defined by the test configuration. For instance, virtual displacement at a stationary grip must be 0. In practice, Eq. (20.3) is discretized by converting the integrals to summations.

While infinitely many kinematically admissible virtual displacements are possible, optimized piecewise virtual fields are often used because specification of boundary conditions is straight-forward and direct solution of Eq. (20.3) is possible, e.g. solving for Q by using virtual fields to specify $\varepsilon_1\varepsilon_1^* + \varepsilon_2\varepsilon_2^* + \frac{1}{2}\varepsilon_6\varepsilon_6^* = 1$ and $\varepsilon_1\varepsilon_2^* + \varepsilon_2\varepsilon_1^* - \frac{1}{2}\varepsilon_6\varepsilon_6^* = 0$, and vice versa to determine ν. An additional advantage of using optimized piecewise virtual fields is the calculation of η, a parameter that represents a type of deviation of the fit of Q or ν. The reader is referred to Pierron and Grédiac [9] for calculation of η.

The virtual mesh used to identify Q and ν is shown in Fig. 20.1a, showing four elements across specimen width and six elements along its length. This mesh density (4 × 6) was used for both the homogeneous and heterogeneous analysis. General practice is to have 2× the # of unconstrained nodes as the number of unknown stiffnesses.

Here, heterogeneity will be examined by dividing the specimen into different subregions. Each subregion is assumed to be homogeneous within itself. Dividing S into three subregions, $S^{(1)}$, $S^{(2)}$, and $S^{(3)}$ gives a new VFM equation as:

$$\begin{aligned} &\int_{S^{(1)}} Q^{(1)}\left[\left(\varepsilon_1\varepsilon_1^* + \varepsilon_2\varepsilon_2^* + \frac{1}{2}\varepsilon_6\varepsilon_6^*\right) + \nu^{(1)}\left(\varepsilon_1\varepsilon_2^* + \varepsilon_2\varepsilon_1^* - \frac{1}{2}\varepsilon_6\varepsilon_6^*\right)\right]dS^{(1)} + \cdots \\ &\int_{S^{(2)}} Q^{(2)}\left[\left(\varepsilon_1\varepsilon_1^* + \varepsilon_2\varepsilon_2^* + \frac{1}{2}\varepsilon_6\varepsilon_6^*\right) + \nu^{(2)}\left(\varepsilon_1\varepsilon_2^* + \varepsilon_2\varepsilon_1^* - \frac{1}{2}\varepsilon_6\varepsilon_6^*\right)\right]dS^{(2)} + \cdots \\ &\int_{S^{(3)}} Q^{(3)}\left[\left(\varepsilon_1\varepsilon_1^* + \varepsilon_2\varepsilon_2^* + \frac{1}{2}\varepsilon_6\varepsilon_6^*\right) + \nu^{(3)}\left(\varepsilon_1\varepsilon_2^* + \varepsilon_2\varepsilon_1^* - \frac{1}{2}\varepsilon_6\varepsilon_6^*\right)\right]dS^{(3)} + \cdots \\ &\cdots = \int_{L_f} \overline{T}_1 u_1^* dl \end{aligned} \tag{20.4}$$

Equation (20.4) allows the identification of six constitutive parameters. Although many subregions could be defined, several practical limitations arise and foremost is the need for sufficient strain information. Problems begin to occur when the subregions are too small or when the subregion contains low magnitude strains as compared to other subregions. Practical usage has shown that special care is needed in the identification process when there are six (or more) subregions.

A further development in VFM [13] comes from the understanding that PVW can also be used to examine local equilibrium. A similar procedure has been used to examine errors in FEM modeling [14]. Discrepancies in local equilibrium are 'equilibrium gaps' and can be used to locate boundaries of regions which are not in equilibrium with each other when material homogeneity is assumed.

Equation (20.3) can be rewritten as:

$$\int_{S'} Q\left[\left(\varepsilon_1\varepsilon_1^* + \varepsilon_2\varepsilon_2^* + \frac{1}{2}\varepsilon_6\varepsilon_6^*\right) + \nu\left(\varepsilon_1\varepsilon_2^* + \varepsilon_2\varepsilon_1^* - \frac{1}{2}\varepsilon_6\varepsilon_6^*\right)\right]dS' - \int_{L_f} \overline{T}_1 u_1^* dl = 0 \tag{20.5}$$

Equation (20.5) describes the equilibrium of a region S', where S' is a portion of a larger region, S. If both Q and ν are known, e.g. by use of Eq. (20.3), and u^* are kinematically admissible then summation of the integrals should be zero; a

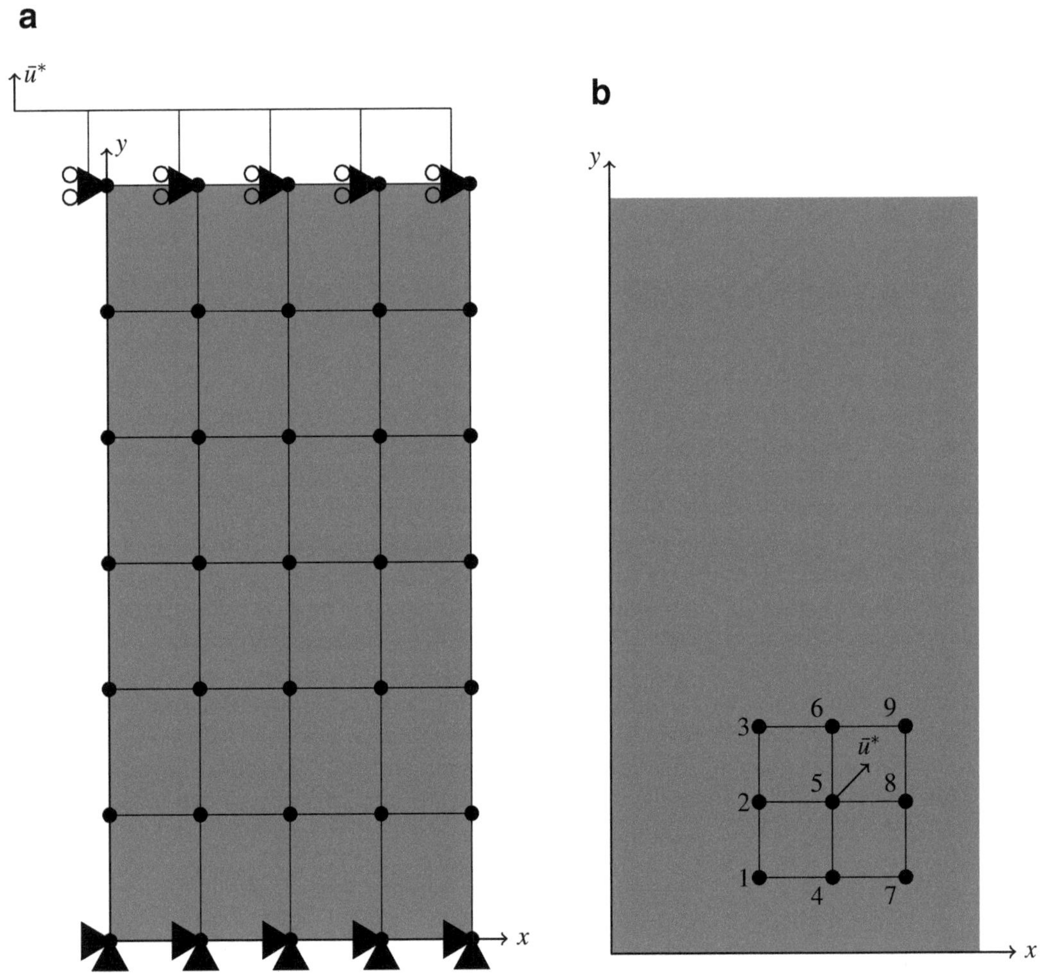

Fig. 20.1 VFM meshes used. (**a**) VFM mesh for analysis, showing kinematic constraints, (**b**) four virtual elements used in EG analysis, not to scale

non-zero result is termed to be an 'equilibrium gap.' In absence of external forces, the final term containing $\overline{T}_i$ is 0. The virtual mesh used for the Equilibrium Gap (EG) analysis is shown in Fig. 20.1b The inspection window, composed of the four virtual elements, is rastered, horizontally and vertically, across the entire specimen. Two parameters can be adjusted to produce different types of EG maps, the number of data points in each element and the movement of the inspection window to the next location. For this work, each virtual element contained 25 DIC data points (5 points × 5 points). The inspection window was moved by a single DIC row or column.

All virtual degrees of freedom for nodes on the edge of the inspection window are set to zero, i.e. nodes 1, 2, 3, 4, 6, 7, 8, 9 are fixed. This process eliminates all work of external forces on the window boundary. Virtual node #5 is subject to a 45° displacement fixed at $|u| = 1$. Equation (20.5) is discretized using summations and the equilibrium gap contour is determined by sliding the window over the entire specimen.

Examination of a simulated homogeneous isotropic material under uniaxial vertical tension provides a clarifying example. In this case, both ε_1 and ε_2 are constants and ε_6 is zero. Displacement of virtual node #5 at 45° with a value of 1, means the bottom two virtual elements have positive ε_2^* and the top two elements have negative ε_2^*. Because the exterior nodes are fixed, every individual ε_2^* within the EG window is balanced by another ε_2^* of opposite sign elsewhere in the window. The same explanation is made for ε_1^*. So, if Q and ν are constant everywhere in the window, then a homogeneous isotropic material under uniaxial tension will indicate that the entire specimen is in equilibrium, i.e. has no gaps. Note that the only requirement is that Q and ν are constants, not that they are the actual values of Q and ν for the material. Therefore, equilibrium gaps indicate only that either Q or ν is not homogeneous within the EG window. Finally, Eq. (20.5) can also be used if the heterogeneous Q and ν are known by using the spatially corresponding value of $Q(x, y)$ and $\nu(x, y)$ for each $\varepsilon_i(x, y)$.

20.4 Results and Discussion

The tensile test examined in this work is shown in Fig. 20.2 where the symbols denote specific data examined with the different VFM analyses discussed in VFM section. Because the reference image for all DIC analyses was at load = 0 kN/m, all identified stiffnesses are secant stiffnesses, i.e. the slope of a line from the origin to the data point. For a linear elastic material, secant stiffness and tangential stiffness are identical but, as illustrated in Fig. 20.2, the paperboard material was nonlinear. The first point selected for analyses was at load = 2.05 kN/m.

Figure 20.3 shows the evolution of strain during the test, where each column corresponds to a symbol in Fig. 20.2 and each row is mapped on the same scale, shown in the far right of the figure. Based on the scale, the strain heterogeneity in all three strains became readily apparent at load = 6.50 kN/m. Especially for ε_2, the strain heterogeneity was similar, but gradients increased, as the test progressed. For this single test, the specimen regions 'activated' by the external stress remained 'activated' as the external stress increased. Alternatively stated, the load bearing regions of the specimen did not vary spatially as the test progressed.

Because paperboard is composed of cellulose fibers, it is geometrically possible for regions of the material to be auxetic as in foam materials [15], but auxetic behavior was not evident. The strain maps of ε_2 demonstrate the importance of

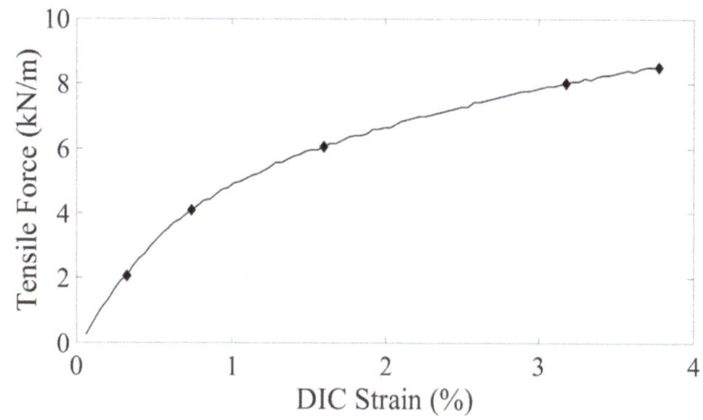

Fig. 20.2 Tensile test for specimen. Symbols denote specific data examined with VFM

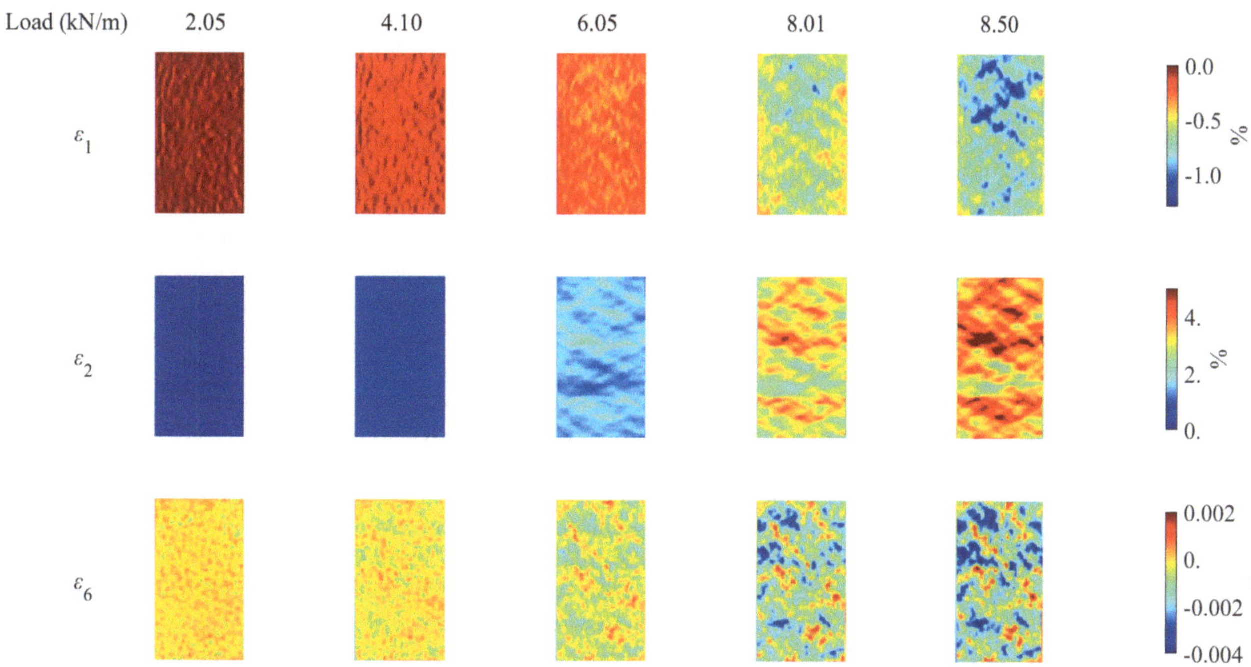

Fig. 20.3 Evolution of strains during tensile test

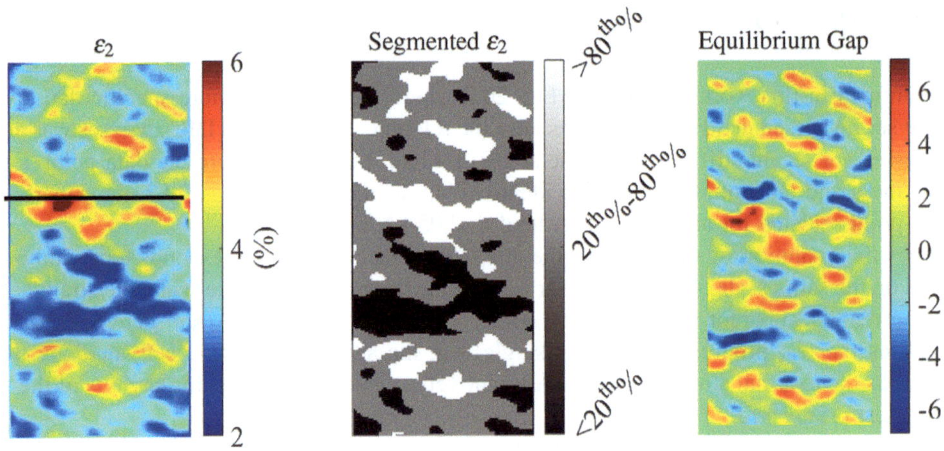

Fig. 20.4 Results from data at load = 8.50 kN/m. Horizontal black line in ε_2 map represents approximate failure location. EG analysis used virtual elements that contained 5 × 5 strain data

requiring the specimen width to be larger than the expected heterogeneity. In this test the regions of high and low ε_2 did not extend completely across the specimen. Of particular interest is the proximity of low and high ε_2 near the specimen mid-section that created a high strain gradient. The magnitude of ε_6 remained small throughout the test, indicating good specimen alignment and that, if present, the heterogeneities were generally aligned in the 1-direction, across the specimen width.

The left map of ε_2 in Fig. 20.4 is the same map as shown in the rightmost column in Fig. 20.3, but on a different scale to demonstrate the high strain gradients. The horizontal black line in the map of ε_2 represents approximate failure location and, perhaps not surprisingly, passes through a region of high strain. The map labeled 'Segmented ε_2' divides ε_2 in a tertiary fashion, with ε_2 divisions of ' < 20th %', '20th %–80th %', and > 80th %'. The segmentation of the material based upon strain was considered reasonable because Ostoja-Starzewski and Castro [7] found a reasonable agreement between local grammage, local strain and local modulus. The EG analysis was performed by assuming material homogeneity and rastering the Equilibrium Gap window shown in Fig. 20.1b throughout the specimen. Zero-value borders are the width of five strain data points, are shown as 0, and indicate areas outside the possible inspection region. Because the EG inspection window started at the lower left of the specimen, the center of the inspection window was located five horizontal and five vertical data points from the corner.

The non-zero regions of the EG map denote location in which the EG inspection window contained non-uniform Q or ν. This can be illustrated by examining the region of low strain in the bottom third of the specimen. When the inspection window was located entirely within that region, the EG map was green, or in equilibrium, even though, as we will subsequently demonstrate, that region has a higher than homogeneous stiffness. An examination of Eq. (20.5) shows that as long as Q and ν vary by the same proportion as compared to regions in equilibrium relative to the homogeneous parameter, the EG window will indicate equilibrium. The value of the EG map shows regions where either Q or ν varies within the inspection window.

Figure 20.5 shows the changes in secant stiffness and ν during test progression. Mean data for secant stiffness was determined by $load(i)/\varepsilon_2(i)/(1-\nu_{21}(i)\nu_{12})$ where i is one of the five examination times. Based on other tests reported in [10], $\nu_{12} = 0.45$. Mean data for ν was determined by $-mean(\varepsilon_1(i))/mean(\varepsilon_2(i))$ and this value was also used for $\nu_{21}(i)$ on the previous calculation. The homogeneous VFM identification was performed using Eq. (20.3). The segmented VFM identification was performed using Eq. (20.4), using the regions denoted in the segmented ε_2 map of Fig. 20.4. This division of the specimen is arbitrary. The ε_2 map suggests that a higher portion of 'stiff' material could have been reasonably allocated.

The secant stiffnesses of the 80th % region and those of the 20th % region appear to lie equally below and above the secant stiffness of the median strain region, denoted by the 20th %–80th %. The median strain region, the mean data stiffnesses and those of the homogeneous identification are nearly identical. If the papermaker were given the binary choice of eliminating low grammage regions (high strain) or high grammage regions (low strain) they would recognize the benefit of reducing the high grammage regions, which cost more for material (cellulose fiber) and require greater drying energy. By reducing the high grammage regions the papermaker reduces costs with only a marginal reduction in overall stiffness.

The evolution of ν was different than Q. For ν, the identification was similar for the mean data and homogeneous analysis. For the segmented analysis, the 20th %–80th % segment had a lower ν than for the high and low strain regions and the high

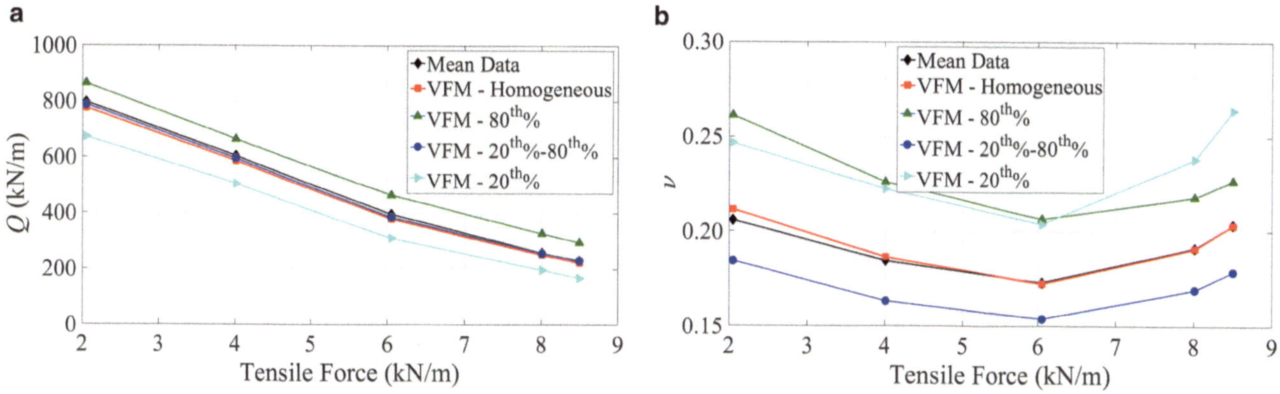

Fig. 20.5 Evolution of Q and ν with tensile force. (**a**) Evolution of secant stiffness for each analysis, (**b**) evolution of Poisson's ratio for each analysis

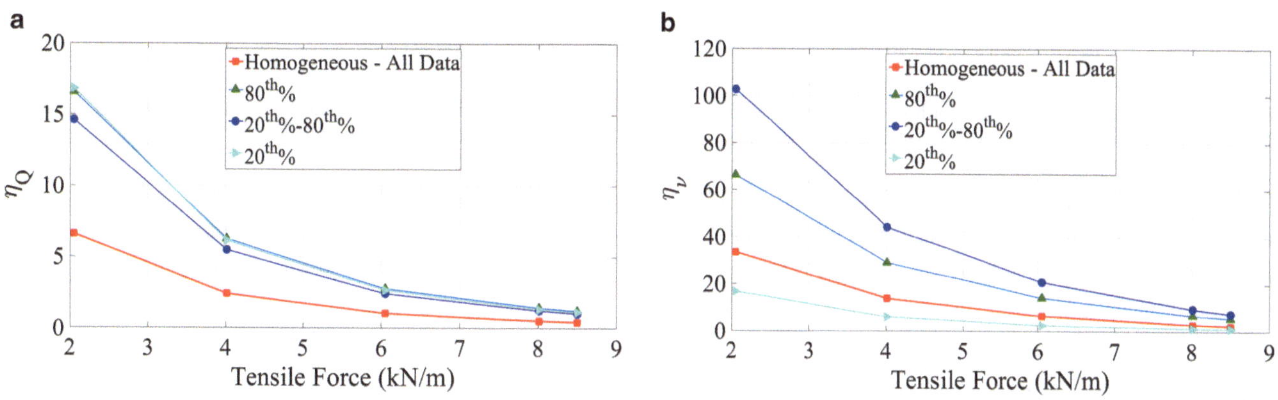

Fig. 20.6 Evolution of η's with increasing tensile force. (**a**) Evolution of η_Q, (**b**) evolution of η_ν

and low strain regions had very similar ν to each other, until near failure, where our linear elastic definition of ν does not apply. While ν did change during the test, the magnitude of the change was small.

The use of optimized, piecewise virtual fields also provide the opportunity to examine the quality of the identification using the η parameter. The derivation of this parameter is beyond the scope of this work; the reader is referred to Chap. 3.7 in *The Virtual Fields Method* [9]. η is a kind of standard deviation of the identified parameter across the specimen. Here, η is normalized by the identified parameter, either Q or ν, and the $\sqrt{\text{number of data points}}$. The calculated η's are shown in Fig. 20.6.

Figure 20.6a shows that the segmented analysis did not reduce the variation in the identification of Q, suggesting that segmentation based on ε_2 may not be the best choice, but the low value of η_Q suggests it was still a good choice. As the test progressed, and load and strain increased, the signal-to-noise ratio in the strain measurement increased and deviation was reduced. The generally higher values of η_ν as compared to η_Q indicate a less uniform identification, not surprisingly, because that identification relies on strain values about an order of magnitude less that those for Q. Important here is the result that the high and low strain regions had more uniform identification than the mean strain region. The high strain region even had a lower η than homogeneous identification.

The value of segmenting the specimen can be shown by comparing the EG map before and after segmentation in Fig. 20.7. While segmentation by ε_2 may not be the best choice, based upon η_Q, the EG map shows that it was still a good one, the local equilibrium was greatly improved by using six constitutive parameters instead of two. Though no qualitative assessment is provided here, the EG maps in Fig. 20.7 suggest that the non-contiguous high stiffness regions had similar high stiffnesses and the non-contiguous low stiffness regions had similar low stiffnesses.

Comparing how the stiffness changed during the test, the low strain region as related to the mean strain had stiffness that was initially 10 % higher and ended 28 % higher. Comparing the high strain region to the mean strain region, its stiffness

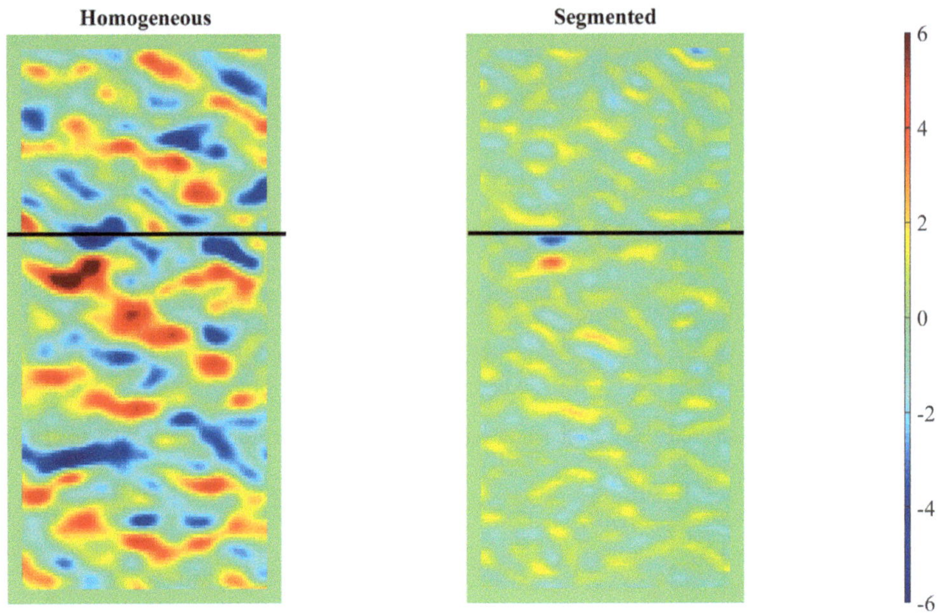

Fig. 20.7 Comparison of EG maps for homogeneous and segmented analysis. Left map is identical to map in Fig. 20.4. Units are dimensionless. Black horizontal lines represent approximate failure location

began 14 % lower and ended 27 % lower than the stiffness of the mean strain region. The ±30 % stiffness variation within this particular specimen was large, especially when a ±10 % variation is assumed when stiffness is related proportionally to grammage [5]. However, the larger than expected variation may also be created during multiply forming process.

20.5 Conclusion

This work re-examined previously published work [10] for the purpose of identifying heterogeneous stiffness of a paperboard specimen using VFM. For the purposes of stiffness identification, the material was segmented into three non-contiguous regions based on magnitude of local axial strain. The division and resulting stiffness identification improved the local equilibrium of the specimen. The confidence of the segmented stiffness identification was slightly less than that of the homogeneous identification, while the quality of the Poisson's ratio identification was similar for the segmented and homogeneous identification. The stiffness variation increased with tensile force from an initial ±10–±30 % just prior to failure.

References

1. Axelrad, D.R., Rezai, K., Atack, D.: Probabilistic mechanics of fibrous structures. J. Appl. Math. Phys. (ZAMP) **35**(4), 497–513 (1984)
2. de Oliveira, R., Mark, R.E., Perkins, R.W.: Evaluation of the effects of heterogeneous structure on strain distribution in low density papers. In: Perkins, R.W. (ed.) Mechanics of Wood and Paper Materials, AMD Vol. 112, MD Vol. 23, pp. 37–61. American Society of Mechanical Engineers, New York (1990)
3. Hanada, A., Onabe, F.: Analysis of physical properties of paper by interference holography (I) Construction of measuring system for local strain distribution of paper sheets. Japan Tappi J. **47**(5), 645–652 (1993)
4. Hanada, A., Onabe, F.: Analysis of physical properties of paper by interference holography (II) Visualization of local strain distribution of paper sheets through tensile testing process. Japan Tappi J. **47**(6), 761–766 (1993)
5. Korteoja, M.J., Lukkarinen, A., Kaski, K., Gunderson, D.E., Dahlke, J.L., Niskanen, K.J.: Local strain fields in paper. Tappi J. **79**, 217–224 (1996)
6. Korteoja, M., Salminen, L.I., Niskanen, K.J., Alava, M.J.: Strength distribution in paper. Mater. Sci. Eng. A-Struct. **248**(1), 173–180 (1998)
7. Ostoja-Starzewski, M., Castro, J.: Random formation, inelastic response and scale effects in paper. Phil. Trans. R. Soc. London, Ser. A **361** (1806), 965–985 (2003)

8. Yamauchi, T., Murakami, K.: Observation of deforming and fracturing processes of paper by use of thermography. In: Products of Papermaking, Transactions of the Xth Fundamental Research Symposium, Oxford, UK, vol. 10, pp. 825–847. FRC, Oxford (1993)
9. Pierron, F., Grédiac, M.: The Virtual Fields Method: Extracting Constitutive Mechanical Parameters from Full-Field Deformation Measurements. Springer, New York (2012)
10. Hagman, A., Nygårds, M.: Investigation of sample-size effects on in-plane tensile testing of paperboard. Nord. Pulp Pap. Res. J. **27**(2), 295–304 (2012)
11. Grédiac, M., Vautrin, A.: A new method for determination of bending rigidities of thin anisotropic plates. J. Appl. Mech. **57**(4), 964–968 (1990)
12. Moulart, R., Avril, S., Pierron, F.: Identification of the through-thickness rigidities of a thick laminated composite tube. Compos. Part A-Appl. Sci. **37**(2), 326–336 (2006)
13. Devivier, C., Pierron, F., Wisnom, M.R.: Impact damage detection in composite plates using deflectometry and the Virtual Fields Method. Compos. Part A-Appl. Sci. **48**, 201–218 (2013)
14. Ladeveze, P., Leguillon, D.: Error estimate procedure in the finite element method and applications. SIAM J. Numer. Anal. **20**(3), 485–509 (1983)
15. Lakes, R.: Foam structures with a negative Poisson's Ratio. Science **235**(4792), 1038–1040 (1987)

Chapter 21
Rigid-Body Motion Tolerance for Industrial Helical CT Measurements of Logs

Edward Angus and Gary S. Schajer

Abstract The major cost in sawmills is the log raw material. It is therefore important to maximize the value of the product yielded from each log. Computed Tomography (CT) has been explored as a sensing solution for determining log defects and making data-driven choices in product breakdown. However, the harsh conditions of the sawmill environment lead to limitations in data acquisition and log manipulation. This paper presents an iterative-solver CT reconstruction scheme that includes rigid-body motion compensation, greatly increasing reconstruction robustness for misalignments in the radiographic data. The motion compensation is carried out by using the known nominal distribution of density in softwood logs to approximate the geometric center of the log and its radius from the radiographs. This is then applied to an iterative reconstruction methodology based on a log-specific voxel geometry previously developed. The method is validated for synthetic phantoms, a physical phantom, as well as real log samples. Results indicate that the rigid body compensation effectively ameliorates motion blur for movement within in the detector field of view.

Keywords Tomography • Sawmill • Lumber • CT • Wood

21.1 Introduction

The major operating cost of modern sawmills is held in the uncut logs before they are sawed into products. In most mills, this may be in excess of 75 % of total costs [1]. As such, there is a major incentive to optimize log breakdown so that the maximum value is obtained for each tree.

The large natural variation in log properties has driven the development and implementation of a wide variety of sensing technologies for data-driven breakdown, from optical systems to ultrasound and radiographic systems. Among these implementations have been X-ray computed tomography (CT) systems offering both full and partial internal reconstructions of log features [2]. Adoption of CT by the wood industry has been stunted however due to the considerable requirements demanded on such a system from the harsh sawmill environment. Any such system must be capable of reconstruction from data collected at an accelerated rate (logs move within a mill at a rate up to 3 m/s), with data compromised by motion blur and limited sampling [2]. The methods proposed here seek to account for these difficulties in the form of a robust reconstruction system suitable for continuous use in sawmills. The method is an extension of the feature-based voxel computed tomography models of Schajer and An [3, 4], improved upon for continuous helical scanning and movement compensation within the reconstruction space.

21.2 Feature-Tailored Voxel CT Scanning

The recent work of Schajer and An takes into account the known cylindrical shape of logs and features by considering a reconstruction space with a polar discretization. The cylindrical field of view of the detector within an X-ray cone beam is divided in to axial slices whose normals are parallel to the detector surface. Each slice is then divided up by a superposition of annular and sector shapes within each slice. The cone beam is comprised of pixel-rays that penetrate the field of view on

E. Angus (✉)
FPInnovations, Inc., Vancouver, Canada
e-mail: tedangus@gmail.com

G.S. Schajer
Department of Mechanical Engineering, University of British Columbia, Vancouver, Canada

S. Quinn, X. Balandraud (eds.), *Residual Stress, Thermomechanics & Infrared Imaging, Hybrid Techniques and Inverse Problems, Volume 9*, Conference Proceedings of the Society for Experimental Mechanics Series, DOI 10.1007/978-3-319-42255-8_21

the way to the detector. Figure 21.1 shows an overview of the source detector geometry as well as the cylindrical field of view with distinct slices. The polar voxel pattern is shown in 2, while 3 demonstrates the slice placement within the cone beam. The slices in one frame of capture are delineated by the furthest axial (z axis) extents of the cone beam's intersection with the field of view.

The method used here is an implementation of iterative CT reconstruction also known as the Algebraic Reconstruction Technique (ART). Compared to the backprojection family of tomographic reconstructions, ART is more robust for making as good reconstructions from fewer frame captures and lower quality data. Physical laws of X-ray attenuation underpin the inverse problem of Eq. (21.1), in which the pixel ray radiographic data vector $\vec{y}$ (samples of the two dimensional Radon transform) are shown to be equivalent to the densities in the discretized voxel space $\vec{\rho}$ times some basis G whose column space acts as weights for the relation of each pixel ray with each voxel. The physical laws of X-ray optics show that an exact basis may be formed when the correlation is the length of transmission of each pixel ray through each voxel (here on called the 'path length').

$$G\vec{\rho} = \vec{y} \tag{21.1}$$

Conventional ART operates on a field of view discretized into Cartesian space, making each frame capture for the detector have a unique contribution to the basis. The axisymmetric voxel arrangement of Fig. 21.2 has an advantage in

Fig. 21.1 Source detector geometry

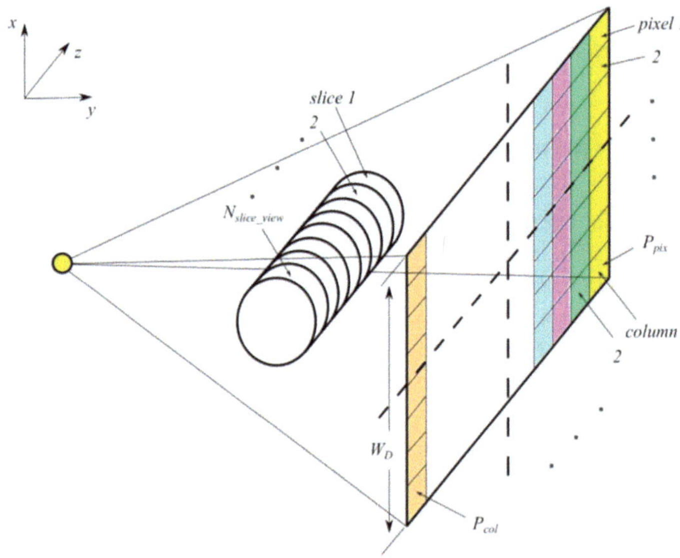

Fig. 21.2 Polar voxel pattern

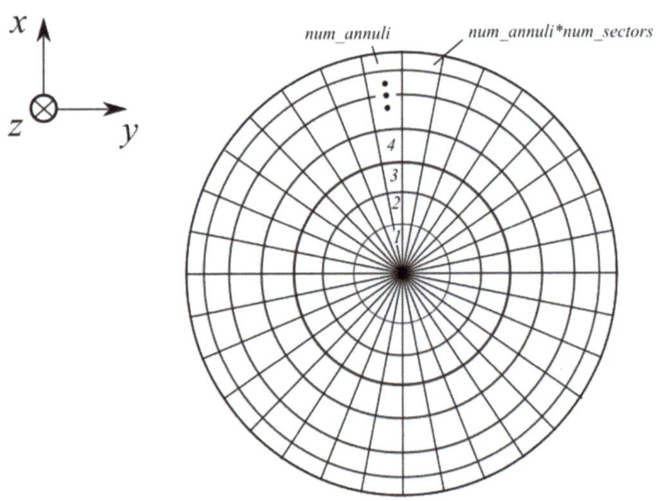

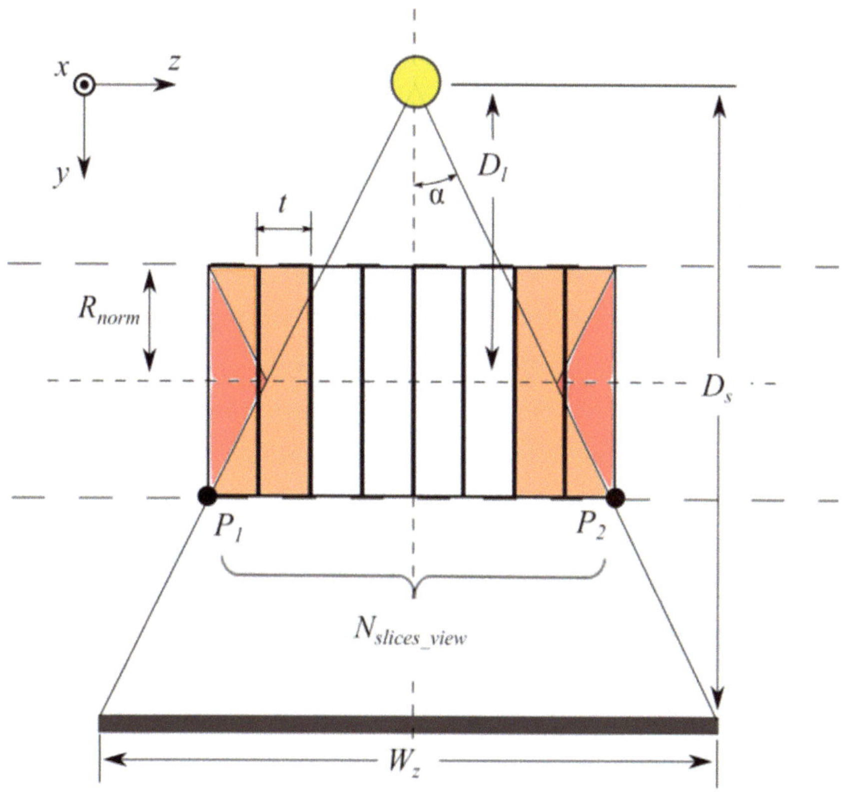

Fig. 21.3 Log placement in cone beam

implementation due to the symmetry of the voxel pattern with respect to the origin. Thus the contribution of each frame to the total basis G is a manipulation of the values from just one view for rotations equal to an integer number of sectors. This approach may be extended into helical scanning by ensuring that the object scanned translates axially by a distance equal to the physical thickness of one reconstruction slice as is shown in Fig. 21.1c. This makes it so that path lengths of one view are usable for frame series having both translation and rotation (Fig. 21.3).

The feature-tailored has been explored more thoroughly by Jian et al. and Thibaudeau et al. under the name 'polar ART'. They showed that this is axisymmetric voxel scheme when applied to volumetric purely rotational reconstructions led to significant savings in memory, as well as some increased efficiencies in the iterative solution methods common with ART [5, 6]. The work here makes use of the simultaneous algebraic reconstruction technique or SART, introduced by Anderson and Kak and proven to be an effective solver for the polar ART reconstruction concept [7].

21.3 Log Motion Compensation

A conventional CT reconstruction system is tied to the absolute space determined by the geometry of the X-ray sources and detectors. This arrangement ties the reconstruction to real coordinates in space, meaning that motion blur will occur if the position of the log relative to the X-ray sources and detectors is not precisely controlled. In sawmills, precise control of the log as it moves along the breakdown line is impractical, so it is important to create a CT system tolerant of extraneous rigid-body motions of the log within the field of view.

Tracking the log's motion in space at the time of frame capture and compensating for the observed displacement is a practical solution to the problem of motion artefact. It makes sense to use the radiographic signal from the detector to do this, as this will automatically correlate the frame with the amount by which the data must be manipulated to counteract the effects of motion. Standard boundary detection methods could be used to follow the signal, but the irregular surfaces of logs due to knot growth, loose bark and detritus would make such methods difficult to implement. A better solution may be arrived at by seeking to use the entire radiographic signal, not just the edge details. This can be done by recognizing that the centroid of the radiographic signal's pixel columns are highly correlated with the geometric center location of the log. If the motion of the log is viewed in the fan beam as being either radial (towards or away from the source) or circumferential (transverse to the source) it makes sense to re-bin the pixel columns into equal-angular values using the relation of Eq. (21.2)

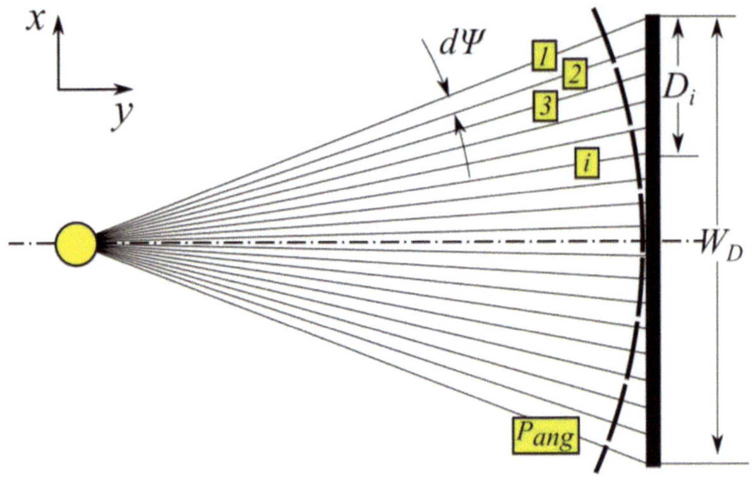

Fig. 21.4 Cylindrical re-binning to equal angular pixel rays

and demonstrated in Fig. 21.4. This give the result of a cylindrical detector centered at the source, and effectively linearizes circumferential shifting of the log's Radon transform. In the geometry of a cone beam projecting radiographic data onto the re-binned cylindrical detector, movement of the log radially by a distance e to the source will cause the data to scale, while movement circumferentially by e will lead translation of the radiographic data (Fig. 21.5a, b).

$$D_i = \frac{W_D}{2} - D_s \sin(\Psi - id\Psi) \tag{21.2}$$

Looking at a uniform circular cylinder with density function $\rho(r, \phi)$ in the xy plane of Fig. 21.6, it may be shown that it's Radon transform R along some line t for frame orientation θ will have the form of a half ellipse, in which the density function is represented in polar coordinates and 2 * R_{pix} is the number of pixel-rays occluded by the cylinder. The elliptical Radon Transform is simply the value of the parallel line integrals across the density function of the circle in one direction. This may be taken to be true as well for divergent rays in an X-ray cone beam if the cone beam angle is sufficiently small. This the case for which the cone beam apex angle was $2\Psi = 15$ degrees. The center of the log may then be approximated in its radiograph as the lateral centroid of this semi-ellipse along the detector array C_x, shown in Fig. 21.7. Although for real logs the radon transform will not exactly have the half-ellipse shape due to non-uniform features, the nominal density distribution means that it will closely resemble this form.

Tracking the center of the radiograph helps with circumferential movement within the fan beam, but the scaling associated with radial error (Fig. 21.5a) is still present. This is compensated for by truncating the data set outside of the log radiograph and scaling the signal for a path length basis that has been previously computed. In order to determine the extents of the array containing the useful data, the radius of the log must be estimated. This can be done by relating the known major axis dimension of a semi-ellipse h and its centroidal height C_y relative to its minor axis dimension R_{pix}. It is half of this minor axis that gives the log's radius in terms of equal angularly spaced pixel-ray detectors. Manipulating the semi-ellipse geometric formula, the expression of Eq. (21.3) may be developed for the log's radius in terms of angular pixels.

$$R_{pix} = \frac{16}{3\pi^2} \frac{\left(\sum b_i\right)^2}{\sum b_i^2} \tag{21.3}$$

This angular radius in terms of pixel-rays may be used to approximate the physical size of the log by using the geometry of the source-detector setup to solve for the real log diameter R_l, using Eq. (21.4).

$$R_l = D_l \sin\left(R_{pix} d\Psi\right) \tag{21.4}$$

It is now necessary to scale the data to fit into the number of pixel-rays contained within the previously calculated path length matrix. This will cause the data to be distorted as the number of pixel-rays used in the calculation will either be fewer than or

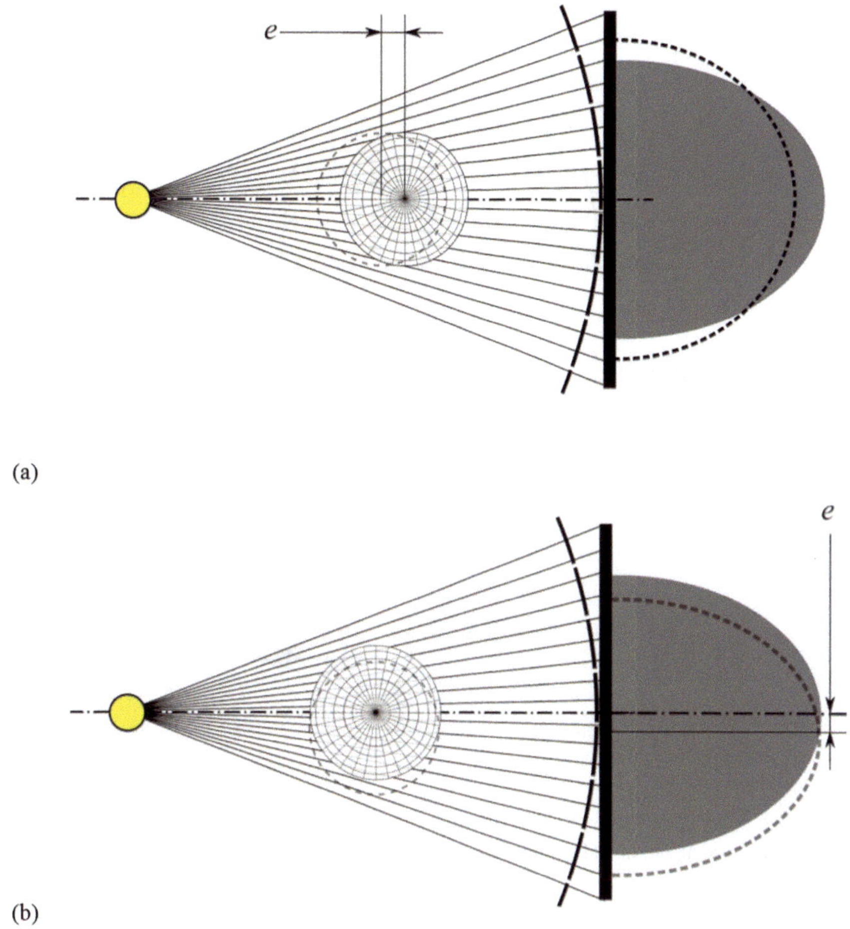

Fig. 21.5 Log shift effect on radiograph

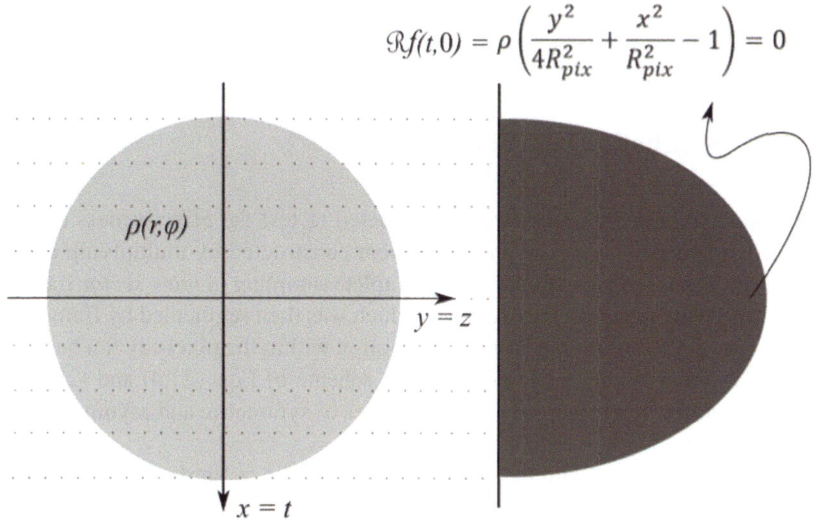

Fig. 21.6 Radon Transform of a uniform cylinder

greater than the actual number of pixel-rays subtended by the log in each frame. This effectively leads to an apparent loss of mass within the log, necessitating that the data be scaled to retain the true mass. This may be accomplished by scaling the data according to the ratio of the real log radius as estimated from the radiograph to the radius used to calculate the path length basis. A benefit to this method is that the entirety of the basis is being used for solving non-empty voxels, and no

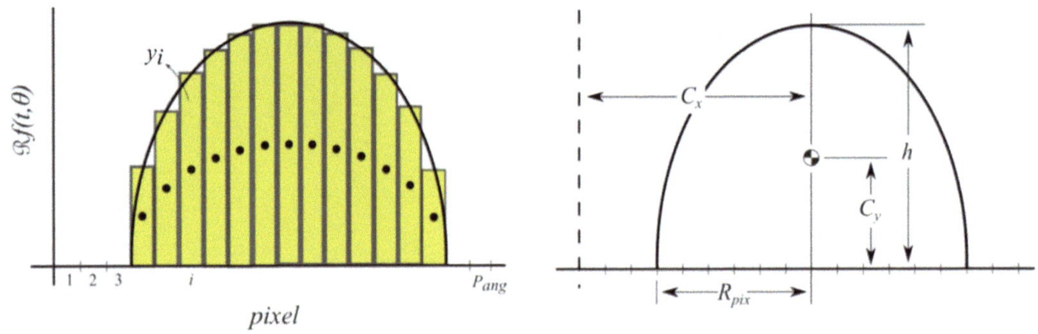

Fig. 21.7 Semi-elliptical Radon Transform centroids

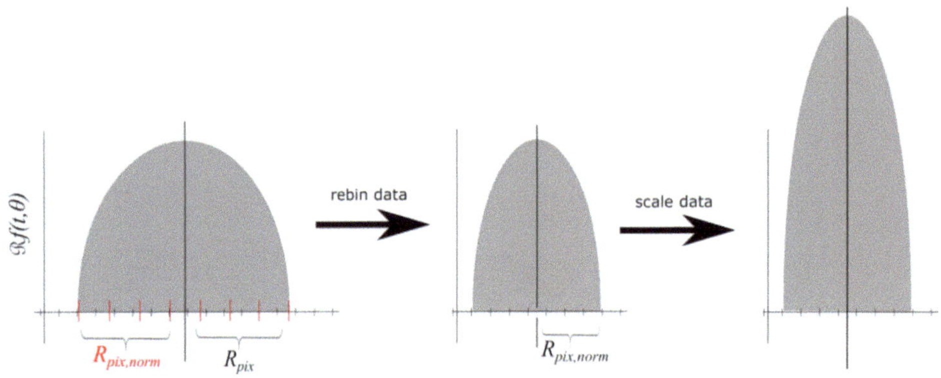

Fig. 21.8 Normalization and scaling of Radon transform

computation waste occurs working with empty exterior voxels. Figure 21.8 illustrates the normalization and scaling steps; Eq. (21.5) gives and the scaling formula.

$$\Re f(t,\theta)_{scaled} = \Re f(t,\theta)\frac{R_l}{R_{norm}} \tag{21.5}$$

21.4 Synthetic Data Validation

Several synthetic phantoms were constructed to test the effectiveness of this scheme for locating the center of a log and approximating its size. The phantoms were constructed by multiplying a known density distribution $\vec{\rho}$ by the path length basis for a single slice consisting of a complete sampling of *num_sector* frames evenly distributed about the circumference of the slice. This yielded a data vector $\vec{y}$, which was then segmented by frame and each of these inserted at a shifted location in an empty frame to simulate rigid body motion within the pixel-ray fan beam. These synthetic frames could then be fed back into the center finding and normalization scheme of Eqs. (21.4) and (21.5) and the density vectors examined to see if the reconstruction was effectively made. A series of symmetric and asymmetric phantoms were tested, their forms are presented in Fig. 21.9.

It was found that with the asymmetric phantoms, a 'starburst' artifact was exhibited in the reconstructions around the centermost voxels. This was determined to be due to small errors in estimating the true geometric center of the Radon transform from the centroid of the signal, as hard-coding in the correct center made the artifact disappear. This was a very small error, only on the order of 1 % in the worst cases. To counteract this difference between geometric and centroid pixel-ray indices, the centroid finding of the data signal was instead performed on the square root of the data. This 'biased' the centroid towards the true geometric center by minimizing the disparity between large density spikes and valleys that led to the error in the first place. This was coupled with a selective low pass filter that provided inter-voxel regularization between iterates of the SART solver per Herman [8]. The end effect was to greatly reduce the effect of the artifact, Fig. 21.10 shoes the original artefact as well as the results of center biasing and regularization on an asymmetric synthetic phantom.

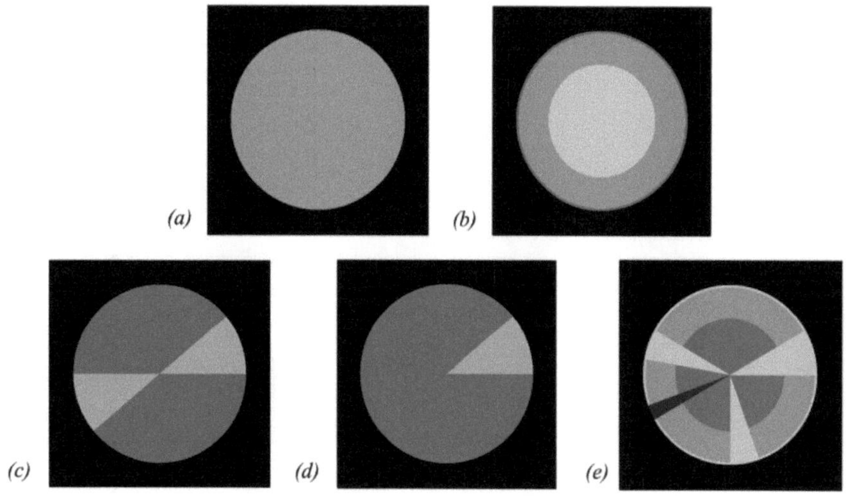

Fig. 21.9 Synthetic phantoms tested (**a**–**c**) symmetrical (**d**, **e**) asymmetrical

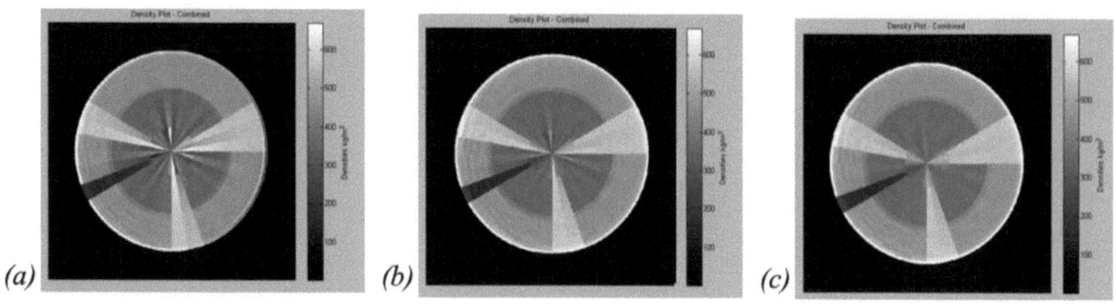

Fig. 21.10 Elimination of artifact (**a**) Artifact present (**b**) Square root signal center finding (**c**) Updated center finding with regularization

21.5 Real Data Results and Discussion

A series of experiments was run to evaluate the center finding helical CT algorithm on physical data. The experimental setup is a modified version of the one used by An to validate the altered voxel geometry [3]. Figure 21.11 shows a schematic plan view. It consists of a log carriage that is capable of both translating and rotating a ~1 m long log sample of approximately 30 cm in diameter through an X-ray cone beam. The cone beam is produced by a Comet MXR 160kVp X-ray tube and Phillips MG-160 X-ray generator. This system is capable of producing X-rays up to 160kVp with and anode current up to 10 mA. The detector is a custom design and consists of a gadolinium oxysulfide scintillator screen contained within a light-proof box and viewed by a highly sensitive Andor iXon 897 EMCCD camera. This allows for a very large scanning area when compared to commercial flat panel detectors and at much less cost. In addition, the arrangement is easily scalable to accommodate larger logs. Figure 21.12 shows an image and schematic of the detector box. The camera views the detector screen through a mirror so that it can be positioned outside of the cone beam to avoid damage to the sensor.

The cart consists of two wheeled platforms and a driveshaft attached to each. These shafts may be directly attached to a log specimen using lag screws. The log motion is driven by two size 34 stepper motors driving a 1 in. HTD timing belt that is attached to the carts. This translation is mechanically linked to rotation by a pulley over which a separate fixed belt runs, thereby causing helical rotation. The diameter of this pulley determines the pitch of the helix that the motion of the log follows. For a given voxel pattern containing *num_sectors*, this pulley size can be used to determine the slice thickness on capture within the cone beam.

Two physical samples were considered in studying the reconstruction algorithm and center correcting scheme, a real log sample and a physical phantom. The physical phantom was constructed by first making an empty shell out of 12″ diameter concrete foundation tube and sealing one end. An inner cavity of 6″ diameter foundation tube was centered within the larger tube, and a sector-shaped prism of housing insulation inserted to simulate a crack. The void in the inner tube was then filled with compressed grain seed, and the outer annular void with compacted cedar sawdust to make large constant density

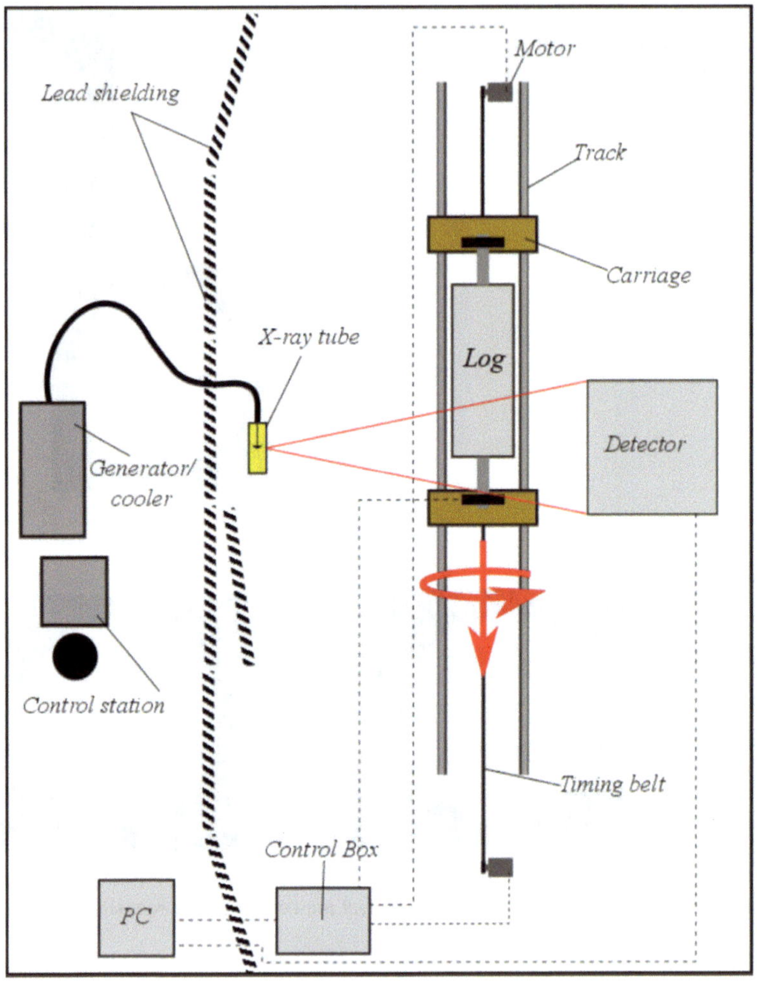

Fig. 21.11 Schematic of CT scanner system

(a)

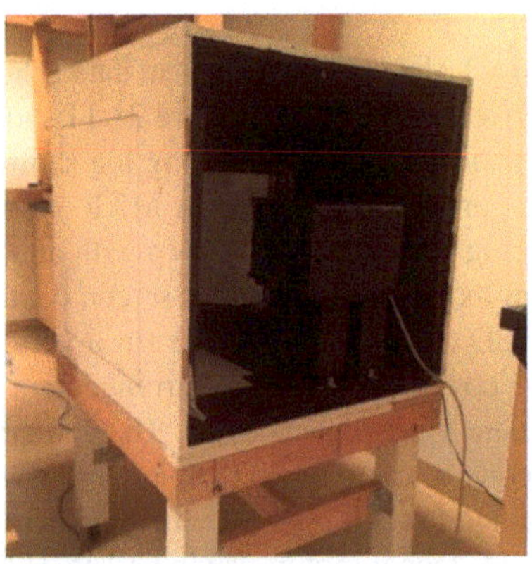

(b)

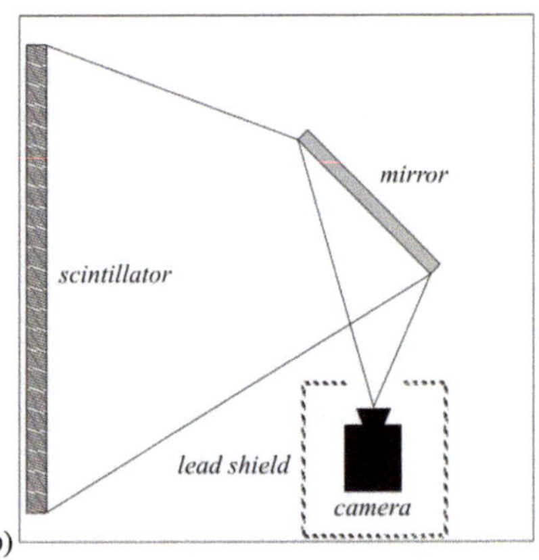

Fig. 21.12 Detector and schematic of detector

volumes. At three planar locations in the phantom, 'knot' clusters were inserted in the form of sector shaped slices of 1″ medium density fiberboard. Figure 21.13 presents a schematic of the physical phantom. Table 21.1 the three knot cross sections and their orientations. Figure 21.14 shows photographs of the assembled physical phantom.

Figure 21.15 presents a sample of radiographic data for the physical phantom. These data is for the case when the center of the cylindrical phantom is aligned with the shafts on the carts, so that no rigid body motion of the phantom occurs during data captures. Figure 21.16 shows reconstructions of resolution 60 sectors by 15 annuli of the three knot slices. Table 21.2 gives the knot positions as determined from the reconstructions.

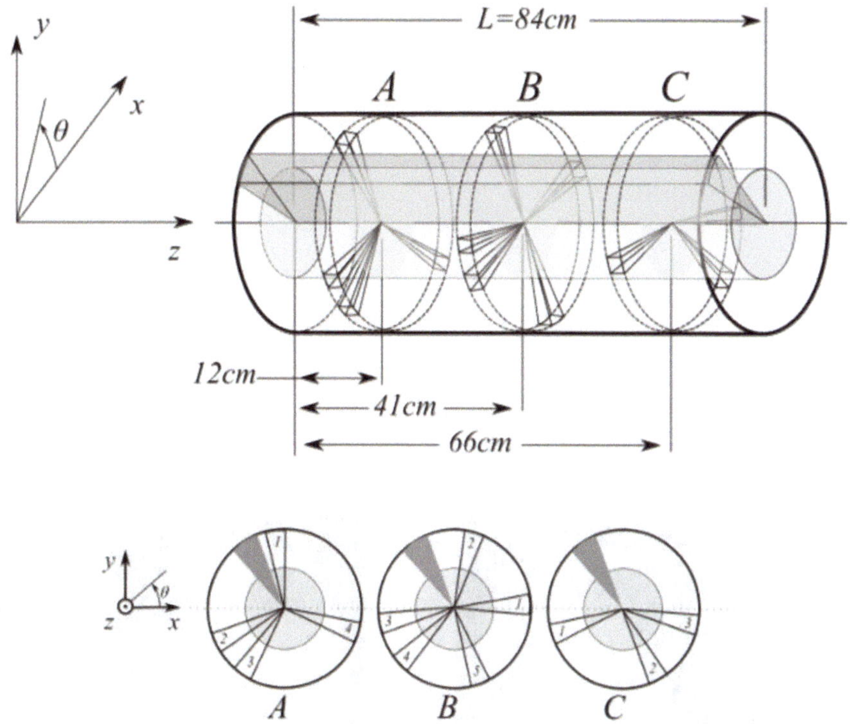

Fig. 21.13 Physical phantom schematic

Table 21.1 Table of knot inclinations in degrees

Knot	Slice A	Slice B	Slice C
1	99	0	212
2	200	78	318
3	240	198	353
4	332	229	//
5	//	300	//

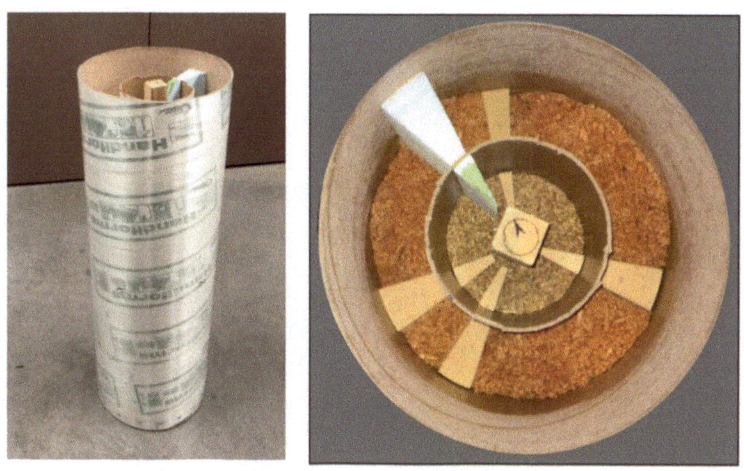

Fig. 21.14 Log phantom

To test the data normalization methods, the log was mounted with a deliberate eccentricity of 1 in. from the mounting shafts, which was the maximum allowed by the geometry of the cone beam. This feature is shown in the data captures of Fig. 21.17. The individual knot-cluster reconstructions were again identified and the position of the knots determined (Fig. 21.18). It is clear that the centering and normalization methods used here are effective at removing motion blur from the reconstruction to the extent that can be tested with this phantom.

Fig. 21.15 Phantom radiograph

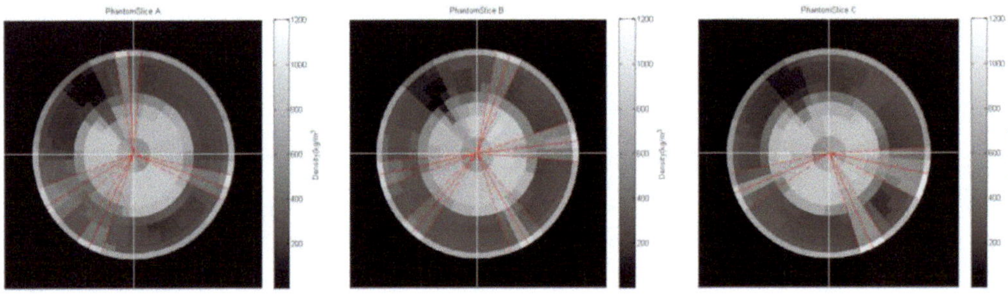

Slice A *Slice B* *Slice C*

Fig. 21.16 Phantom reconstructed slices

Table 21.2 Knot locations in degrees determined from reconstruction

Knot	Slice A	Slice B	Slice C
1	93	6	203
2	210	70	300
3	243	191	349
4	341	229	//
5	//	300	//

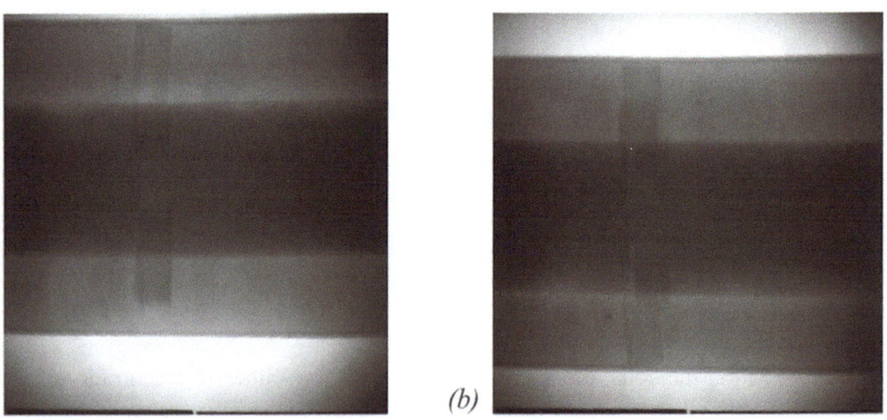

(a) *(b)*

Fig. 21.17 Phantom frames showing eccentric rotation

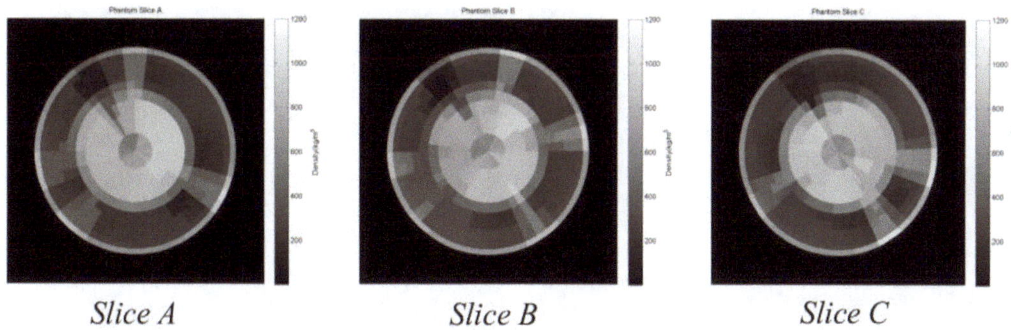

Fig. 21.18 Eccentric knot cluster reconstructions

Fig. 21.19 Dried Douglas fir sample

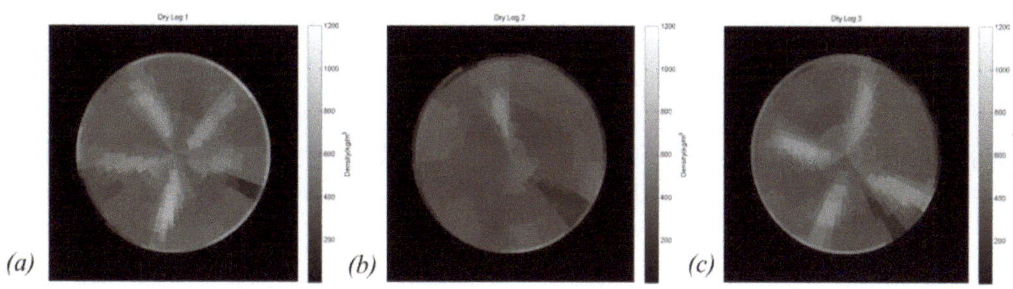

Fig. 21.20 Reconstruction of dried logs (**a**) Five knot cluster (**b**) single knot (**c**) Four knot cluster

The real log scanned consisted of a dry sample of Douglas fir measuring 27 cm in diameter and 90 cm long. The dried log exhibited numerous knotty sections along its length in addition to a multi-stepped radial crack running the length of the log. Figure 21.19 clearly shows the log along with the radial crack.

The log was mounted at a 1 in. eccentricity from the cart shafts and scanned. A voxel pattern of 72 sectors by 15 annuli was used to reconstruct 180 slices of thickness 0.48 cm each. Figure 21.20 shows cross cut planes of this coarse reconstruction from which the surfaces of the features of interest were identified using the open-source segmentation software ITK-SNAP [9]. Figure 21.21 shows the segmentation. It is clear that the knots are easily differentiable from the surrounding xylem, as well as their inclinations. The multi-stepped crack is also clearly defined. This shows the algorithm's robustness in normalizing and scaling the reconstruction from data that are subject to rigid body motion.

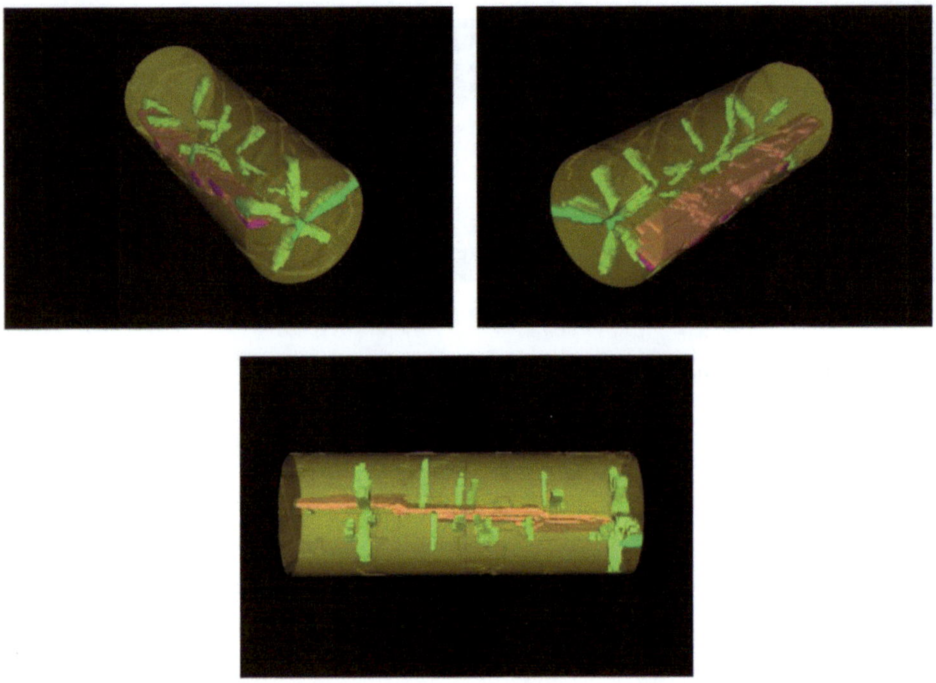

Fig. 21.21 Surface segmentation of log

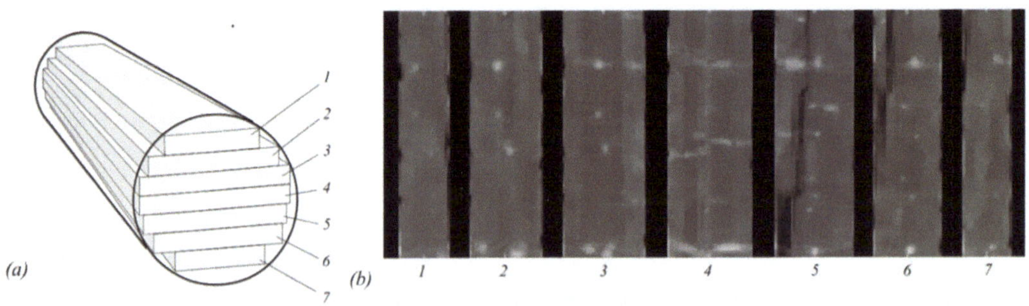

Fig. 21.22 Plane sawn saw pattern and reconstructed board surfaces

These data may then be used to make quality control decisions about the product from various cut patterns. Figure 21.22 shows how the volume may be used to consider the open lumber faces for a plane sawn cut pattern.

21.6 Conclusion

The high value of logs in modern sawmills has made internal sensing solutions an attractive potentiality in recovering maximum profits from raw material. CT scanning has received attention by researchers for several years as a potential solution to the problem of determining natural log features and damage that directly affects recoverable product. The problems associated with a harsh industrial environment (speed of sampling and reconstruction, incomplete or compromised data frames, and lack of precision in material manipulation) have thus far made CT systems in sawmills costly and ungainly to implement. This work uses an ART coarse resolution reconstruction scheme with techniques to maximize the reconstruction efficiency of the image space and to accommodate rigid body displacements of the log in the field of view. In total, the system is more resilient to bad data and lack of data (allowing for increased temporal resolution due to decreased sampling requirements) as well as less sensitive to displacements of the reconstructed trunk. Further optimization will make for a system capable of detecting major features of the lumber that is anticipated to be made.

Acknowledgements The authors sincerely thank the Natural Sciences and Engineering Research Council of Canada (NSERC) and FPInnovations, Inc., Vancouver, Canada, for their financial support of this work.

References

1. Lundahl, C.G.: Optimized processes in sawmills. PhD thesis, Lulea University of Technology (2007)
2. Varas, M.: Scheduling production for a Sawmill: A robust optimization approach. Int. J. Prod. Econ. **150**, 35–51 (2014)
3. An, Y., Schajer, G.S.: Geometry-based CT scanner for measuring logs in sawmills. Comput. Electron. Agric. **105**, 66–73 (2014)
4. Schajer, G.S.: Low resolution CT scanning for aspen log sorting. FPInnovations Report (2006)
5. Jian, L., Litao, L., Peng, C., Qi, S., Zhifang, W.: Rotating polar-coordinate ART applied in industrial CT image reconstruction. NDT E Int. **40**(4), 333–336 (2007)
6. Thibaudeau, C.: Fully 3D iterative CT reconstruction using polar coordinates. Med. Phys. **40**(11), 111904 (2013)
7. Andersen, A.H., Kak, A.C.: Simultaneous algebraic reconstruction technique (SART): a superior implementation of the ART algorithm. Ultrason. Imaging **6**(1), 81–94 (1984)
8. Herman, G.T.: Fundamentals of computerized tomography—image reconstruction from projections, 2nd edn. Springer, New York (2009)
9. Yushkevich, P.A.: User-guided 3D active contour segmentation of anatomical structures: significantly improved efficiency and reliability. Neuroimage **31**(3), 1116–1128 (2006)
10. Rihnnhofer, A., Petutschnigg, A., Andreu, J.-P.: Internal log scanning for optimizing breakdown. Comput. Electron. Agric. **41**, 7–21 (2003)

Chapter 22
Development and Experimental Validation of Thermally Stable Unimorph SMP Actuators Incorporating Transverse Curvature

Jason T. Cantrell and Peter G. Ifju

Abstract Shape memory polymers (SMP) have the potential to be utilized as a lightweight, solid state actuator in modern reconfigurable structures including as deployment systems for satellite solar panels or morphing aircraft wings. This paper is predominantly focused on the use of Veriflex-S®, a thermally activated SMP, and bi-directional carbon-fiber-reinforced polymer (CFRP) in a flexural unimorph actuator configuration. The disadvantage of a unimorph composite actuator (UCA) as opposed to an actuator with a SMP matrix or a SMP composite sandwich structure is that UCA behaves like a bimaterial strip when heated or cooled. This means that large temperature swings, like those seen in space environments, will result in large out-of-plane curvature. These deformations can greatly affect the effectiveness of reconfigurable structures. This paper explains the development and experimental validation of a closed-form solution for a thermally stable unimorph actuator which exhibits minimal out-of-plane deformation when subjected to a thermal stimulus. A closed-form solution of the SMP actuator was developed and a set of UCA actuators were experimentally evaluated utilizing digital image correlation (DIC) to validate the conceptual model created. The experimental results indicate that the closed-form solution appears to be accurate as the maximum out-of-plane deformations for several non-ideal thermally stable actuators were less than 0.6 mm for a 65 °C temperature change.

Keywords Digital image correlation • Shape memory polymer • Unimorph • Transverse curvature • Composite • Thermal stability

Nomenclature

c	Polymer width
CF	Carbon fiber
CFRP	Carbon-fiber-reinforced polymer
COV	Coefficient of variation
CTE	Coefficient of thermal expansion
L	Actuator length
MAV	Micro air vehicle
NA	Neutral axis
s	Substrate width
SMP	Shape memory polymer
t	Polymer thickness
T_g	Glass transition temperature
u, v, w	Lengthwise, widthwise, and vertical displacements
UCA	Unimorph composite actuator
x, y, z	Lengthwise, widthwise, and vertical coordinates
ρ	Substrate radius of curvature

J.T. Cantrell (✉) • P.G. Ifju
Mechanical and Aerospace Engineering Department, University of Florida,
571 Gale Lemerand Dr., MAE-C 134, Gainesville, FL 32611, USA
e-mail: jasontcantrell@gmail.com

S. Quinn, X. Balandraud (eds.), *Residual Stress, Thermomechanics & Infrared Imaging, Hybrid Techniques and Inverse Problems, Volume 9*, Conference Proceedings of the Society for Experimental Mechanics Series, DOI 10.1007/978-3-319-42255-8_22

22.1 Introduction

Shape memory polymers (SMPs) are a subset of smart materials with the ability to change their shape from a programmed state to a temporary state when activated via an external stimulus [1, 2]. Veriflex-S®, the SMP used in this study, uses a thermal external stimulus to allow for reconfiguration from the original/programmed state (flat beams in this study) to a deformed/stored shape (a U-shaped configuration for this study) and then return to a deployed shape (ideally the same as the original/programmed shape). Veriflex-S® and its higher glass transition temperature (T_g) counterpart Veriflex-E® have been used in applications including: deployment of satellite solar panels, active disassembly for electronics recycling, and morphing of aircraft structures [3–6]. Veriflex®, like other SMPs has two distinct classifications of stiffness and material behavior: the high modulus "glassy state" and the low modulus "rubbery state" [7–10]. When the SMP is below the T_g of 62 °C it is stiff, brittle, and has a high elastic modulus (~1.1 GPa). When Veriflex® is heated above T_g the modulus drops by several orders of magnitude (~1 MPa) and becomes pliable. Veriflex® can take and hold temporary shapes indefinitely if heated above its T_g, mechanically deformed, and held in place as it cools below T_g. The sample can return to its unconstrained original/programmed state with the reapplication of heat. Ideally the sample would return to 100 % of the original shape seen before the thermos-mechanical cycle; however, in reality Veriflex® can achieve a final shape that is only close to the original shape. A typical thermo-mechanical cycle is illustrated in Fig. 22.1 which shows the ideal versus actual stress–strain-temperature behavior for a SMP. The actual and idealized cycles match until the load is released in the stored position. Ideally, the SMP retains its current configuration; however, some relaxation occurs during this step after the load is released. This is shown as shape retention or fixity loss in Fig. 22.1. Once reheated and allowed to cool the SMP will try to return to its original/programmed shape. Unfortunately, as previously stated the SMP is often unable to recover fully and experiences some residual deformation or shape recovery loss. Researchers have studied both fixity and recoverability of SMPs extensively to determine their practicality in reconfigurable structures [9, 11–15].

Despite the lack of shape recovery and fixity in certain settings, SMPs are still seen as beneficial materials for adaptive structures. The current authors have successfully utilized SMPs in the past to provide a feasible solution for the folding of micro air vehicle (MAVs) wings and as a multipurpose morphing actuator with nearly 100 % recoverability without affecting fixity [5, 15–17]. This was accomplished with the incorporation of transverse curvature (similar to an extendable tape measure) into the carbon-fiber-reinforced polymer (CFRP) substrate of a unimorph composite actuator to increase the bending stiffness of the unimorph actuator to compensate for the lack of recovery of the SMP. A unimorph composite actuator (UCA) is described as an element capable of bi-stable configuration when supplied with an external stimulus (heat in this instance) consisting of one active layer (SMP) to which the stimulus is applied and one inactive layer (CFRP) that supports the active layer. The three types of UCAs that were tested are shown in Fig. 22.2 in the stored configuration while the variables of interest for the UCAs studied in this paper and previous experiments are shown in Fig. 22.3 on a concave transversely curved actuator in the programmed configuration. In previous studies a simple flat CFRP

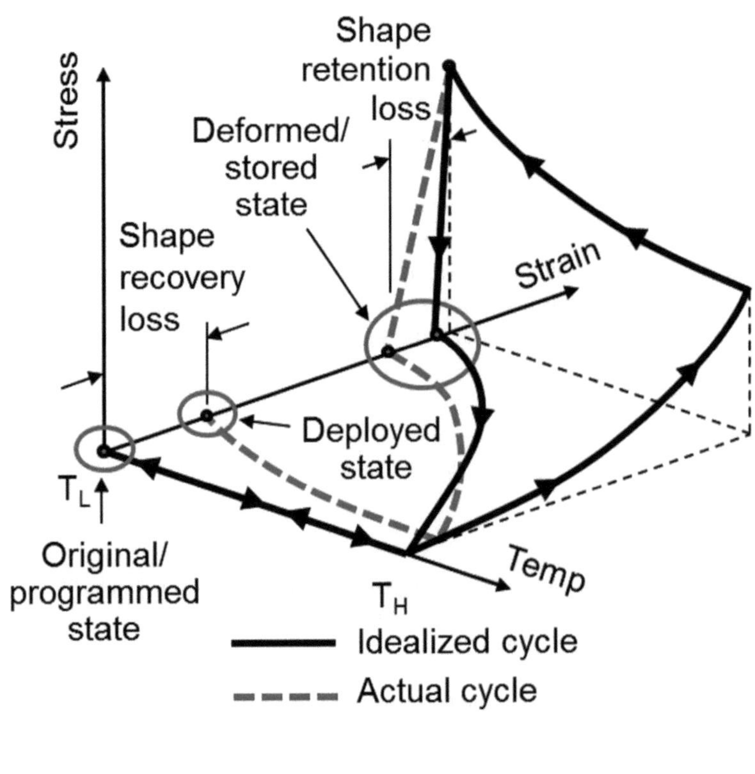

Fig. 22.1 Schematic of the stress–strain-temperature behavior of a SMP during a thermo-mechanical cycle

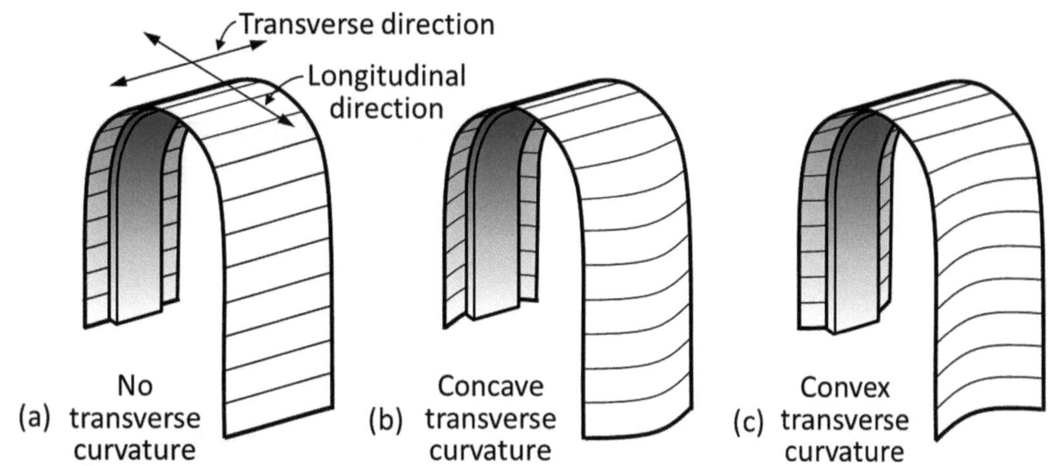

Fig. 22.2 Comparison of a UCA (**a**) without curvature (flat substrate), (**b**) with concave transverse curvature, and (**c**) with convex transverse curvature in the stored configuration

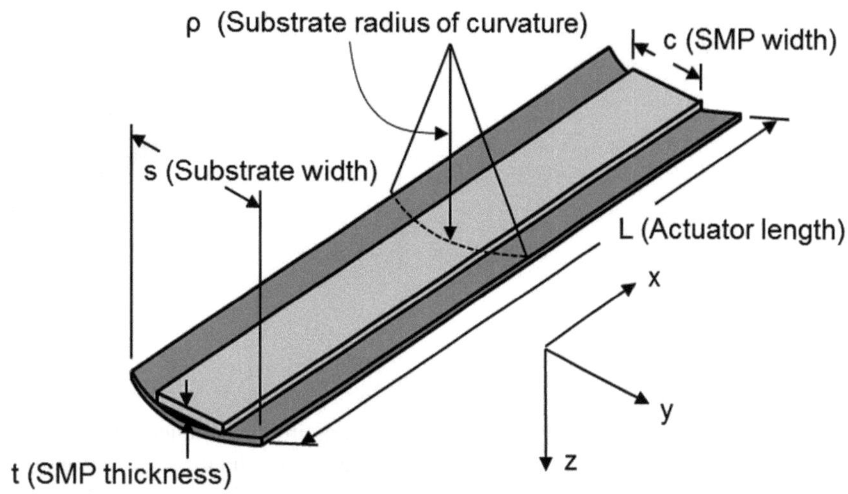

Fig. 22.3 Illustration of the variables of interest for this study on an original/programmed concave UCA

beam with SMP adhered to its surface was compared against a transversely curved CFRP beam to determine the effect of transverse curvature on the recovery and residual deformation of UCAs [5, 16]. A study on the effects of transverse curvature on the shape retention loss (fixity) and deflection angles over time was also performed [17]. These studies have shown that concave transverse curvature increases recoverability while not hindering shape fixity when compared to an actuator without transverse curvature. Additionally, the concave transverse curvature actuator outperforms the convex actuator with respect to both fixity and recoverability with all other variables held constant.

The concept of using transverse curvature in morphing structures is not a new one and has been studied extensively in a tape spring actuator setting without SMP [18–30]. Primarily, transversely curved composite structures have been used to create tape spring actuators for use in self-deploying space structures like antennas [22, 23, 31, 32] and deployable booms [25, 33–36]. One of the design requirements of these actuators is that they be thermally stable, meaning that when the actuators are exposed to environmental conditions the actuators do not warp, expand, or contract such that the out-of-plane shape changes. A thermally stable actuator is especially critical in space environments where large swings in temperature are common and can greatly affect the deployment and effectiveness of reconfigurable structures. A tape spring actuator without SMPs is thermally stable; however, deployable structures without SMPs often utilize either electric motors or stored energy like mechanical springs to actuate reconfigurable structures. Therefore, these devices often require locking mechanisms which increases launch weight or in the case of stored energy mechanisms rapidly deploy within seconds which greatly affects the dynamics of the spacecraft. Using SMPs is advantageous over the current state-of-the-art technologies because SMPs can provide an inexpensive and lightweight solid state actuator to slowly deploy antennae, solar panels, or other deployable structures. Some researchers have begun using SMPs in deployable actuators by either using SMP as the composite matrix material or creating a unimorph like the current authors [3, 37–40]. The use of SMPs as a composite matrix material allows for the actuator to behave like a tape spring but also maintain a stored shape without the need for locking

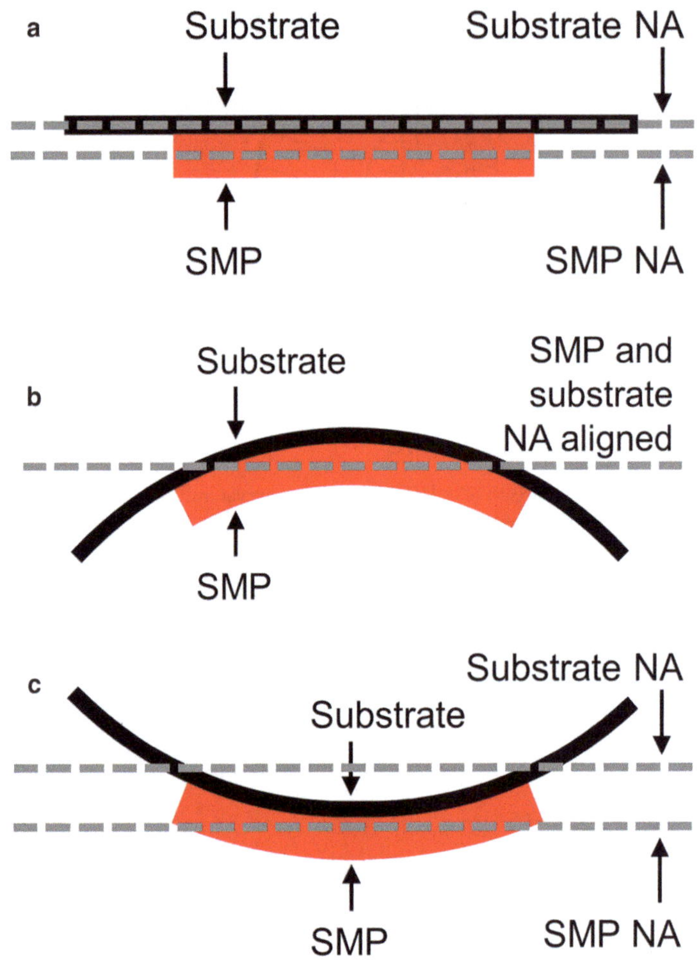

Fig. 22.4 Comparison of the neutral axis (NA) locations for the UCAs (**a**) without curvature (flat substrate), (**b**) with concave transverse curvature, and (**c**) with convex transverse curvature

mechanisms or motors. However, in high temperature environments these SMP matrix actuators become weak due to the drop in SMPs modulus. The solution to this is the unimorph tape spring or unimorph composite actuator which combines the stiffness of the composite substrate with the reconfigurable nature of SMP. Unfortunately, the disadvantage of a UCA as opposed to an actuator with a SMP matrix or a SMP composite sandwich structure is that UCA behaves like a bimaterial strip when heated or cooled. This means that large out-of-plane curvature or deflections may occur in space environments due to the large difference in the coefficients of thermal expansion. The solution to eliminating this out-of-plane curvature or deflection is to create a UCA such that the neutral axis (NA) of the CFRP substrate is at the same position as the SMP. If the neutral axes are aligned the UCA should behave similar to a composite sandwich structure or a symmetric and balanced CFRP layup where out-of-plane deformation due to thermal expansion is minimal. In order to accomplish this task a closed form solution for the alignment of the UCAs NAs must be found and evaluated experimentally to determine if the closed form solution is valid. Figure 22.4 shows a comparison of the three UCA geometries available to develop a closed form solution for. Figure 22.4 shows that NA alignment is only possible for a concave transversely curved actuator and thus a closed form solution for this geometry will be developed. The details of this thermal stability closed form solution for a concave transverse curvature UCA as well as the experimental procedures and results used to validate the closed form solution are described in the following sections.

22.2 Fabrication Methods, Experimental Set-Ups, and Procedures

22.2.1 Materials and Specimen Fabrication

The individual materials used to form the UCAs are Veriflex-S® shape memory polymer and [±45°] oriented, plain weave, bi-directional carbon-fiber-reinforced polymer. The properties for the CFRP or simply carbon fiber (CF) were gathered in the

Table 22.1 Material properties for [±45°] oriented, plain weave, bi-directional carbon fiber and Veriflex-S® shape memory polymer

	CFRP [±45°]		Veriflex-S®	
Property	Value	COV	Value	COV
Coefficient of thermal expansion (α)	$2.5 \pm 0.2E^{-6}\,K^{-1}$	8 %	$160 \pm 15E^{-6}\,K^{-1}$	9.6 %
Young's modulus (E)	11.8 ± 0.3 GPa	2.8 %	1.1 ± 0.05 GPa	4.5 %
Poisson's ratio (ν)	0.79 ± 0.02	2.5 %	0.39 ± 0.01	2.6 %
Glass transition temperature[a] (T_g)	121 °C	–	62 °C	–

[a]Manufacturer published values

[±45°] orientation as this was the direction of actuation. The materials were characterized to establish the modulus of elasticity, coefficient of thermal expansion (CTE), and Poisson's ratio of both materials below their glass transition temperatures. These values are provided below in Table 22.1.

Each unimorph composite actuator is comprised of a carbon-fiber-reinforce polymer substrate bonded to a thin sheet of SMP. Both unimorphs incorporating transverse curvature in the substrate and flat CF unimorphs followed the same fabrication procedures. A single layer of [±45°] oriented, plain weave, bi-directional CF, pre-impregnated with epoxy, was cut and placed on a Teflon® covered flat plate or curved tooling board. The [±45°] configuration was used versus a [0°/90°] because it allowed for the CF to be rolled to a smaller diameter in the stored state and was more stable during storage. The entire assembly was covered in an additional layer of Teflon® film, wrapped with breather fabric, vacuum bagged, and cured in a convection oven at 130 °C for 4 h. After curing, the CF was cut to the proper dimensions, and then a Veriflex-S® shape memory polymer panel was bonded to it using Araldite® 2011 two-part epoxy at room temperature to create a unimorph composite beam.

22.2.2 *Digital Image Correlation and Experimental Set-up*

The primary objective of this research was to validate the closed-form solution of a thermally stabile actuator by determining the deformation and shape of actuator samples. This was done through the use of DIC, a non-contact, full-field shape and deformation measurement technique developed at the University of South Carolina [41–43]. The DIC system utilized two Point Grey Research, 1.5-megapixel, grayscale cameras that simultaneously captured images of the random speckle pattern applied to the samples. Images were captured via VIC-Snap™ 2009 and later processed via VIC-3D™ 2009 to determine the deformations and strains. A pair of reference images (one image per camera) of the CF substrate side of the actuators were taken after the substrate was brought to an initial temperature of 25 °C and held for 20 min to allow it to reach steady state. These reference images were contrasted against pairs of images taken at each temperature hold to determine the deflection as the sample was heated. Ten images were taken at each temperature and averaged to account for any vibrations from the environmental chamber. Figure 22.5 shows the DIC experimental set-up used to measure the out-of-plane position of the UCA specimens.

The UCAs were placed in a Sun Systems® Model EC12 environmental chamber with a glass window which was used to regulate the temperature to the desired point in the temperature cycle. Each sample was placed in the chamber and heated to an initial steady-state temperature of 25 °C so that all samples would have a common starting point. Each successive temperature change was made in increments of five degrees up to the final temperature of 90 °C. The specimens were held at each temperature for 20 min to ensure the samples reached steady-state before an image was taken for DIC as previous research has confirmed this is an adequate hold time for the UCA specimens [16]. The temperature was monitored via a thermocouple inside of the chamber and confirmed via a Fluke® 561 series infrared thermometer.

22.2.3 *Experimental Procedure*

The step-by-step procedure for measuring the out-of-plane residual deformation of the UCA using DIC during a temperature cycle is enumerated below.

Step 1. After applying a speckle pattern to the sample, place the undeformed UCA in the in the environmental chamber for 20 min at 25 °C.

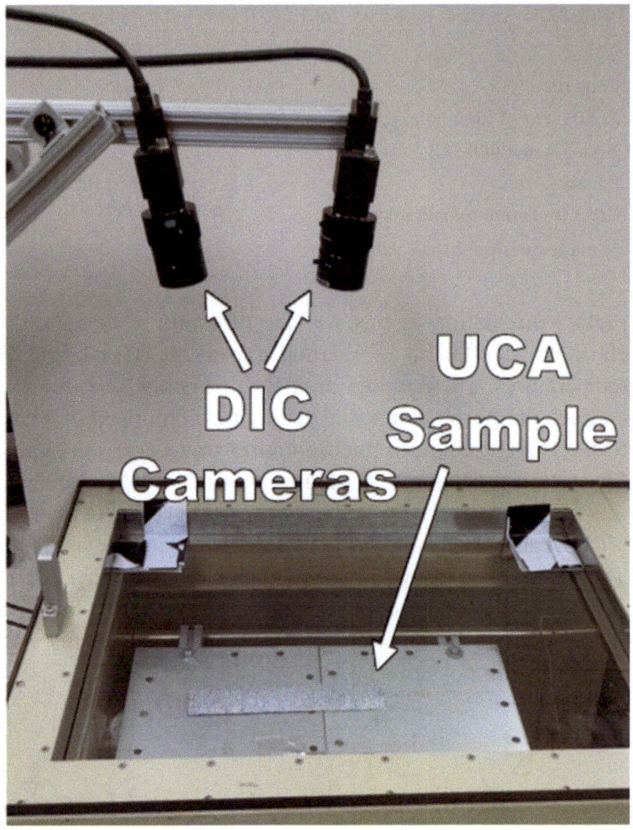

Fig. 22.5 Experimental set-up used to photograph the UCAs through the glass window of the environmental chamber during the heating cycle

Step 2. Take an initial (reference) image of the UCA in the environmental chamber using the DIC set-up shown in Fig. 22.5.
Step 3. Increase the temperature of the environmental chamber by 5 °C and allow the sample to heat for 20 min.
Step 4. Take ten images of the UCA at the new temperature.

Measuring out-of-plane deformation at each additional temperature step was completed by following steps three and four until the maximum observation temperature of 90 °C was reached.

22.3 Theoretical Calculations

The primary assumption for this closed-form solution is that both the SMP and CF substrate have a curved geometry with a negligible glue layer between the two. This means that by assuming the radius of curvature is much greater than the thickness of the substrate the geometry can be assumed to be the arc of a circle. The equations for the arc and segments of a circle are taken from Beer, et al. and Zwillinger respectively [44, 45]. An illustration of the arc of a circle centroid/neutral axis definition is provided for reader clarity in Fig. 22.6.

The UCAs used in this study have a radius of curvature approximately 180 times greater than the thickness of the substrate so the arc approximation holds and the centroid of the carbon fiber substrate can be written as:

$$\overline{X}_{\mathrm{Substrate}} = \frac{r_{\mathrm{Substrate}} \sin \alpha_{\mathrm{Substrate}}}{\alpha_{\mathrm{Substrate}}} \tag{22.1}$$

Where the substrate radius of curvature is given as

$$r_{\mathrm{Substrate}} = \text{Actuator Radius of Curvature} - \frac{\text{Thickness of substrate}}{2}$$

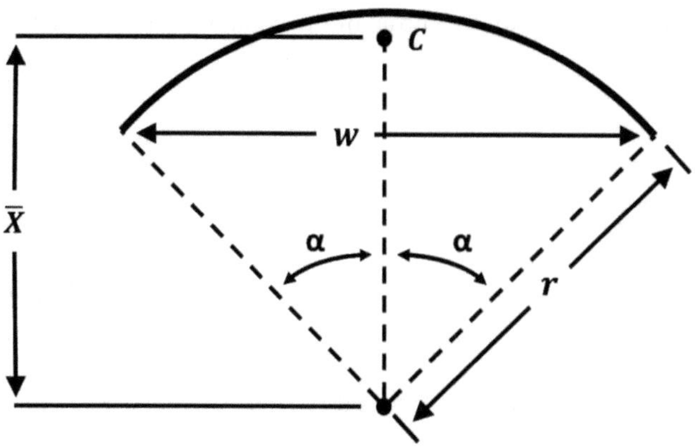

Fig. 22.6 Illustration of the arc of a circle centroid definition and the variables used in the derivation of a thermally stable UCA

$$r_{Substrate} = R - \frac{T_{Substrate}}{2} \tag{22.2}$$

And the angle from the centroid to the arc edge is

$$\alpha_{Substrate} = \sin^{-1}\left(\frac{\text{Width of substrate}}{2^*\text{Radius of substate}}\right) = \sin^{-1}\left(\frac{w_{Substrate}}{2^* r_{Substrate}}\right) \tag{22.3}$$

The area of the carbon fiber substrate is given as

$$A_{Substrate} = 2^*\alpha_{Substrate}{}^* r_{Substrate} \tag{22.4}$$

So with shortened subscripts

$$\overline{X}_{Sub} = \frac{r_{Sub}\sin\alpha_{Sub}}{\alpha_{Sub}} \quad \text{And} \quad A_{Sub} = 2^*\alpha_{Sub}{}^* r_{Sub}$$

The centroid of the SMP layer is also given as the arc of a circle as the radius of curvature is approximately 80 times greater than the thickness of the substrate therefore

$$\overline{X}_{SMP} = \frac{r_{SMP}\sin\alpha_{SMP}}{\alpha_{SMP}} \tag{22.5}$$

Where the radius of the SMP is

$$r_{SMP} = \text{Actuator Radius of Curvature} - \text{Thickness of substrate} - \frac{\text{Thickness of SMP}}{2}$$

$$r_{SMP} = R - T_{Sub} - \frac{T_{SMP}}{2} \tag{22.6}$$

And

$$\alpha_{SMP} = \sin^{-1}\left(\frac{\text{Width of SMP}}{2^*\text{Radius of SMP}}\right) = \sin^{-1}\left(\frac{w_{SMP}}{2^* r_{SMP}}\right) \tag{22.7}$$

The area of the SMP like the substrate is

$$A_{SMP} = 2^*\alpha_{SMP}{}^* r_{SMP} \tag{22.8}$$

So the neutral axis/centroid of the composite actuator is:

$$\overline{X}_{\text{Composite Actuator}} = \overline{X}_{CA} = \frac{\sum \overline{x}A}{\sum A} = \frac{n\overline{X}_{SMP}A_{SMP} + \overline{X}_{Sub}A_{Sub}}{nA_{SMP} + A_{Sub}} \tag{22.9}$$

Where

$$n = \frac{\text{Elastic Modulus of Substrate}}{\text{Elastic Modulus of SMP}} = \frac{E_{Sub}}{E_{SMP}} \tag{22.10}$$

For equilibrium to be satisfied

$$\overline{X}_{CA} = \overline{X}_{SMP}$$

So

$$\overline{X}_{SMP} = \frac{n\overline{X}_{SMP}A_{SMP} + \overline{X}_{Sub}A_{Sub}}{nA_{SMP} + A_{Sub}}$$

Rearranging the equation gives

$$\overline{X}_{SMP}(nA_{SMP} + A_{Sub}) = n\overline{X}_{SMP}A_{SMP} + \overline{X}_{Sub}A_{Sub}$$

$$n\overline{X}_{SMP}A_{SMP} + \overline{X}_{SMP}A_{Sub} = n\overline{X}_{SMP}A_{SMP} + \overline{X}_{Sub}A_{Sub}$$

$$n\overline{X}_{SMP}A_{SMP} + \overline{X}_{SMP}A_{Sub} = n\overline{X}_{SMP}A_{SMP} + \overline{X}_{Sub}A_{Sub}$$

$$\overline{X}_{SMP}A_{Sub} = \overline{X}_{Sub}A_{Sub}$$

$$\overline{X}_{Sub} = \overline{X}_{SMP} \tag{22.11}$$

Therefore to have the neutral axis of the actuator at the neutral axis of the SMP the substrate and SMP neutral axes should be equal. This gives the following:

$$\overline{X}_{Sub} = \overline{X}_{SMP} = \frac{r_{Sub}\sin\alpha_{Sub}}{\alpha_{Sub}} = \frac{r_{SMP}\sin\alpha_{SMP}}{\alpha_{SMP}}$$

Where

$$r_{Sub} = R - \frac{T_{Sub}}{2}, \quad \alpha_{Sub} = \sin^{-1}\left(\frac{w_{Sub}}{2^{*}r_{Sub}}\right),$$

$$r_{SMP} = R - T_{Sub} - \frac{T_{SMP}}{2}, \quad \alpha_{SMP} = \sin^{-1}\left(\frac{w_{SMP}}{2^{*}r_{SMP}}\right)$$

Using these equations with a prescribed actuator radius of curvature, substrate thickness, and SMP thickness will give a range of values for which the neutral axis of the substrate and SMP align resulting in a thermally stable actuator. An example case is shown in Fig. 22.7 with an actuator radius of curvature of 63.5 mm, a substrate thickness of 0.35 mm, and a SMP thickness of 0.8 mm. Figure 22.7 illustrates the combination of widths at each NA location that would result in a thermal stable actuator, which when heated or cooled would result in no out-of-plane deformation.

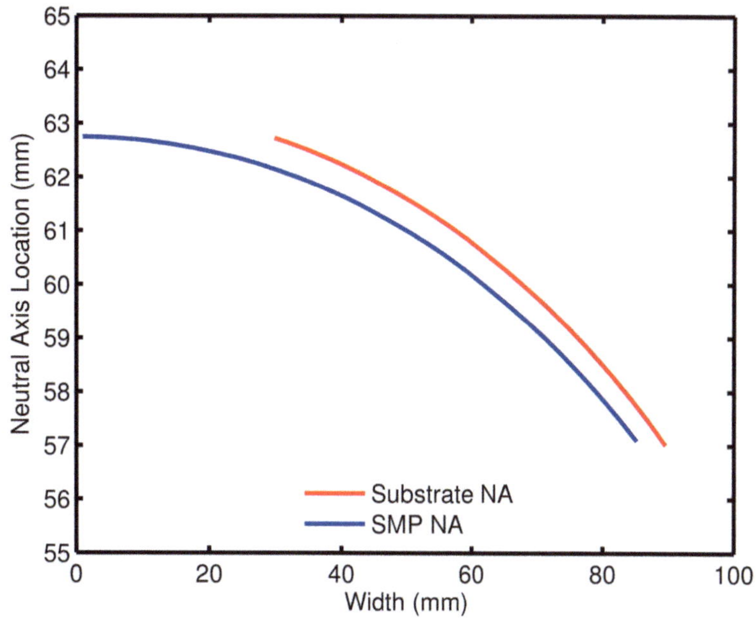

Fig. 22.7 Substrate and SMP width versus NA location for an actuator radius of curvature of 63.5 mm, a substrate thickness of 0.35 mm, and a SMP thickness of 0.8 mm showing the combinations of SMP and substrate widths which result in a thermally stable actuator

Table 22.2 Radius of curvature (ρ), SMP thickness (t) and substrate width (s) possibilities and their respective designations for the UCAs studied

Nomenclature and supporting values			
Transverse curvature	Flat	Low	High
Radius of curvature (ρ)	∞	127 mm	62.5 mm
SMP thickness (t)	**Thin**	**Thick**	–
Thickness values	0.8 mm	1.6 mm	–
Substrate width (s)	**Narrow**	**Intermediate**	**Wide**
Width values	25.4 mm	38.1 mm	50.8 mm

22.4 Experimental Results

Out-of-plane deflection (w) was the main focus of this study as it was used to determine the susceptibility of UCAs to deformation during temperature swings and to determine if the theoretical calculations for a thermal stability were accurate. The goal was to determine how the alignment of the substrate and SMP neutral axes affected the out-of-plane deflection with increasing temperature. Ideally, new concave UCAs with aligned NAs would be created and used for this experiment; however, a lack of SMP material led to the reuse of UCAs from previous research. Therefore, the samples used in this study did not have ideal thermal stability properties, but would allow general trends to be established about thermal stability when the NA was aligned either above or below the ideal location. As shown previously in Fig. 22.3 there were five variables: actuator length, substrate radius of curvature, substrate width, SMP width, and SMP thickness investigated. It was determined that the SMP width and actuator length were to be held constant which left substrate radius of curvature, SMP thickness, and substrate width as the variables that were altered. Three transverse curvatures were investigated and given designations; **flat** (infinite radius of curvature), **low transverse curvature** (127 mm radius of curvature), and **high transverse curvature** (63.5 mm radius of curvature). Two SMP thicknesses were investigated; **thin** (0.8 mm) and **thick** (1.6 mm). Finally, three substrate widths were tested; **narrow** (25.4 mm), **intermediate** (38.1 mm), and **wide** (50.8 mm). Table 22.2 shows the variations of curvature, thickness, and width investigated and their respective designations for convenience of the reader for the remainder of the paper.

Using the theoretical results explained previously, a graphical representation of thermally stable UCAs was created and is shown in Fig. 22.8 along with the UCAs available for testing. Using a constant SMP width of 12.7 mm and a substrate thickness of 0.35 mm the design space of acceptable substrate width and curvature combinations was created. The dashed line in Fig. 22.8 indicates the acceptable width and curvature combinations for a SMP thickness of 1.6 mm while the solid line indicates NA alignment for a SMP thickness of 0.8 mm. As expected, thicker SMPs warranted higher radius of curvatures and wider substrate widths to ensure the alignment of the SMP and substrate NAs. The experimental UCAs above

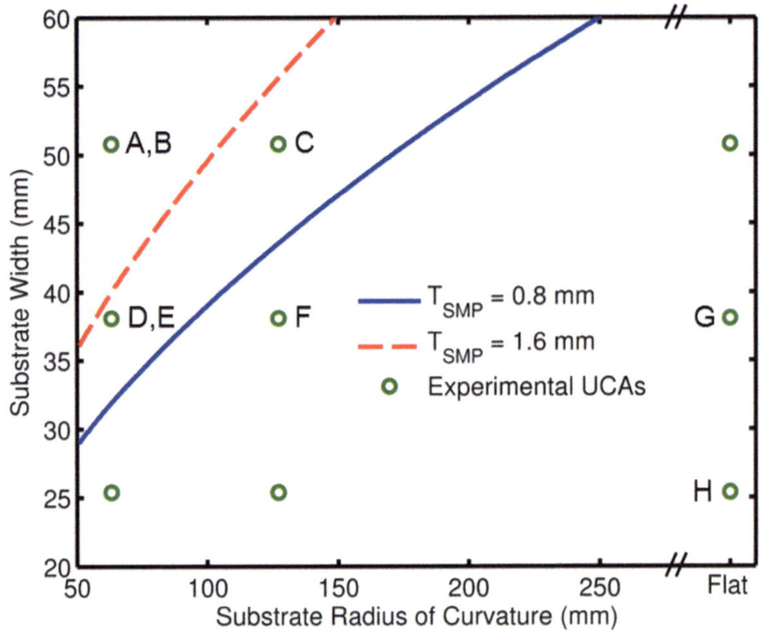

Fig. 22.8 Graph illustrating the combinations of thermally stable UCAs for various substrate radius of curvatures and widths with a substrate thickness of 0.35 mm and SMP width of 12.7 mm. Each line indicates the combinations which result in the alignment of the SMP and substrate neutral axes at specified SMP thickness. The UCAs available to be used for experimental validation of the closed-form solution are also shown for reference

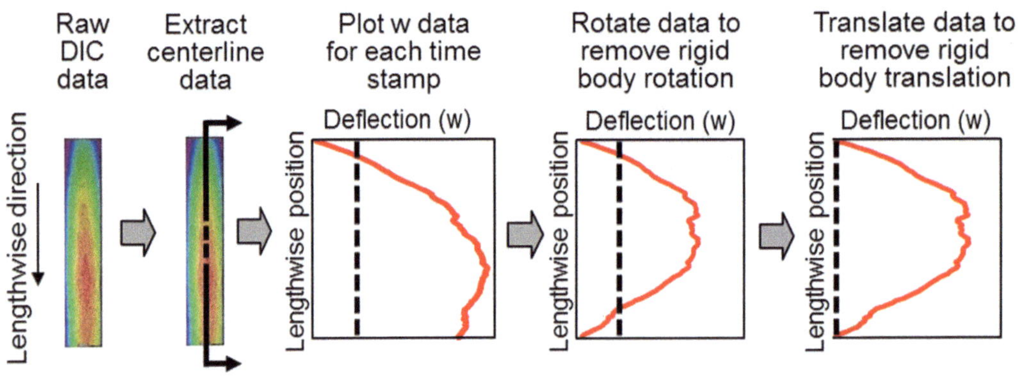

Fig. 22.9 Illustration of the process of converting the DIC data to the desired coordinate system and removing rigid body motion in the longitudinal direction

the lines indicate UCAs with the substrate NA below the SMP and vice versa for the UCAs below the respective lines. Two thick SMP, concave transverse curvature UCAs, one with a substrate NA above the SMP NA and one below were selected for study. Six thin SMP UCAs, four with concave transverse curvature and two flat specimens were also evaluated.

Before evaluating the results of thermal stability experiments post-processing of the DIC data was required to properly determine the out-of-plane deflection of the UCA at each temperature. Post-processing was done by extracting the xyz coordinates and uvw displacements for the centerline of each sample at the desired temperature. Next, the data was written to an Excel file, the deformation (w) data sorted by timestamp, and shifted to the desired coordinate system via MATLAB. Once in the desired x-z plane, the data was rotated to eliminate rigid body motion making sure to rotate the sample in the x or lengthwise direction to maintain the correct displacement directions. After rotation the data underwent a final vertical translation to the x-axis ensuring all images can be compared in the same coordinate system. This process is illustrated in Fig. 22.9.

Data for the UCA samples was collected in five degree intervals from 25 to 90 °C. The centerline shape was measured for the reference (at 25 °C) and at various times during the temperature cycle. To obtain the deformation (w), the reference shape at 25 °C was subtracted from the shape at each time in the temperature cycle. The lengthwise position versus out-of-plane deflection was obtained for each of the eight UCAs as well as the maximum out-of-plane deflection throughout the heating cycle. The maximum out-of-plane deflection for the UCAs is shown in Table 22.3 while the maximum out-of-plane deflections for the thin SMP, intermediate substrate UCAs of various curvatures are shown in Fig. 22.10a, b. Additionally, the lengthwise position versus out-of-plane deflection for these flat, low, and high transverse curvature UCAs are shown in

Table 22.3 Maximum out-of-plane deflection during the heating cycle for the UCA samples tested

Specimen code	UCA combination Substrate curvature (ρ), SMP thickness (t), substrate width (s)	Maximum deflection (mm)
A	High, Thin, Wide	0.27
B	High, Thick, Wide	0.32
C	Low, Thin, Wide	0.38
D	High, Thin, Intermediate	0.16
E	High, Thick, Intermediate	0.58
F	Low, Thin, Intermediate	0.33
G	Flat, Thin, Intermediate	33.15
H	Flat, Thin, Narrow	31.96

Fig. 22.10c–e respectively to give an idea of the deformations along the length of the UCA as temperature increases over time. The results in Table 22.3 confirm that flat substrate UCAs perform poorly and experience the largest amount of out-of-plane deflections with increasing temperature as the NAs are the farthest from alignment and these UCAs have the lowest overall stiffness without transverse curvature. The high curvature, thin SMP, intermediate substrate width UCA has the least amount of deflection and appears to confirm that NA alignment does increase thermal stability as temperature increases. All of the actuators tested have less than 0.6 mm of deflection throughout the temperature cycle suggesting that the combination of increased stiffness due to transverse curvature and being closer to NA alignment aids in reducing out-of-plane deflections with temperature fluctuations. Looking at Fig. 22.10 reveals additional details about NA alignment and out-of-plane deflection. In Fig. 22.10a the flat UCA has nearly of its out-of-plane deflection occur before the glass transition temperature. After going above T_g and transitioning to the low stiffness rubber phase the actuator only deflects an additional 50 μm after deflecting nearly 33 mm in the high stiffness phase. Fig. 22.10b reveals two sets of behavior as the two actuators are on either side of the NA alignment. The low curvature UCA whose substrate NA is above the SMP NA reaches T_g and continues to deform with increasing temperature. Additionally, the out-of-plane deflection increases above T_g suggesting that while the modulus of the SMP maybe low, the nearly threefold increase in CTE in the rubbery phase which has been noted by other researchers in Veriflex® SMPs allows the UCA to continue to deform as the temperature increases [46, 47]. When the substrate NA is below the SMP NA, but near alignment the out-of-plane deflection appears to be hindered as deflection peaks at 0.16 mm and then regresses after the SMP enters the rubbery phase. After a slight regression there is an increase in out-of-plane deflection again which may be caused by the large difference in CTEs between the substrate and SMP. Further testing at higher temperatures will be needed to determine if deflection is increasing or remaining approximately constant. Additional testing will also be required to determine the repeatability of the heating cycle results.

22.5 Conclusion

A closed form solution for a thermally stable unimorph composite actuator with aligned neutral axes was created and tested experimentally using carbon-fiber-reinforced polymer as a rigid substrate for Veriflex-S® shape memory polymer. Existing unimorph composite actuators with varying neutral axes were evaluated versus increasing temperatures to determine the effect of transverse curvature, carbon-fiber-reinforced polymer substrate width, and shape memory polymer thickness on thermal stability. Images of the actuators were captured at various temperatures and then processed to measure the out-of-plane deflection at each temperature. In the research conducted, a unimorph composite actuator with a 63.5 mm radius of curvature, 0.8 mm thick shape memory polymer layer, and 38.1 mm substrate width had a maximum out-of-plane deformation of 160 μm over a 65 °C temperature change while a flat actuator with the same substrate width and shape memory polymer thickness had a maximum deflection of 33.15 mm. All actuators near the thermally stable region where the neutral axes of the shape memory polymer and substrate are aligned exhibited out-of-plane deflections of less than 0.6 mm. These results indicate that increased transverse curvature and alignment of the neutral axes reduce the deformation seen with changing temperatures and can be considered a first step in the confirmation of the closed form solution for thermally stable unimorph composite actuators.

Future research will include developing a finite element model of a thermally stable unimorph composite actuator as well as the actuators used in the experiments in this paper. The results of the finite element model will be compared to the results presented here as well as the closed form solution. Additionally, further testing of the present actuators will be conducted to determine the repeatability of the out-of-plane deflection with increasing temperature.

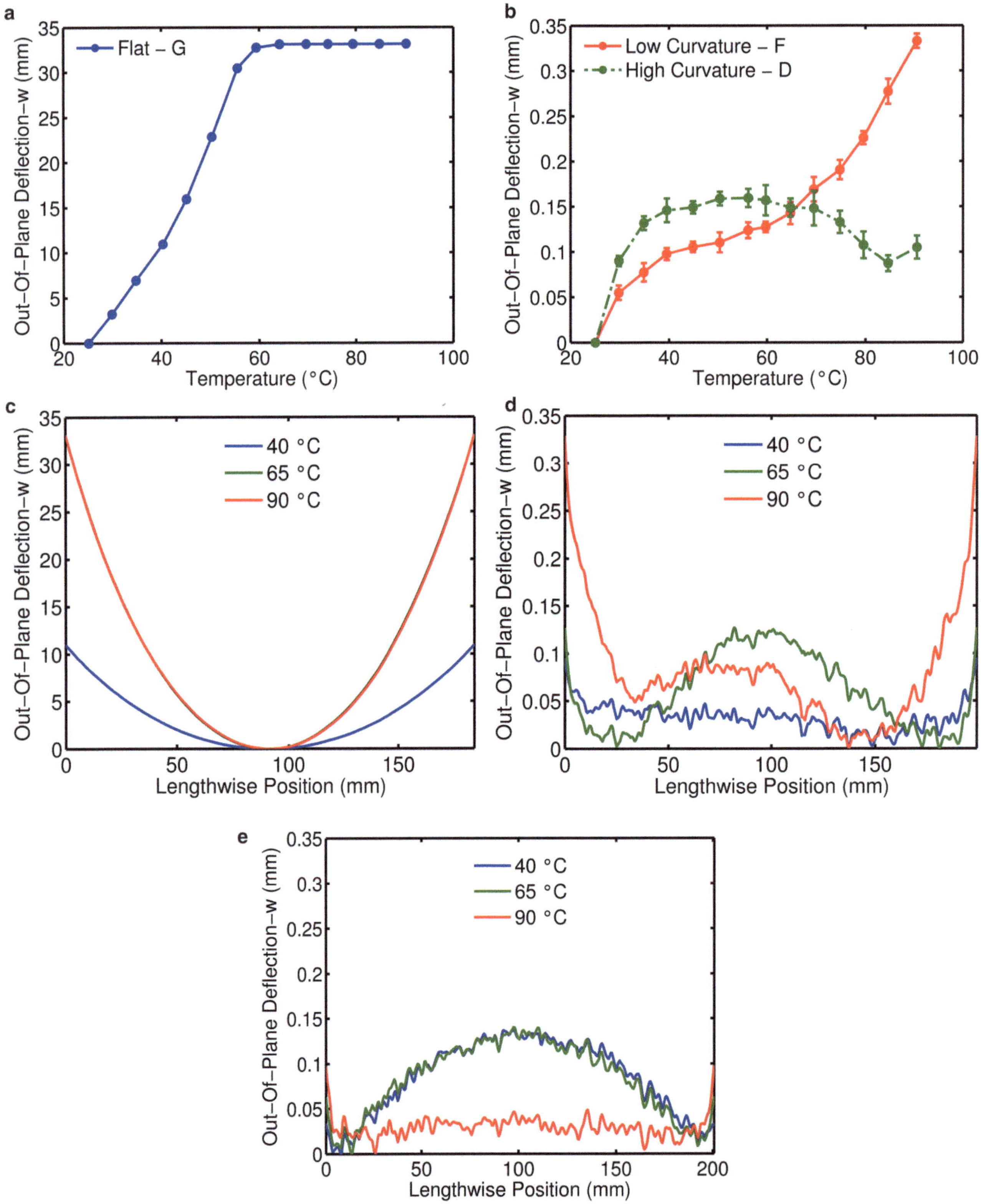

Fig. 22.10 Out-of-plane deflection versus temperature for the thin SMP, intermediate substrate, (**a**) flat (G), (**b**) low transverse curvature (F), and high transverse curvature UCAs (D). Out-of-plane deflection versus lengthwise position for the thin SMP, intermediate substrate, (**c**) flat (G), (**d**) low transverse curvature (F), and (**e**) high transverse curvature (D) UCAs at select temperatures

References

1. Beloshenko, V.A., Varyukhin, V.N., Voznyak, Y.V.: The shape memory effect in polymers. Russ. Chem. Rev. **74**, 265–283 (2005). doi:10.1070/RC2005v074n03ABEH000876
2. Leng, J., Lu, H., Liu, Y., et al.: Shape-memory polymers—a class of novel smart materials. MRS Bull. **34**(11), 848–855 (2009)
3. Lan, X., Liu, Y., Lu, H., et al.: Fiber reinforced shape-memory polymer composite and its application in a deployable hinge. Smart Mater. Struct. **18**, 024002 (2009). doi:10.1088/0964-1726/18/2/024002
4. Carrell, J., Tate, D., Wang, S., Zhang, H.-C.: Shape memory polymer snap-fits for active disassembly. J. Clean. Prod. **19**, 2066–2074 (2011). doi:10.1016/j.jclepro.2011.06.027
5. Cantrell, J.T., Ifju, P.G.: Novel design concepts for micro Air vehicles and unimorph composite actuators incorporating transverse curvature. In: Tandon, G.P., Baur, J.W., McClung, A.J.W. (eds.) Shape-Memory Polymers for Aerospace Applications: Novel Synthesis, Modeling, Characterization and Design, pp. 1–22. DEStech Publications, Lancaster (2015)
6. Joo, J., Smyers, B., Beblo, R., et al.: Load-bearing multi-functional structure with direct thermal harvesting for thermally activated reconfigurable wing design. In: International Conference on Composite Materials, pp. 1–6. Society of Composite Materals, Jeju (2011)
7. McClung, A.J.W., Tandon, G.P., Baur, J.W.: Strain rate- and temperature-dependent tensile properties of an epoxy-based, thermosetting, shape memory polymer (Veriflex-E). Mech. Time-Depend. Mater. **16**, 205–221 (2011). doi:10.1007/s11043-011-9148-7
8. McClung, A.J.W., Tandon, G.P., Goecke, K.E., Baur, J.W.: Non-contact technique for characterizing full-field surface deformation of shape memory polymers at elevated and room temperatures. Polym. Test. **30**, 140–149 (2011). doi:10.1016/j.polymertesting.2010.11.010
9. McClung, A.J.W., Tandon, G.P., Baur, J.W.: Fatigue cycling of shape memory polymer resin. In: Proulx, T. (ed.) Mechanics of Time-Dependent Materials and Processes in Conventional and Multifunctional Materials, pp. 119–127. Springer, New York (2011)
10. Klesa, J., Placet, V., Foltête, E., Collet, M.: Experimental evaluation of the rheological properties of Veriflex® shape memory polymer. ESOMAT 2009—8th European Symposium on Martensitic Transformations (2009). doi:10.1051/esomat/200904006
11. Tandon, G.P., Goecke, K., Cable, K., Baur, J.: Durability assessment of styrene- and epoxy-based shape-memory polymer resins. J. Intell. Mater. Syst. Struct. **20**, 2127–2143 (2009). doi:10.1177/1045389X09348255
12. Nahid, M.N.H., Wahab, M.A.A., Lian, K.: Degradation of shape memory polymer due to water and diesel fuels. In: Proulx, T. (ed.) Mechanics of Time-Dependent Materials and Processes in Conventional and Multifunctional Materials, vol. 3, pp. 37–48. Society of Experimental Mechanics, New York (2011)
13. McClung, A.J.W., Tandon, G.P., Baur, J.W.: Deformation rate-, hold time-, and cycle-dependent shape-memory performance of Veriflex-E resin. Mech. Time-Depend. Mater. **17**, 39–52 (2011). doi:10.1007/s11043-011-9157-6
14. Fulcher, J.T., Karaca, H.E., Tandon, G.P., Lu, Y.C.: Thermomechanical and shape memory properties of thermosetting shape memory polymer under compressive loadings. J. Appl. Polym. Sci. **129**, 1096–1103 (2012). doi:10.1002/app.38791
15. Cantrell, J.T., Ifju, P.G.: Unimorph shape memory polymer actuators incorporating transverse curvature in the substrate. In: 2014 Annual Conference on Experimental and Applied Mechanics, pp. 1–10. Springer International Publishing, Greenville, (2014)
16. Cantrell, J.T., Ifju, P.G.: Experimental characterization of unimorph shape memory polymer actuators incorporating transverse curvature in the substrate. Exp. Mech. **55**, 1395–1409 (2015). doi:10.1007/s11340-015-0035-z
17. Cantrell, J.T., Van Hall, J.R., Young, A.J., Ifju, P.G.: Experimental characterization of shape fixity in transversely curved unimorph shape memory polymer actuators. In: 2015 Annual Conference on Experimental and Applied Mechanics, pp. 1–9. Springer International Publishing, Costa Mesa (2015)
18. Seffen, K.A.: On the behavior of folded tape-springs. J. Appl. Mech. **68**, 369 (2001). doi:10.1115/1.1365153
19. Seffen, K.A., You, Z., Pellegrino, S.: Folding and deployment of curved tape springs. Int. J. Mech. Sci. **42**, 2055–2073 (2000). doi:10.1016/S0020-7403(99)00056-9
20. Seffen, K.A., Pellegrino, S.: Deployment dynamics of tape springs. Proc. R Soc. A Math. Phys. Eng. Sci. **455**, 1003–1048 (1999). doi:10.1098/rspa.1999.0347
21. Pellegrino, S.: Bistable shell structures. In: 46th AIAA/ASME/ASCE/AHS/ASC Structures, Structural Dynamics, and Materials Conference, pp. 1–8. AIAA, Austin (2005)
22. Soykasap, O., Pellegrino, S., Howard, P., Notter, M.: Folding large antenna tape spring. J. Spacecr. Rocket. **45**, 560–567 (2008). doi:10.2514/1.28421
23. Soykasap, O., Pellegrino, S., Howard, P., Notter, M.: Tape spring large deployable antenna. In: 47th AIAA/ASME/ASCE/AHS/ASC Structures, Structural Dynamics, and Materials Conference, 1–4 May 2006, pp. 1–12. Newport, Rhode Isl. Newport, Rhode Island (2006)
24. Watt, A.M., Pellegrino, S.: Tape-spring rolling hinges. In: 36th Aerospace Mechanisms Symposium, pp. 1–14. Cleveland (2002)
25. Mallikarachchi, H.M.Y.C., Pellegrino, S.: Optimized designs of composite booms with integral tape-spring hinges. In: Proceeding of the 51st AIAA/ASME/ASCE/AHS/ASC Structures, Structural Dynamics, and Materials Conference, pp. 1–16. Orlando (2010)
26. Tan, L.T., Soykasap, O., Pellegrino, S.: Design & manufacture of stiffened spring-back reflector demonstrator. In: 46th AIAA/ASME/ASCE/AHS/ASC Structures, Structural Dynamics, and Materials Conference, pp. 1–12. American Institute of Aeronautics and Astronautics, Austin (2005)
27. Mallikarachchi, H.M.Y.C., Pellegrino, S.: Quasi-static folding and deployment of ultrathin composite tape-spring hinges. J. Spacecr. Rocket. **48**, 187–198 (2011). doi:10.2514/1.47321
28. Yee, J.C.H., Soykasap, Ö., Pellegrino, S.: Carbon fibre reinforced plastic tape springs. In: 45th AIAA/ASME/ASCE/AHS Structures, Structural Dynamics, and Materials Conference, pp. 1–9. American Institute of Aeronautics and Astronautics, Palm Springs (2004)
29. Kwok, K., Pellegrino, S.: Folding, stowage, and deployment of viscoelastic tape springs. AIAA J. **51**, 1908–1918 (2013). doi:10.2514/1.J052269
30. Yee, J.C.H., Pellegrino, S.: Composite tube hinges. J. Aerosp. Eng. **18**, 224–231 (2005)
31. Murphey, T.W., Jeon, S., Biskner, A., Sanford, G.: Deployable booms and antennas using bi-stable tape-springs. In: 24th Annual Conference on Small Satellites, pp. 1–7. Logan (2010)

32. Soykasap, O., Karakaya, S., Turkmen, D.: Curved large tape springs for an ultra-thin shell deployable reflector. J. Reinf. Plast. Compos. **31**, 691–703 (2012). doi:10.1177/0731684412444018
33. Sickinger, C., Herbeck, L., Breitbach, E.: Structural engineering on deployable CFRP booms for a solar propelled sailcraft. Acta Astronaut. **58**, 185–196 (2006). doi:10.1016/j.actaastro.2005.09.011
34. Block, J., Straubel, M., Wiedemann, M.: Ultralight deployable booms for solar sails and other large gossamer structures in space. Acta Astronaut. **68**, 984–992 (2011). doi:10.1016/j.actaastro.2010.09.005
35. Soykasap, Ö.: Analysis of tape spring hinges. Int. J. Mech. Sci. **49**, 853–860 (2007). doi:10.1016/j.ijmecsci.2006.11.013
36. Soykasap, Ö.: Deployment analysis of a self-deployable composite boom. Compos. Struct. **89**, 374–381 (2009). doi:10.1016/j.compstruct.2008.08.012
37. Roh, J.-H., Kim, H.-J., Bae, J.-S.: Shape memory polymer composites with woven fabric reinforcement for self-deployable booms. J. Intell. Mater. Syst. Struct. **25**, 2256–2266 (2014). doi:10.1177/1045389X14544148
38. Shi, G.-H., Yang, Q.-S., He, X.Q.: Analysis of intelligent hinged shell structures: deployable deformation and shape memory effect. Smart Mater. Struct. **22**, 125018 (2013). doi:10.1088/0964-1726/22/12/125018
39. Yang, H., Wang, L.Y.: Thermomechanical analysis of shape-memory composite tape spring. Appl. Compos. Mater. **20**, 287–301 (2012). doi:10.1007/s10443-012-9271-x
40. Li, Z., Song, W., Wang, Z.: Numerical and theoretical studies of the buckling of shape memory tape spring. Tech. Sci. Press **17**, 133–143 (2011)
41. Sutton, M.A., Turner, J.L., Bruck, H.A., Chae, T.A.: Full-field representation of discretely sampled surface deformation for displacement and strain analysis. Exp. Mech. **31**, 168–177 (1991)
42. Sutton, M.A.: Springer Handbook of Experimental Solid Mechanics. Department of Mechanical Engineering, The Johns Hopkins University, Baltimore (2008)
43. Sutton, M.A., Orteu, J.J., Schreier, H.: Image Correlation for Shape, Motion and Deformation Measurements: Basic Concepts, Theory and Applications. Springer, New York (2009)
44. Zwillinger, D.: CRC Standard Mathematical Tables and Formulae, 31st ed, pp. 341–343. CRC Press Company, Boca Raton (2003)
45. Beer, F.P., Johnston Jr., E.R., DeWolf, J.T.: Mechanics of Materials, 4th edn. McGraw-Hill, New York (2006)
46. Terzak, J.C.: Modeling of microvascular shape memory composites. Youngstown State University (2013)
47. Volk, B.L.: Thermomechanical Characterization and modeling of shape memory polymers. Texas A&M University (2009)

Chapter 23
Identification of Constitutive Model Parameters in Hopkinson Bar Tests

M. Fardmoshiri, M. Sasso, E. Mancini, G. Chiappini, and M. Rossi

Abstract In this work, tension and compression tests have been carried out on aluminium samples at low and high strain rate, the latter performed by means of a direct tension Hopkinson bar equipment. The parameters of the Johnson-Cook constitutive model have been identified using different approaches; the first method consists in the classical Finite Element Model Updating, where numerical simulations are repeated with different material parameters until the mismatch between the experimental and numerical load–displacement curves falls below an acceptable threshold.

The second method is based on the analysis of the digital images acquired by a fast camera during the tests; this permitted to calibrate the JC model by an analytical minimization procedure, without any FE simulation. A third inverse technique was also implemented, consisting in applying the FE model updating but using an enriched cost function, where also the mismatch between the numerical and acquired specimen shape profiles is included and minimized.

The advantages and drawbacks of the different techniques are assessed and compared.

Keywords Hopkinson bar • FastCam • FEMU • Johnson-Cook • Image processing • Optimization

23.1 Introduction

Nowadays Split Hopkinson Pressure Bar (SHPB) has been vastly used in order to determine the stress–strain curves at strain rates in range of 10^2–10^4 s^{-1}. The Hopkinson Pressure Bar was first suggested by Hopkinson [1] as an apparatus to measure stress pulse propagation in a long metal and it was extended by Davies [2] and Kolsky [3]. More recently, SHPB is focused not only on compression tests but also on tensile [4–6] and torsion tests [7] for different classes of materials such as metals, polymers, ceramics, composites and foams.

In this paper, tension and compression tests on AA6061-T6 aluminium specimens of different dimensions are considered. The collected data has been proceeded with aim of an optimization algorithm based on finite element model Updating (FEMU) in order to estimate Johnson-Cook constitutive material constants. In this method, using scripted code simulation, different input parameters such as material properties, geometry dimensions, contact coefficient, interaction types, simulation types, and solver methods can be parameterized [8, 9]. Numerical results are compared with experimental results and using an optimization method, the inputs are updated for the next simulation. Furthermore, nowadays ultra-fast cinematography [10, 11] method is employed in order to capture sample profile change and estimate the necking radius; despite of costs of fast cameras, they are efficient in capturing deformation changes in high strain rate applications. Therefore a fast camera with a resolution of 192 × 128 pixels and a capture rate of 100,000 fps, captured images of the deformation process undergone by the samples during the tests. The profile of the sample, obtained after the image binarization and processing,

This paper was presented during the SEM Annual Meeting in Costa Mesa, CA, June 2015

M. Fardmoshiri
Dipartimento di Ingegneria, Industriale e Scienze Matematiche Università Politecnica delle Marche, via Brecce Bianche 1, Ancona 60131, Italy

M. Sasso (✉) • G. Chiappini • M. Rossi
Dipartimento di Ingegneria, Industriale e Scienze Matematiche, Università Politecnica delle Marche, via Brecce Bianche 1, Ancona 60131, Italy
e-mail: m.sasso@univpm.it

E. Mancini
Università degli Studi eCampus, Via Isimbardi, Novedrate (CO) 22060, Italy

S. Quinn, X. Balandraud (eds.), *Residual Stress, Thermomechanics & Infrared Imaging, Hybrid Techniques and Inverse Problems, Volume 9*, Conference Proceedings of the Society for Experimental Mechanics Series, DOI 10.1007/978-3-319-42255-8_23

has been used to derive the real strain; this method (hereinafter named FastCam) provides accurate neck profiles of the deforming sample, from which the minimum radius and the radius of curvature of the neck profile may be measured and can be used to approximate the stress and strain history of the sample through the well-known Bridgman analysis [12–14]. Finally, accuracy and time cost of FEMU and analytical FastCam methods [15] are compared.

23.2 Theoretical Background

The Kolsky bar or Split Hopkinson Pressure Bar, consists of a pretension bar, input bar and output bar that sandwich a cylindrical specimen. In this paper a direct tension/compression Hopkinson bar has been used, which working principle is schematically shown in Figs. 23.1 and 23.2.

Instantaneous generated wave at the pretension bar propagates through the input bar, specimen and output bar respectively and partially reflected back to the input bar. These pulses are measured by strain gauges placed on input and output bars. Let the incident, reflected and transmitted pulses be denoted respectively by $\varepsilon_I(t)$, $\varepsilon_R(t)$ and $\varepsilon_T(t)$. All of them are function of time and can be regarded also as displacement, strain and stress perturbations travelling along the bars within the sound speed of material C_0. The nominal strain e, strain rate $\dot{e}$ and stress S in the specimen can be derived from equations [16]:

$$\dot{e}(t) = -\frac{2C_0}{L_S}\varepsilon_R(t) \tag{23.1}$$

$$e(t) = -\frac{2C_0}{L_S}\int_0^t \varepsilon_R(t)dt \tag{23.2}$$

$$S(t) = \frac{A_B E_B}{A_S}\varepsilon_T(t) \tag{23.3}$$

where A_B and E_B are the cross-sectional area and Young modulus of bar and L_S and A_S denote the specimen original gauge length and cross sectional area, respectively. It noticeable that the strain rate is not constant during the test and the gained stress and strain values are the engineering (nominal) values, since the initial length and cross-section area of specimen are considered [8].

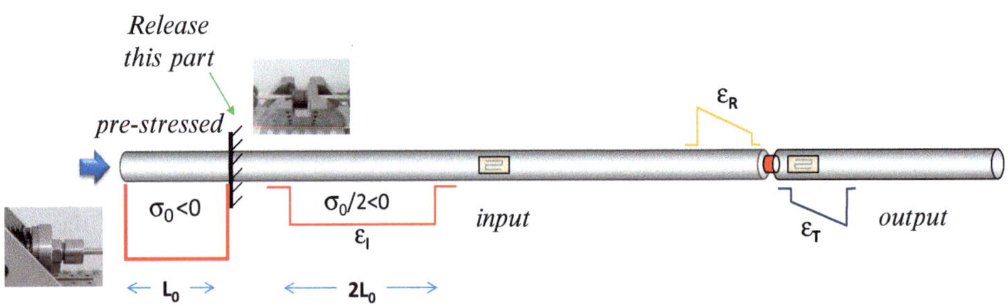

Fig. 23.1 Schematic representation of Kolsky bar (Compression test)

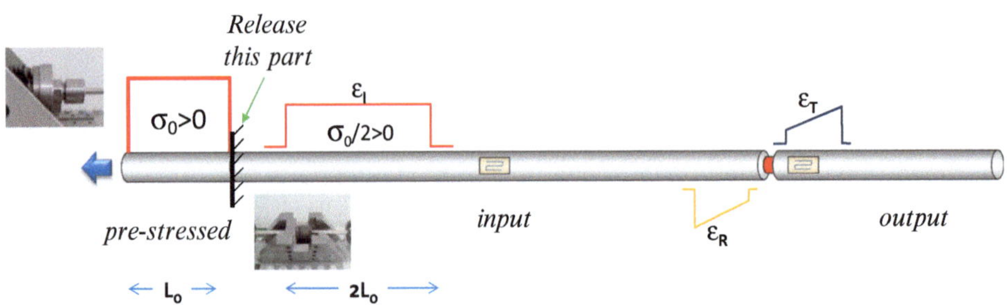

Fig. 23.2 Schematic representation of Kolsky bar (Tension test)

23.3 Experimental Setup and Results

Experimental tests have been carried out, both in tension and compression on aluminium samples AA6061-T6. To facilitate the repeatability and comparison of the tests, all the specimens shown in Fig. 23.3, were made from the same material bar. It should also be pointed out that the experimental apparatus used is virtually identical to the tensile tests and compression, having the same loading mechanism to generate the wave and the same strain gauges. The principal characteristics of the equipment are summarized in Table 23.1.

In order to have different strain rates, the geometry dimensions of the specimens have to be varied. Furthermore, a quasi-static tensile test is performed (Tables 23.2 and 23.3).

The resulting engineering curves are reported in Figs. 23.4 and 23.5. The average engineering strain rates of each test are indicated in the legend.

Figures 23.6, 23.7, 23.8 and 23.9 show an exemplifying images acquired at the beginning and at the end of compression and tension tests. Camera exposure and illumination were set in order to obtain high contrast black/white images. This simplified the subsequent procedures of image binarization and edge detection for the profile identification.

The axial strain can be computed by direct measurement of the distance variation between the bar extremities, which are highlighted by the blue lines in Figs. 23.6, 23.7, 23.8 and 23.9. On the other hand, the profile (highlighted by red curve) can be used to evaluate the diameter variation along the length (7 and 8). In compression tests, the average of the detected edges are used for the strain computation, whilst in tension tests, the profile was fitted by a spline curve [15].

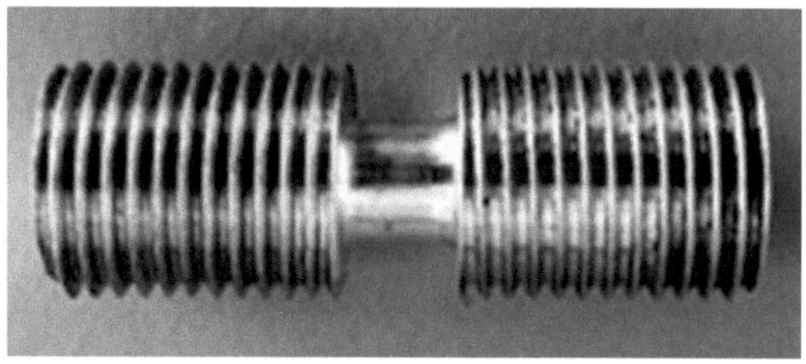

Fig. 23.3 Tension sample AA6061-T6

Table 23.1 Principal Features of the used SHB

	Length (mm)	Diameter (mm)	Material
Pre-tension bar	3000	18	Steel: 17–4 PH
Input bar	7500	18	E = 204 GPa; Density = 7800 kg/m3
Output bar	4000	18	Speed sound = 5114 m/s

Table 23.2 Compression tests geometries

ID	Height (mm)	Width (mm)
C1	7.98	8.12
C2	5.04	6.12
C3	3.04	4.98
C4	2.94	4.94

Table 23.3 Tension test geometries

ID	Length (mm)	Diameter (mm)	Fillet radius (mm)
T1	9	5.94	1.5
T2	6	6	1.5
T3	4.5	5.98	1
QS	12	8	1.5

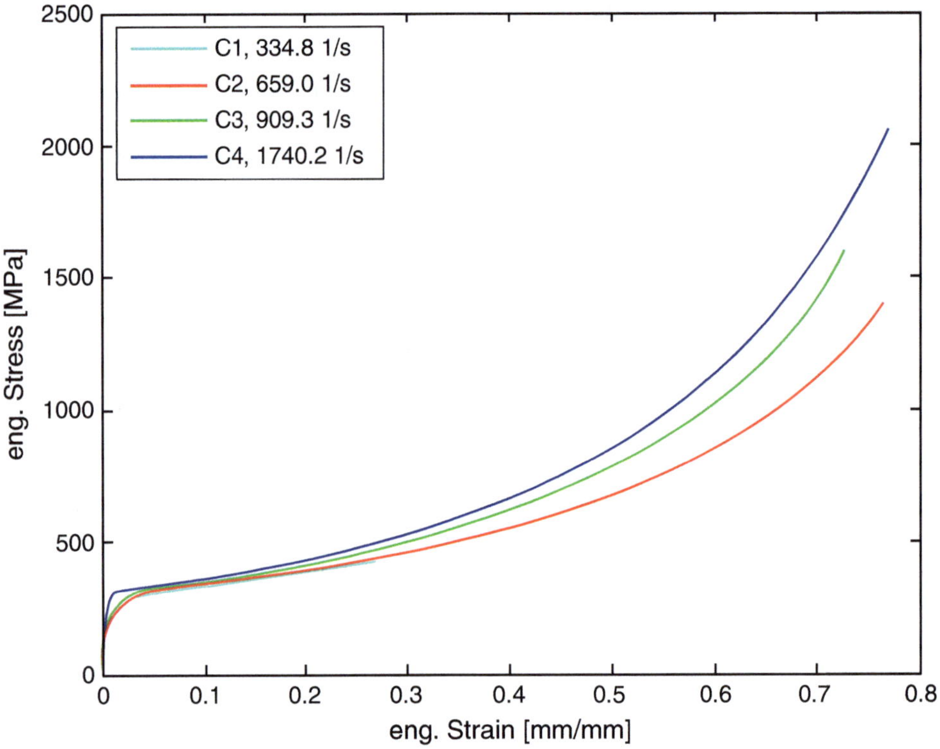

Fig. 23.4 Compression test

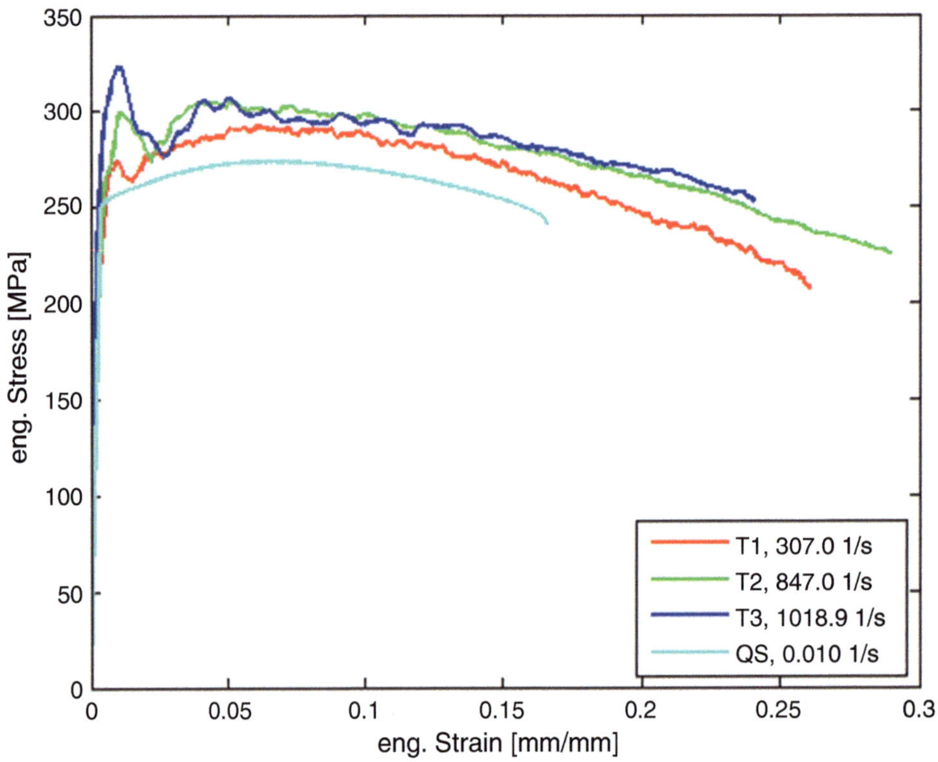

Fig. 23.5 Tension test

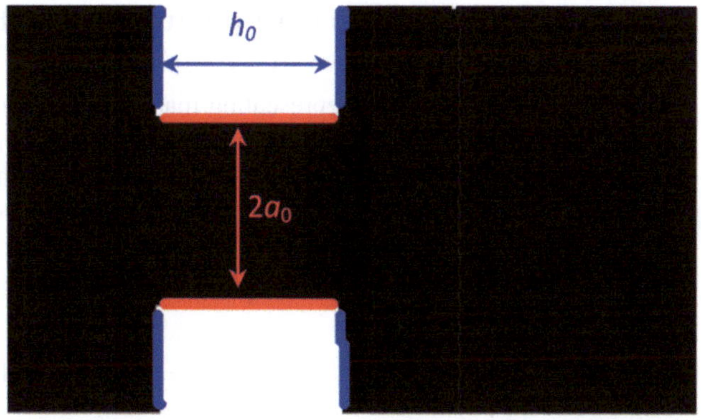

Fig. 23.6 Image acquired at the initial stage of compression a test

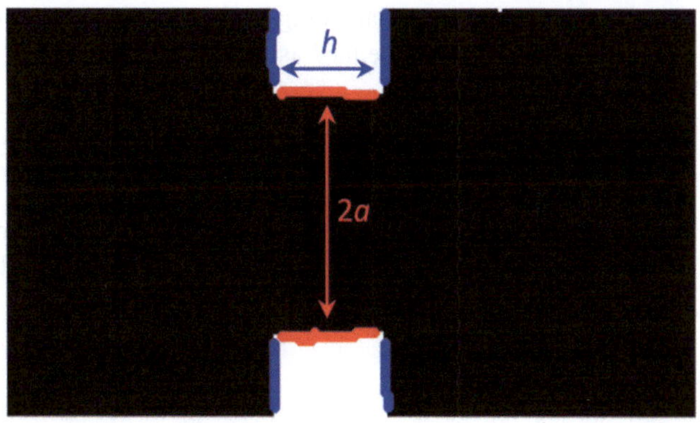

Fig. 23.7 Binarization and edge detection of image acquired in compression test

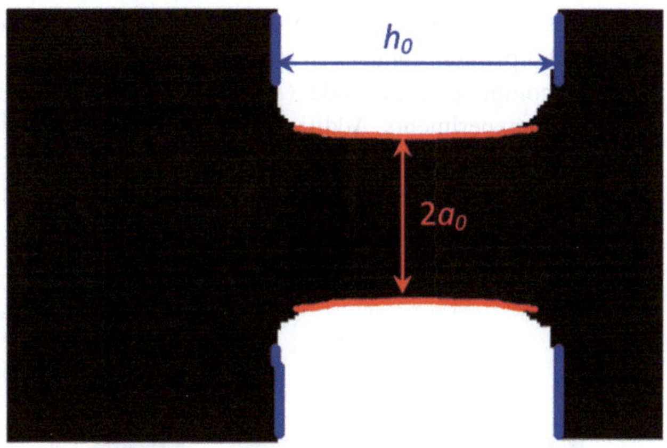

Fig. 23.8 Images acquired at the initial stage of a tension test

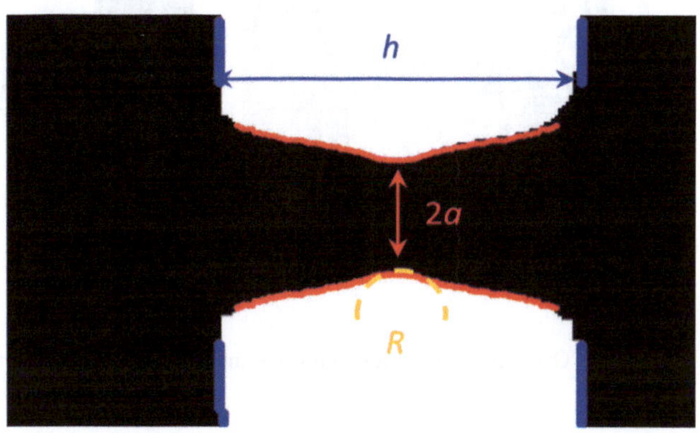

Fig. 23.9 Binarization and edge detection of image acquired in tension test

23.4 Finite Element Model Updating and/or Fitting by FEMU

A variety of material models representing materials behaviour at high strain rate and significant temperature change, have been proposed in literature, e.g. Cowper-Symonds [17], Klopp-Clifton [18], Litonski [19], Zerilli-Armstrong [20]. Johnson-Cook model [21, 22] was practiced as it is well established in the literature and used extensively in most of the finite element computer programs; however, the constitutive constants used in the model are not always available for the specific material under consideration especially as a function of its microstructure and degree of cold working or work hardening. Johnson-Cook Model can be read as following:

$$\sigma_{JC} = (A + B\varepsilon^n)\left(1 + c\log\frac{\dot{\varepsilon}}{\dot{\varepsilon}_0}\right)(1 - T^{*m}) \tag{23.4}$$

In this equation, A, B, and n are yield stress, work hardening coefficient and exponential factor respectively while c and m account for strain rate hardening and thermal softening effects. $\dot{\varepsilon}_0$ is a user defined reference strain rate and T^* is the homologous temperature defined as,

$$T^* = \frac{T - T_{room}}{T_{melt} - T_{room}} \tag{23.5}$$

where T, T_{melt}, T_{room} represent absolute, melting and room temperatures, respectively. Furthermore temperature T is calculated considering the adiabatic heating due to the work done with plastic deformation [23]:

$$T = T_{room} + \frac{\int \sigma d\varepsilon}{\rho C_p} \tag{23.6}$$

The calibration has focused on the coefficients A, B, n, c, m, $\dot{\varepsilon}_0$ and μ. Due to the supposed isotropic behaviour of the material A, B and n were considered equal both in tension and compression. On the contrary, the coefficients c, m, $\dot{\varepsilon}_0$ were evaluated separately for tension and compression tests. Obviously, friction coefficient μ was applied only on compression tests. A commercial FE code (Abaqus® v6.12) [24] was used to simulate tension and compression dynamic and quasi-static experiments. Additionally, gradient base constrained nonlinear optimization programming [25] (MATLAB® fmincon function) [26] is used in order to find constrained minimum of scalar error function of several Johnson-Cook parameters. For better visualization, longest compression, tension and quasi-static sample results are shown in Figs. 23.10 and 23.11.

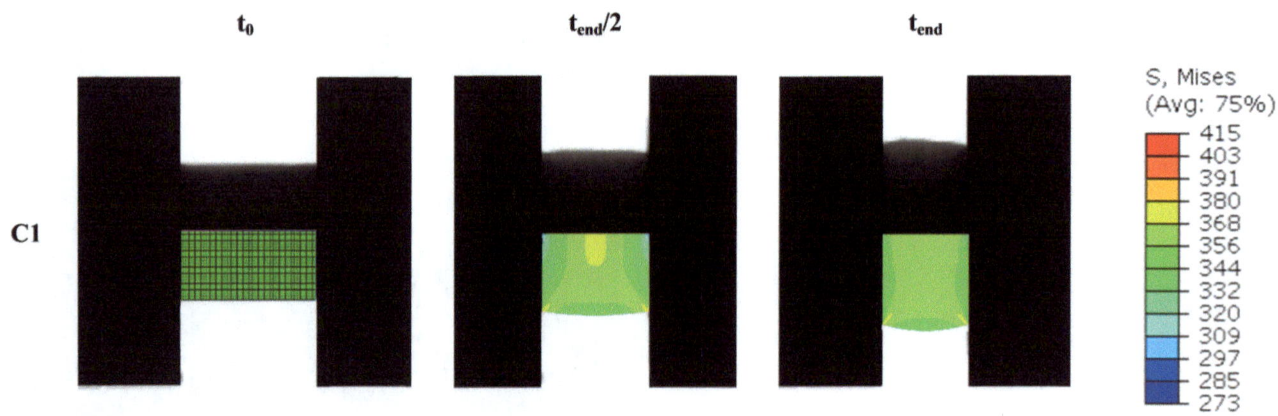

Fig. 23.10 Overlapping of acquired images and FEM contour maps in compression tests

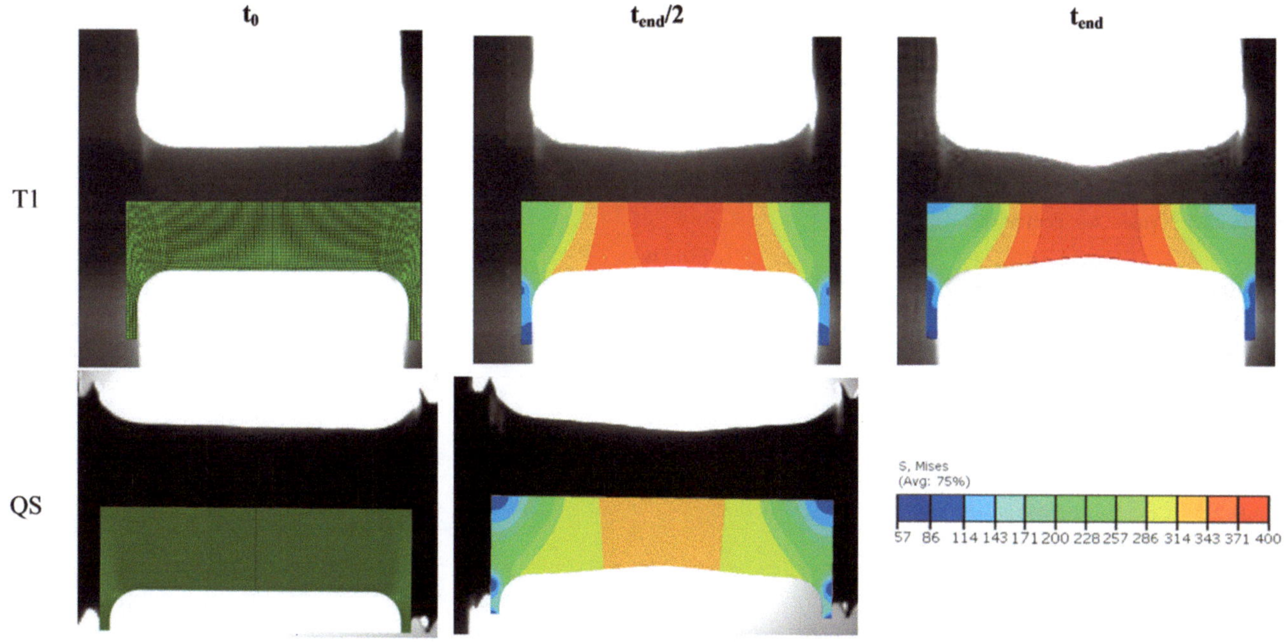

Fig. 23.11 Overlapping of acquired images and FEM contour maps in tension (dynamic T1 and quasi static QS) tests

23.5 Results and Discussion

As aforementioned, the FEMU (force-displacement and x-y profile coordinates) and the analytical procedure were implemented into an optimization routine, in order to derive the best sets of Johnson-Cook coefficients that describe the material behaviour in the experimented conditions.

According to the FEMU method, the cost function to be minimized is represented by the sum of error norms, computed, for each test, in terms of difference between experimental and numerical load histories. The cost function formulation, for force-displacement and profile coordinates cases, is as follows:

$$err = \frac{1}{N}\sum_{k=1}^{N}\frac{1}{\overline{F}_{k,Exp}}\sum_{t=1}^{t_k}\sqrt{\frac{\left[F_{k,Fem}(t) - F_{k,Exp}(t)\right]^2}{t_k}} \tag{23.7}$$

where N is the number of experimental tests, t_k is the number of time increments in the k-th test; $F_{k,Fem}$ and $F_{k,Exp}$ are the numerical and experimental measured entities (force (Newton) or profile coordinates (millimetre)). The term $\overline{F}_{k,Exp}$ represents the average experimental entities (force or coordinates) in the k-th test, used for normalization purposes.

With respect to the analytical/FastCam method, the cost function to be minimized is represented by the sum of error norms computed in terms of differences between true stress histories. The former stress history $\sigma_{k,JC}(t)$ is predicted by the J-C Eq. (23.4), and the latter $\sigma'_k(t)$ is represented by the experimental data modified according to well-known Bridgman formula [12]:

$$err = \frac{1}{N}\sum_{k=1}^{N}\frac{1}{\overline{\sigma'}_k}\sum_{t=1}^{t_k}\sqrt{\frac{\left[\sigma_{k,JC}(t) - \sigma'_k(t)\right]^2}{t_k}} \tag{23.8}$$

where N is the number of experimental tests, t_k is the number of time increments in the k-th test..

The Johnson-Cook parameter sets are shown in Tables 23.4, 23.5 and 23.6, together with the average errors in tension and compression. The experimental and the numerical reconstructed curves are reported in Figs. 23.12, 23.13 and 23.14. All the curves in Figs. 23.12 and 23.13, even including those obtained by the analytical/FastCam method, were converted from true stress–strain to load–displacement and using the correspondent initial area and length. This conversion was done for a better

Table 23.4 Work-hardening coefficients

	A [MPa]	B [MPa]	N
FastCam	218.2	129.5	0.249
FEMU force-displacement	203.7	127.7	0.280
FEMU profile	230.5	109.9	0.249

Table 23.5 Tensile test coefficients

	C	m	$\log\dot{\varepsilon}_0$ [1/*s*]	RMS err
FastCam	0.031	1.28	1.704	5.7 [MPa], 1.8 %
FEMU force-displacement	0.036	1.05	1.531	611.3 [N], 6.0 %
FEMU profile	0.049	0.91	1.699	0.225 [mm], 5.7 %

Table 23.6 Compression test coefficients

	C	M	$\dot{\varepsilon}_0$ [1/s]	μ	RMS err
FastCam	0.165	0.99	3.183	0.100	5.4 [MPa], 1.8 %
FEMU force-displacement	0.070	1.16	2.117	0.110	2552 [N], 9.7 %
FEMU profile	0.049	0.934	2.0	0.087	0.195 [mm], 1.9 %

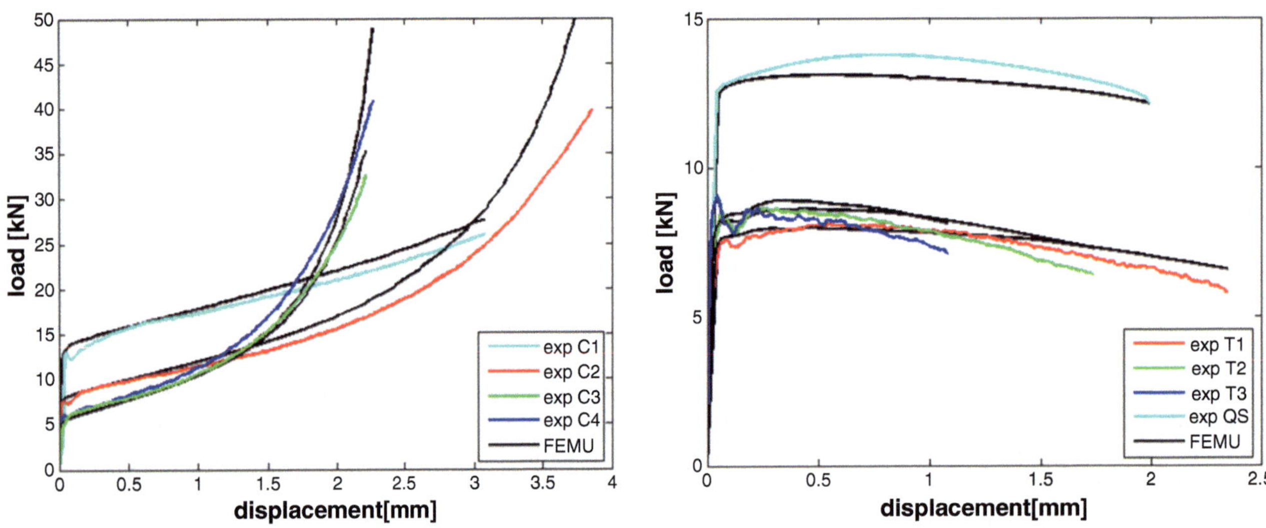

Fig. 23.12 Fitting of experimental load-displacements by FEMU procedure of compression (*left*) and tension (*right*) tests

comparison, since with the present data the load–displacement curves are more distinguishable than the corresponding stress–strain ones. On the other hand, curves reported in Fig. 23.14 represent profile coordinates obtained from FEMU and FastCam methods.

Observing the curves and the average percentage error of the tables, a general good matching is found for the analytical/FastCam method. The FEMU method provided a slightly poorer data fitting; nevertheless, the obtained coefficients are close to each other in tension, with a higher discrepancy in compression. Basically, this originates from the difficulty of applying exact boundary conditions.

Finally, it must be pointed out that to calculate the error with respect to all simulations simultaneously, all dynamic and quasi-static simulations need to be performed every time the parameters change. For the automated FEMU method (force-displacement and profile), the computational time is averagely 4.5 days for each method, taking advantage of a 6-core Intel

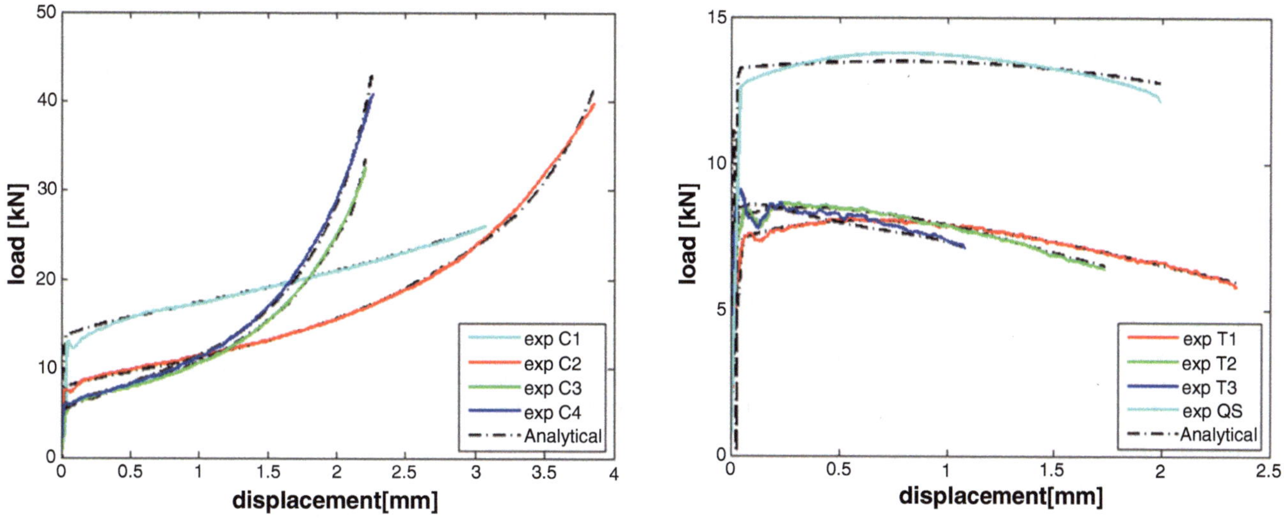

Fig. 23.13 Fitting of experimental load-displacements by FastCam procedure of compression (*left*) and tension (*right*) tests

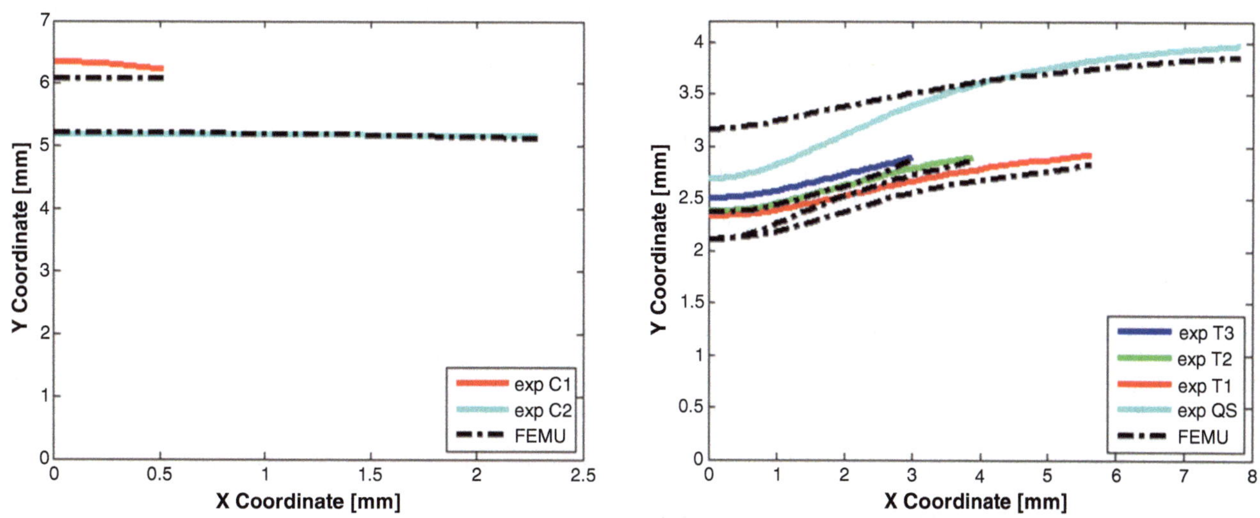

Fig. 23.14 Fitting of experimental profile by FEMU procedure of compression (*left*) and tension (*right*) tests

Xeon® workstation. On the contrary, analytical/FastCam methods require a small programming effort in image processing and synchronization with SHB data. Besides the evident time cost advantage, it should also be considered that constitutive model can be changed and error and sensitivity analysis can be performed in simpler and faster way.

23.6 Conclusion

The study involved the Johnson-Cook constitutive model calibration through a series of experimental tests both in tension and compression at high strain rate using the Hopkinson bar. The popular method of Finite Element Model Updating as well as an analytical method based on the analysis of the images acquired during the tests by means of a fast camera has been implemented. FastCam method analysis had the possibility to evaluate the deformation experienced by the specimen during the tests. It validated the measurements acquired by strain gauges and on the other hand, permitted to measure true strain in the necking section. Analytically, various effects which influence the behaviour of the specimens were considered such as the friction in the compression tests, the triaxiality and non-uniformity of deformation in tension tests, and finally the adiabatic heating due to plastic deformation. The coefficients found with the proposed analytical method resulted in good agreement with those obtained through finite element simulations, with an even better matching with experimental data.

References

1. Hopkinson, B.: A method of measuring the pressure produced in the detonation of explosives or by the impact of bullets. Phil. Trans. Roy. Soc. Lond. **213**, 437–456 (1914)
2. Davies, R.: A critical study of the hopkinson pressure bar. Phil. Trans. Roy. Soc. Lond. **240**, 375–457 (1948)
3. Kolsky, H.: An investigation of the mechanical properties of materials at very high rates of load. Proc. R. Soc. Lond. B **62**(11), 676–700 (1949)
4. Harding, J., Wood, E., Campbell, J.: Tensile testing of materials at impact rates of strain. J. Mech. Eng. Sci. **2**(2), 88–96 (1960)
5. Staab, G., Gilat, A.M.O.S.: Exp. Mech. **31**(3), 232–235 (1991)
6. Nicholas, T.: Tensile testing at high rates of strain. Exp. Mech. **21**(5), 177–185 (1981)
7. Lewis, J.L., Campbell, J.: The development and use of a torsional Hopkinson-bar apparatus. Exp. Mech. **12**(11), 520–524 (1972)
8. Sasso, M., Newaz, G., Amodio, D.: Material characterization at high strain rate by Hopkinson bar tests and finite element optimization. Mater. Sci. Eng. **A487**, 289–300 (2008)
9. Sedighi, M., Khandaei, M., Shokrollahi, H.: An approach in parametric identification of high strain rate constitutive model using Hopkinson pressure bar test results. Mater. Sci. Eng. A **527**(15), 3521–3528 (2010)
10. Rossi, M., Pierron, F.: Identication of plastic constitutive parameters at large deformations from three dimensional displacement fields. Comput. Mech. **49**(1), 53–71 (2012)
11. Rossi, M., Pierron, F., Forquin, P.: Assessment of the metrological performance of an in situ storage image sensor ultra-high speed camera for full-field deformation measurements. Meas. Sci. Technol. **25**(1), 025401 (2014)
12. Chakrabarty, J.: Applied Plasticity, 2nd edn, pp. 161–163. Springer, New York (2010)
13. Cross, L., Bless, A., Rajendran, E., Strader, E., Dawicke, D.: New technique to investigate necking in a tensile Hopkinson Bar. Exp. Mech. **24** (3), 184–186 (1984)
14. Dunnett, T., Balint, D., MacGillivray, H., Church, P., Gould, P.: Scale effects in necking. EPJ Web Conf. **26**, 01008 (2012)
15. Sasso, M., Fardmoshiri, M., Mancini, E., Rossi, M., Cortese, L.: High speed imaging for material parameters calibration at high strain rate. Eur Phys J Special Topics, **225**(2), 295–309 (2016)
16. Chen, W., Song, B.: Split Hopkinson (Kolsky) Bar, Design, Testing and Applications. Mechanical engineering series, pp. 11–17. Springer, New York (2011)
17. Cowper, P.S.G.: Strain hardening and strain rate effect in the impact loading of Cantilever beams. Brown University, Applied Mathematics Report (1958)
18. Klopp, R., Clifton, R.: Pressure-shear impact and the dynamic viscoplastic response of metals. Mech. Mater. **4**(3), 375–385 (1985)
19. Litonski, J.: Plastic flow of a tube under adiabatic torsion. Bull. Acad. Pol. Sci. Ser. Sci. Tech. **XXV**, 7 (1977)
20. Zerilli, F.J., Armstrong, R.W.: Deformation twinning and grain size aspects of numerical simulations of plastic flow. MRS Proc. **362** (1994)
21. Johnson, R., Cook, W.H.: A constitutive model and data for metals subjected to large strains, high strain rates and high temperatures. Int. Symp. Ballistics **7**, 541–547 (1983)
22. Johnson, J., Cook, W.: Fracture characteristics of three metals subjected to various strains, strain rates, temperatures and pressures. Eng. Fract. Mech. **21**(1), 31–48 (1985)
23. Meyers, M.A.: Dynamic Behavior of Materials. Wiley, New York (1994)
24. "Abaqus FEA," D S Simulia. © Dassault Systemes, 2004, 2012.
25. Han, S.: A globally convergent method for nonlinear programming. J. Optim. Theory Appl. **22**, 297 (1977)
26. MATLAB version 8.2. The MathWorks Inc., Natick (2013)

MIX
Papier aus verantwortungsvollen Quellen
Paper from responsible sources
FSC® C105338
FSC
www.fsc.org

If you have any concerns about our products,
you can contact us on
ProductSafety@springernature.com

In case Publisher is established outside the EU,
the EU authorized representative is:
Springer Nature Customer Service Center GmbH
Europaplatz 3, 69115 Heidelberg, Germany

Printed by Libri Plureos GmbH
in Hamburg, Germany